Topics in Current Chemistry Collections

Aims and Scope

The series *Topics in Current Chemistry Collections* presents critical reviews from the journal *Topics in Current Chemistry* organized in topical volumes. The scope of coverage is all areas of chemical science including the interfaces with related disciplines such as biology, medicine and materials science.

The goal of each thematic volume is to give the non-specialist reader, whether in academia or industry, a comprehensive insight into an area where new research is emerging which is of interest to a larger scientific audience.

Each review within the volume critically surveys one aspect of that topic and places it within the context of the volume as a whole. The most significant developments of the last 5 to 10 years are presented using selected examples to illustrate the principles discussed. The coverage is not intended to be an exhaustive summary of the field or include large quantities of data, but should rather be conceptual, concentrating on the methodological thinking that will allow the non-specialist reader to understand the information presented.

Contributions also offer an outlook on potential future developments in the field.

More information about this series at http://www.springer.com/series/14181

Luis Serrano • Rafael Luque • Bert F. Sels
Editors

Lignin Chemistry

With contributions from

Ion Agirre • Iker Agirrezabal-Telleria • V. Laura Barrio
Kepa Bizkarra • Jose F. Cambra • Juan Antonio Cecilia
Juan Carlos Colmenares • Walter Eevers • Elias Feghali
Araceli García • Cristina García-Sancho • Roger Gläser
Aitziber Iriondo • Ayesha Khan • S.-F. Koelewijn • Yuhe Liao
Jing Liu • Alexander Lopez-Urionabarrenechea • Wei Lv
Longlong Ma • Vaishakh Nair • Mikel Oregui-Bengoechea
Pablo Ortiz • Olga I. Poddubnaya • Alexander M. Puziy
Tom Renders • Jesus M. Requies • Wouter Schutyser • Bert F. Sels
Luis Serrano • Olena Sevastyanova • Kirk M. Torr
Daniel J. van de Pas • S. Van den Bosch • G. Van den Bossche
Karolien Vanbroekhoven • Thijs Vangeel • Richard Vendamme
Chenguang Wang • Yuting Zhu

Springer

Editors
Luis Serrano
Departamento de Química Inorgánica
e Ingeniería Química
Universidad de Córdoba
Córdoba, Spain

Rafael Luque
Departamento de Química Orgánica
Universidad de Córdoba
Córdoba, Spain

Bert F. Sels
Center for Sustainable Catalysis
and Engineering
KU Leuven
Leuven, Belgium

Partly previously published in Topics in Current Chemistry Volume 376 (2018); Topics in Current Chemistry Volume 377 (2019).

ISSN 2367-4067
Topics in Current Chemistry Collections
ISBN 978-3-030-00589-4

This Springer imprint is published by the registered company Springer Nature Switzerland AG
The registered company address is: Gewerbestrasse 11, 6330 Cham, Switzerland

Contents

Preface

We are living in exciting times. Given the finiteness of cheap fossil resources and their association with climate change, utilization of fossil resources is debated and the search for alternatives is pertinent and pressing. Among others, lignocellulosics may be able to replace part of the fossil feedstock market. Fractionation of this feedstock and further valorization of the fractions such as cellulose, hemicellulose and lignin into useful products and materials is however challenging. The chemistry is fascinating though it needs further development. The lignocellulosic biorefinery technologies are under development but not yet secure and sufficiently de-risked to readily replace the current fossil refineries, new value chains between feedstock and final usage are required to get the best out of biomass utilization, and life cycle assessments should critically prove lignocellulosics utilization and processing is indeed contributing to sustainability. A lot of work is yet to be done with many academics and industrial researchers, following a multidisciplinary approach, to reach these challenging goals.

Lignin conversion, among a few others, is a crucial research topic in this sustainability adventure. Excellent work has already been done however the urgency came only recently after realizing that biorefineries, to become economically viable, should include valorization of lignin into chemicals and materials. Lignin valorization is complex due to the heterogeneous structure of lignin, as well as due to the many forms of lignin having different structures and compositions. Native lignin is structurally different from lignin side-streams and therefore the chemistries to convert them are different, and therefore also their utilization. Since lignin is aromatic in nature, the groups of chemicals that already form a large part of today's chemicals are likely the preferred target molecules. This view also fits with the potential shortage of aromatics in the future due to the fossil feedstock shifting from oil to gas.

Lignin is not a single molecule, and its refinery will always lead to several different products. It is left to us to discover and explore the best chemistry and catalysis to make these products, chemicals and materials that are most useful in our daily life, and which can be produced and isolated in large amounts in the most sustainable way. A flavor of such endeavors can be read in this topical collection. The reader will hopefully enjoy the overviews of the state of the art and our definitions of the next ways to go.

Dr. Luis Serrano
Universidad de Córdoba

Prof. Rafael Luque
Universidad de Córdoba

Prof. dr. ir. Bert F. Sels
KU Leuven

Top Curr Chem (Z) (2018) 376:20
https://doi.org/10.1007/s41061-018-0198-z

REVIEW

Lignin-Based Composite Materials for Photocatalysis and Photovoltaics

Ayesha Khan[1] · Vaishakh Nair[1] · Juan Carlos Colmenares[1] · Roger Gläser[2]

Received: 14 February 2018 / Accepted: 19 April 2018 / Published online: 2 May 2018

Abstract Depleting conventional fuel reserves has prompted the demand for the exploration of renewable resources. Biomass is a widely available renewable resource that can be valorized to produce fuels, chemicals, and materials. Among all the fractions of biomass, lignin has been underutilized. Due to its complex structure, recalcitrant nature, and heterogeneity, its valorization is relatively challenging. This review focuses on the utilization of lignin for the preparation of composite materials and their application in the field of photocatalysis and photovoltaics. Lignin can be used as a photocatalyst support for its potential application in photodegradation of contaminants. The interaction between the components in hybrid photocatalysts plays a significant role in determining the photocatalytic performance. The application of lignin as a photocatalyst support tends to control the size of the particles and allows uniform distribution of the particles that influence the characteristics of the photocatalyst. Lignin as a semiconductive polymer dopant for photoanodes in photovoltaic cells can improve the photoconversion efficiency of the cell. Recent success in the development of lignosulfonates dopant for hole transport materials

Chapter 1 was originally published as Khan, A., Nair, V., Colmenares, J. C. & Gläser, R. Topics in Current Chemistry (Z) (2018) 376: 20. https://doi.org/10.1007/s41061-018-0198-z.

✉ Ayesha Khan
akhan@ichf.edu.pl

✉ Juan Carlos Colmenares
jcarloscolmenares@ichf.edu.pl

✉ Roger Gläser
roger.glaeser@uni-leipzig.de

1 Institute of Physical Chemistry, Polish Academy of Sciences, Kasprzaka 44/52, 01-224 Warsaw, Poland

2 Institute of Chemical Technology, Leipzig University, Linnéstr. 3, 04103 Leipzig, Germany

in photovoltaics will pave the way for further research in lignin-based high-performance organic electronic devices.

Keywords Lignin · Composite materials · Photocatalysis · Photodegradation · Photovoltaics · Photoactive materials

Abbreviations

UV	Ultraviolet
T_g	Glass transition temperature
T_d	Degradation temperature
RhB	Rhodamine B
ITO	Indium–tin-oxide
HOMO	Highest occupied molecular orbital
LUMO	Lowest unoccupied molecular orbital
PEDOT:PSS	Poly(3,4-ethylenedioxythiophene): poly(styrene sulfonate)
P3HT	Poly(3-hexylthiophene)
PCBM	[6]-Phenyl-C61-butyric acid methyl ester
LA	Lignin-amine
MO	Methyl orange
AL	Aminated lignin
LPQAS	Lignin-phosphate quaternary ammonium salt
SLS	Sodium lignosulphonates
PFI	Perfluorinated ionomer
HIL	Hole injection layer
SL	Lignosulfonate
ASL	Alkyl chain cross-linked lignosulfonate
PCE	Photoconversion efficiency
GSL	Grafted sulfonated acetone–formaldehyde lignin
G	Guaiacyl
S	Syringyl
H	*p*-Hydroxybenzaldehyde
SAF	Sulfonated acetone–formaldehyde
MWNTs	Multi-walled carbon nanotubes
AL	Alkali lignin
HTL	Hole transport layer
HTM	Hole transport material
HEL	Hole extracting layer
$PC_{71}BM$	[6,6]-Phenyl-C_{71}-butyric acid methyl ester
PTB7	Poly[[4,8-bis[(2-ethylhexyl)oxy]benzo[1,2-b:4,5-*b'*]dithiophene-2,6-diyl][3-fluoro-2-[(2-ethylhexyl)carbonyl]thieno[3,4-*b*]thiophenediyl]]
Th	Benzodithiophene
J_{SC}	Short-circuit current density
V_{OC}	Open-circuit voltage
FF	Fill factor

PFN	Poly[(9,9-bis(3′-(*N*,*N*-dimethylamino) propyl)-2,7-fluorene)-alt-2,7-(9,9-dioctylfluorene)]
[Ir(ppy)2(dtbbpy)]PF6	4,4′-Di-*tert*-butyl-2,2′-bipyridine)bis[(2-pyridinyl)phenyl] iridium(III) hexafluorophosphate

1 Introduction

The chemical industry principally depends on fossil resources for the manufacturing of carbon-based compounds. However, dwindling supply of conventional fuels and the search for alternative raw materials for chemical production has made biomass an attractive resource that has significant potential for the production of chemicals, fuels, and materials, paving the way for a sustainable future [1]. Lignin is a major fraction of biomass besides cellulose and hemicellulose that accounts for 40% of the total lignocellulosic biomass energy. However, little attention has been paid to the valorization of lignin due to its complex nature [2]. Recently, works have been reported for the conversion of lignin into value-added chemicals like aromatics [3], low molecular weight hydrocarbons, and fuel [4] through depolymerization reactions and gasification, respectively.

The term “lignin” is devised from the Latin word “lignum”, which means wood [5]. Lignin accounts for about 15–30% of the total biomass content in plants, annually about 150 billion tons of lignin is produced by plants, which make it the most abundantly available natural polymer next to cellulose. It stores about 0.082% (3000 EJ year^{-1}) of all the solar radiation intercepted the earth surface, which accounts for approximately 5.4 times the present global energy consumption rate. With the empirical formula of $C_{31}H_{34}O_{11}$, lignin contains about 95 billion tons of the carbon in the earth crust, which illustrates the unexploited high carbon energy reserve [6].

Lignin has an extensively branched three-dimensional chemical structure with various functional moieties such as carboxyl (COOH), carbonyl (C=O), and methoxy (CH_3O), respectively. It is a macromolecule made up of repeating phenyl propane-based monolignols subunits, which are coniferyl alcohol (G), sinapyl alcohol (S), and low amounts of *p*-coumaryl alcohol (H) that take part in lignin formation (Fig. 1). Common linkages found in heterogeneous, high molecular weight lignin are β-O-4, α-O-4, β-5, β–β, 5-5′, 4-O-5, β-1′ [7]. The percentage content of monolignol subunits varies among different plant species. Similar to the monomers content, there is also a variation in the percentage of linkages with respect to plant species [8]. The variation in linkages and monolignols content with plant species make the actual structure determination of lignin rather difficult [9].

Among all the fractions of biomass, lignin has been comparatively underutilized attributed to its complex structure, recalcitrant nature, and heterogeneity that make its valorization relatively challenging. Lignin comprises 30% of all the organic carbon stockpiled in the biosphere [10, 11]. Moreover, the pulp and paper as well as the bioethanol industry produce copious amounts of lignin as a side product that is mainly exploited for power and heat generation via combustion. Out of 50 million tons of lignin produced by the pulp and paper industry in 2010, only 2% has been

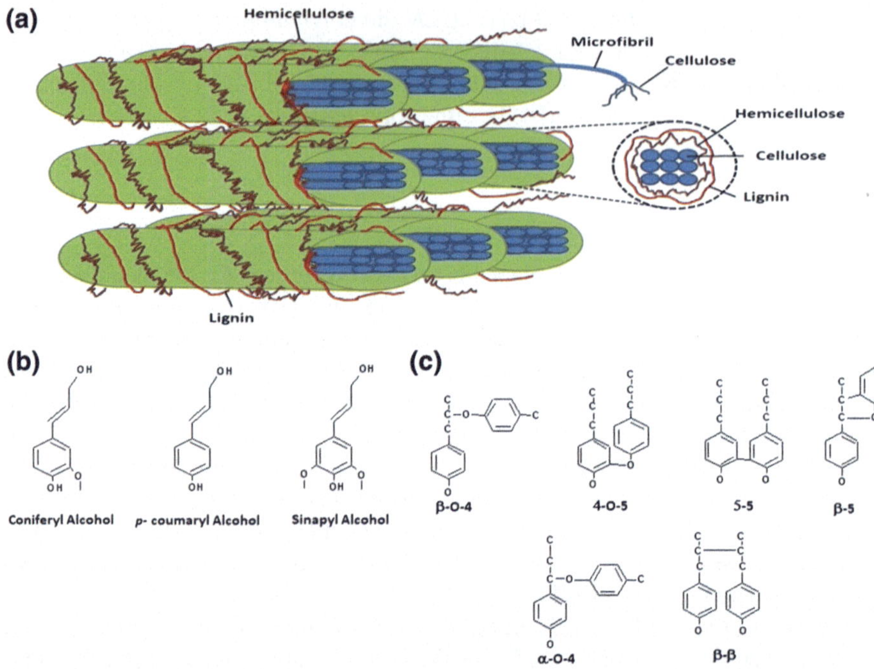

Fig. 1 **a** Plant cell wall structure and microfibril cross section (strands of cellulose molecules embedded in a matrix of hemicellulose and lignin) [12]. **b** Monolignols units in lignin [12]. **c** Common linkages found in lignin [13]

utilized for the production of chemicals, while a high fraction is just burned as low-quality fuel [9]. However, the high aromaticity of lignin makes it a potential precursor for the production of a number of chemicals. In the context of sustainability and economic viability, lignin valorization may play a key role. Therefore, there is a need to concentrate on developing efficient, cost-effective, and green methods for the valorization of lignin [11].

Along with the generation of value-added chemicals, application of lignin for manufacturing of high-performance materials with tunable structure and properties such as high-value polymeric materials or supercapacitors to adsorbents and from biomedical application to electrode materials [14, 15] is progressively becoming a focus of current research [9]. In 2014, the overall lignin market around the globe was valued at approximately US$775 million, and it is anticipated that in 2020 it will reach approximately US$900 million, consistent with an annual average growth rate of 2.5% within 2015 and 2020 [16]. A moderately slow growth rate originated from the impediments in handling of lignin due to its complex nature as well as from the isolation methods employed that utilizes lignin as a fuel [17].

The basic aim of this review is to summarize the recent progress in lignin-based materials for their application in heterogeneous photocatalysis (2013–2017) and photovoltaics (2015–2016). Photocatalysis and photovoltaics are the two major routes for the exploitation of solar energy. Moreover, photocatalysis is a key route, specifically for the degradation of contaminants in water. Recently, some lignin-based composites have shown high activity for the photocatalytic environmental remediation processes such as desulphurization and degradation of organic dyes [18–20]. Organic photovoltaic cells have gained much attention over the last few decades due to their potential application in reducing energy and environmental impact caused by the increasing combustion of fossil fuels. Substantial efforts have been made towards understanding the mechanism of photovoltaic cells together with modifying chemical structural motifs and device structure leading to the enhancement of power conversion efficiency from 4% to over 25% using silicon. Although the energy conversion efficiency of the conventional materials like crystalline Si is high (25%), these materials also have high manufacturing and installation costs [21]. Another critical aspect in photovoltaic devices is the anode interface, where hole extraction and hole injection takes place; a hole conducting polymer is required for this function. PEDOT:PSS is most commonly used conducting polymer applied in anode interfaces. However, it has several drawbacks such as its acidic nature that induces corrosion and variable conductivities due to its microstructural and electrical inhomogeneity. Lignosulphonates have recently been introduced as dopant with tunable conductivities and work function to modify the anode interface [22]. Application of lignin in the field of photovoltaics for the development of organic electronic devices is a sustainable approach towards electricity generation. Nevertheless, several articles have been published on carbon-based hybrid materials for photocatalysis, specifically on the degradation of organic contaminants in water [23], but insufficient information is available on the synthesis of lignin-based composite materials and their limitations and applications in the field of photocatalysis and photovoltaics.

2 Native Versus Processed Lignin

2.1 Sources

Lignocellulosic biomass is the natural source of lignin. Both woody and non-woody biomass resources contain significant amounts of lignin [24]. Lignin along with cellulose and hemicellulose is a principal constituent of a plant's cell wall. The key function of lignin in the cell wall is to provide rigidity by reinforcing strength of crystalline cellulose and middle lamella that enable the erect growth of plants [9]. The content of lignin varies considerably among different plant species, and typically decreases from softwood to hardwood to grasses [24]. Furthermore, approximately 50 million tons of lignin is produced annually by chemical processing of pulp in bioethanol refineries and paper industry [17, 25].

2.2 Classification of Lignin

Lignin is categorized into the following three major classes based on its origin [24]

- Guaiacyl lignin contains significantly high concentrations of coniferyl alcohol with the G:S:H ratio of 90:2:8. It is also named softwood lignin, mainly derived from coniferous trees.
- Guaiacyl–syringyl lignin contains significant amounts of sinapyl alcohol in addition to coniferyl alcohol. It is also known as hardwood lignin, principally found in deciduous trees and shrubs.
- Guaiacyl syringyl *p*-hydroxybenzaldehyde lignin contains a significant proportion of *p*-hydroxybenzaldehyde, approximately 30% in combination with other phenyl propane subunits. Lignin derived from monocotyledons falls into this category.

2.3 Extraction of Lignin

The isolation of lignin from lignocellulosic biomass is carried out under different conditions where polymeric lignin is chemically degraded to low molecular weight fragments, with different physicochemical properties. In addition to the source, composition and properties of the isolated lignin vary depending on the method of extraction. Generally, acid- or base-catalyzed reactions are commonly used in the extraction and depolymerization of lignin [24]. At an industrial scale, four methods are generally employed for lignin extraction that can be further categorized into two classes based on the presence of sulfur. Sulphur-containing lignin is extracted through the sulfite and Kraft processes. On the other hand, soda and organosolv processes are applied for the isolation of non-sulphur-containing lignin [9].

2.3.1 Sulfite Process

Sulfite-pulping process is a commonly used method for the production of commercial lignin. Different concentrations of sulfite or bisulfite salts of ammonium, magnesium, sodium, or calcium are used in an aqueous solution within the pH range of 1–13.5 [26]. The reaction temperature is usually maintained between 140 and 160 °C [27]. Delignification in sulfite pulping process involves the sulfonation of the aliphatic chain of the lignin via cleavage of α-O-4-ether. The lignosulfonates produced through the sulfite process are water soluble and easily dissolved in pulping liquor in an aqueous media [9].

2.3.2 Soda Process

The first pulping process applied for the isolation of lignin was the soda process, introduced by Watt and Burgess in 1854 [28]. The soda process involves heating of the biomass in an alkaline aqueous solution (sodium hydroxide solution) at a temperature of around 160 °C [9]. The reaction proceeds with the protonation of phenolic hydroxyl moieties of lignin with simultaneous cleavage of α-O-4 and β-O-4 bonds [17]. The lignin produced through the soda process is soluble in water and upon acidification it is isolated from pulp liquor through precipitation reaction [9].

2.3.3 Kraft Process

The most commonly used method for pulping is the Kraft process, which produces sulphur-containing lignin. Large fractions (98%) of lignin produced through the Kraft process are utilized for energy purposes through combustion, and merely minor fractions (2%) are used for material or chemical synthesis [9, 17]. The Kraft process is supposed to be an advancement or progression of the soda process, as it involves the heating of pulping liquor with sodium sulfide in addition to sodium hydroxide at a temperature between 150 and 180 °C [9, 26]. The process of lignin depolymerization in the Kraft process is the same as in the soda process; the reaction proceed with the cleavage of α-O-4 and β-O-4 ether linkages and results in soluble fragments of lignin [17]. A minor fraction of resultant lignin is sulphated due to the presence of anions of hydrosulfide. A large fraction of lignin produced is sulphate-free and isolated via acidification and precipitation method [9, 29].

2.3.4 Organosolv Process

The most recent process used for the extraction of lignin at an industrial scale is the organosolv process. This technique involves heating the biomass with a mixture of organic solvents for isolating lignin. Polar organic solvents such as methanol, acetone, ethanol, acetic acid, and formic acid are commonly used for the extraction process [9, 17]. The nature of the solvent used significantly determines the structure and properties of the isolated lignin [9].

2.4 Properties of Lignin

Native lignin is colorless, but acid–alkali treatment changes its color to dark brown [5]. Physical and chemical properties of the lignin vary with the extraction procedure and the monolignols content [9], as shown in Table 1. The functional moieties of lignin like carboxylic, phenolic hydroxyls, methoxy, aliphatic and carbonyl groups depend on the monomeric linkages. Moreover, these moieties significantly contribute to the chemical modification of lignin [9]. The molecular weight of lignin ranges between 1000 and 20,000 g mol^{-1}. As lignin constantly degrades during the extraction process, it is therefore difficult to predict the degree of polymerization attributed to random repetition of subunits [25].

The glass transition temperature (T_g) of lignin differs with moisture content, molecular weight, extraction method, cross-link density, and measurement method. T_g generally increases with the increase in molecular weight, based on the structure and fragments' molecular mass the T_g of lignin falls between 70 and 170 °C [30]. Analogous discrepancies have been observed for the decomposition temperature of lignin, lignin source, extraction method, and measurement techniques influence lignin decomposition processes [31]. Distinct decomposition

Table 1 Summary of characteristic properties of lignin

Type of lignin	Kraft	Lignosulfonate	Soda	Organosolv	References
Structure	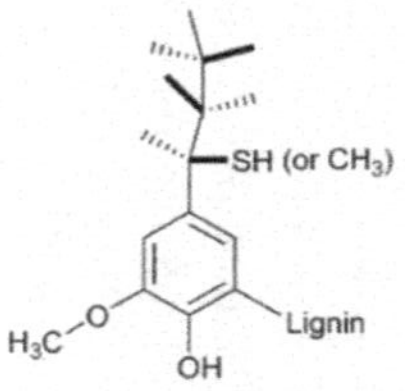	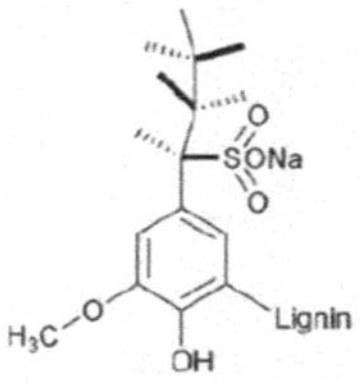	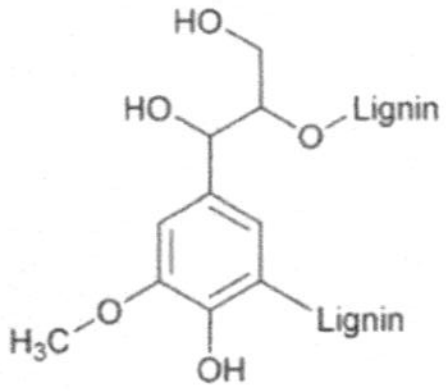	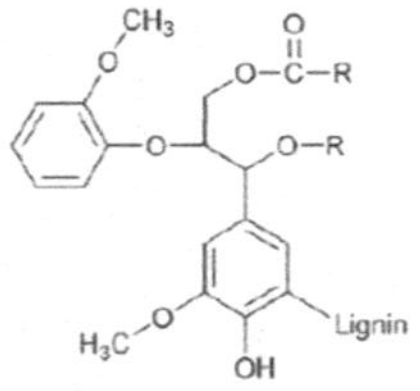	[80]
Separation methods	Precipitation (pH change)–ultrafiltration	Ultrafiltration	Precipitation (pH change)–ultrafiltration	Precipitation (addition of non-solvent) Dissolved air flotation	[80]
Active extracting agent	NaOH, Na_2S	H^+, HSO_3^-	NaOH	40–60 wt% aqueous ethanol	[26, 111]
pH of isolation medium	13–14	1–2 (acid bisulfite) 3–5 (bisulfite)	13–14	–	[26, 111]
Temp. (°C)	155–175	125–145 (acid bisulfite) 150–170 (bisulfite)	155–175	180–210	[26, 111]
Sulphur (%)	1.0–3.0	3.5–8.0	0	0	[80]
Nitrogen (%)	0.05	0.02	0.2–1.0	0–0.3	[80]
Molecular weight ($\times 10^3$ g mol^{-1})	1.5–5 (up to 25)	1–50 (up to 150)	0.8–3 (up to 15)	0.5–5	[80]
Solubility	Alkali and some organic solvents	Water	Alkali	Broad range of organic solvents	[80]
T_g (°C)	140–150	130	140	90–110	[80]
T_d (°C)	340–370	250–260	360–370	390–400	[80]
Scale (ktpa)	60	1000	5–10	~ 3	[112]

of lignin comparable to other biomass components have been observed in the range of 360–480 °C via thermogravimetry [32].

Furthermore, lignin has many valuable properties such as antimicrobial and antioxidant nature [33], mechano-thermal stability [34], and blending properties [5] that make it a potential candidate as a constituent of composites. In addition to its benefits, it has some limitations, such as abrupt changes in its properties with change in moisture content. Bleaching of produced radicals and reaction with atmospheric oxygen is the major drawback of lignin for its application in material synthesis [35]. These limitations may be overcome by chemical modification introduction of lignin into composite materials [36].

3 General Applications of Lignin

3.1 Synthesis of Materials from Lignin

Depletion of fossil resources and associated environmental problems has resulted in exploration of renewable resources for the synthesis of materials. Lignin as a renewable biopolymer has a big role to play in this field due its natural abundance. Lignin is a potential candidate for the fabrication of composite materials credited to it several remarkable features such as biodegradability, antioxidant activity, antimicrobial activity, and reinforcing properties etc. [37]. Recently composites based on lignin gained a lot of attention due to its property to be used as a reinforcing material for the synthesis of high-performance composites. Morandim-Giannetti et al. investigated the application of lignin as additive for the fabrication of polypropylene–coir composites [38]. The tensile strength of the composite has not been affected to a large extent by the addition of lignin, though initial degradation temperature and oxidation induction time of the composite has been increased due to the presence of lignin. Moreover, the multifunctional nature of lignin makes it a reactive component for the manufacturing of resins and polymer-based materials. This goal is achieved either through chemical modification of lignin through esterification, phenolation and oxypropylation reaction, or through partial substitution of traditional materials by lignin. Similarly, lignin has shown promising application for the preparation of phenol–formaldehyde resins, epoxy resins, polyurethanes, and graft copolymers [17].

Lignosulfonate and kraft lignin in combination with activated carbon have been used in direct carbon fuel cells for the production of electricity [39]. Lignosulfonate showed better performance than Kraft lignin in direct carbon fuel cells due to high hydrophilicity of lignosulfonate. High wettability of lignosulfonate also enhances electrochemical reactivity and electrical conductivity in fuel cells. Using lignosulfonate-activated carbon, the maximum power density reaches 25 mW cm^{-2}, while Kraft lignin-activated carbon shows a power density of 12 mW cm^{-2} [40]. Furthermore, a study reported the production of highly nanoporous carbon for supercapacitors application using low-cost renewable lignin as precursor [41]. The better control over the surface functional groups, pore structure, and electrical conductivity of lignin-based carbon materials enhances the electrocapacitive performance of electrode for supercapacitors [14]. However, there are certain challenges associated with the lignin for its use in energy storage devices such as low electrical conductivity that make the active sites of lignin electrochemically inaccessible. The other factor is the solubility of some types of lignin like lignosulfonates in an aqueous media that may result in degradation of the active material in an electrochemical device [42]. Designing hybrid capacitors using lignin in combination with metal oxides and conductive polymers is a futuristic approach to improve the electrode capacitance of supercapacitors.

4 Overview of Carbon-Based Materials in Photocatalysis and Photovoltaics

Recently, tremendous attention has been paid to the development of porous carbon-based materials derived from environmentally friendly renewable biomass resources [43]. Carbon-based materials derived from biomass such as wood, cellulose, lignin, hemicellulose, and biochar widely used as template for semiconductors in photocatalytic applications [44]. Introducing carbon-based materials as doping agent plays a significant role in the modification of photocatalyst by improving visible light responsive performance of the photocatalyst [45, 46]. During photocatalysis, the high surface area, electrical conductivity, and porosity of carbon materials may increase the adsorption and modify the mechanism of photochemical reaction. This synergistic effect induced by carbon materials enhances the photocatalytic degradation of environmental pollutants, photocatalytic production of H_2, and photocurrent generation attributed to their high electroconductivity [47].

Various efforts have been made for the utilization of carbon-based materials like activated carbon and biochar for fabrication of composites. Activated carbon derived from biomass is considered to be a potential support for the photocatalytic material attributed to its ability to improve the interface charge transfer rate and reduce the electron hole recombination rate [48]. It has shown promising properties as support for TiO_2 in case of gas and water remediation. Moreover, the heterojunction formed between the components leads to inoculation of electrons from activated carbon to TiO_2 [49]. Similarly, TiO_2 in combination with biochar (a porous solid-rich byproduct of thermal decomposition of organic waste) derived from Miscanthus straw pellets and soft wood pellets has shown enhanced photocatalytic activity for selective oxidation of methanol to methyl methanoate as well as for phenol degradation [50].

Among various biomass fractions, cellulose has been extensively utilized for the fabrication of hybrid photocatalyst [20, 51]. Composite films of TiO_2 show high photocatalytic degradation efficiency for concentrated phenol accredited to the void formation in TiO_2 assembly and their immobilization by hydroxyl groups [52]. Carbon-based photocatalyst composites with well-defined physicochemical characteristics such as specific surface area, pore volume, microstructure, and solubility etc., may enhance the photocatalytic system that ultimately pave the way to understand the reaction mechanism of the material synthesis based on its structure and composition [44]. Still, further investigation is required to understand the reaction kinetics and mechanism, interphase interaction, and leaching of components from composites.

Photovoltaics is another foremost mode for the utilization of solar energy for power generation. Due to high energy demand, there is a dire need to fabricate energy conversion devices utilizing renewable resource in accordance with the principles of green chemistry. About 1.8×10^{11} MW power is intercepted by the earth from the sun, which is quite greater than the current rate of overall energy consumption [53]. The field of photovoltaics offers great potential for the utilization of renewable solar power by converting it into electricity. Steady progress has been achieved in the field of photovoltaics in order to increase the power conversion efficiencies and lower the cost of production by using organic molecule or polymers in chemical design [54].

Photovoltaic technologies are required to be economically viable and environmentally friendly. The current photovoltaic devices are mostly centered on the use of toxic and expensive inorganic chemicals [55] such as CdTe, GaAs, $CuIn_xGa_{1-x}Se_2$ [56]. Semiconductors derived from organic polymer are potential alternative for inorganic semiconductive materials in the field of photovoltaics. Low cost, renewability, and conjugated structure are the most important features of organic polymers for their application in photovoltaics [57].

The importance of natural polymers for the fabrication of photovoltaic devices can be understood by the working principle of photovoltaics. A photovoltaic cell is composed of a layered structure in which the layer that absorbs light is packed between two different types of electrode, as depicted in Fig. 2. One of the electrodes is made up of indium–tin-oxide ITO, while the other electrode is often composed of metals like aluminum, calcium-magnesium, and gold, etc. Upon exposure to light, electrons residing in the highest occupied molecular orbital (HOMO) absorbing a certain wavelength of radiation and shifted to the lowest unoccupied molecular orbital (LUMO) result in the formation of exciton. Free electrons and holes generated through exciton dissociation move towards Al and ITO respectively, as depicted in Fig. 3. Movement of electrons in the external circuit generates an electric current. Asymmetrical electrodes' ionization energy or work functions provide an electric field required to avoid recombination. This asymmetry is responsible for the flow of electrons from the region of low work function to the region of high work function, the process is known as rectification. In case the semiconductors are based

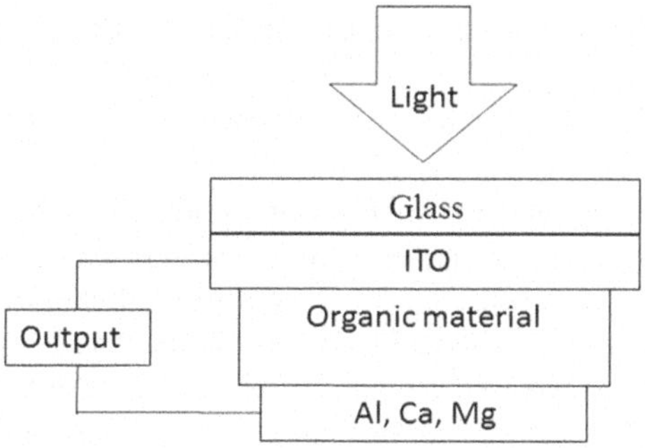

Fig. 2 Basic construction of a photovoltaic cell with typical electrode materials. An organic material is sandwiched between two electrodes [57]

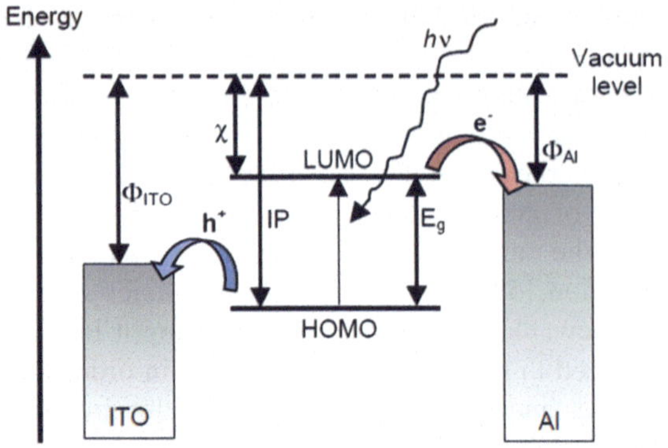

Fig. 3 Energy levels and light harvesting. Upon irradiation, an electron is promoted to the LUMO leaving a hole behind in the HOMO. Electrons are collected at the Al electrode and holes at the ITO electrode. Φ, workfunction; χ, electron affinity; IP, ionization potential; E_g, optical band gap [57]

on inorganic materials, interaction of HOMOs and LUMOs of adjacent molecules result in a conduction and valence band. While in case of semiconductors based on organic dyes instead of bands, charge transfer takes place between localized states through hopping. Polymers have a conjugated structure placed in between organic dyes and inorganic semiconductors [57].

Nanocellulose has been used as substrate for the fabrication of photovoltaic and solar cells [58]. Carboxymethylated nanocellulose paper has been applied as a substrate in a device consisting of poly(3,4-ethylenedioxythiophene):poly(styrene sulfonate) (PEDOT:PSS), poly(3-hexylthiophene) (P3HT), and [6]-phenyl-C61-butyric acid methyl ester (PCBM). The photo conversion efficiency of the device was not very high (0.2%), which was attributed to the high resistance of ITO [59].

The dye-sensitized solar cell is a type of photovoltaic cell that works on the principle of oxidation–reduction reaction with the capability of maximizing the power conversion efficiency up to 11% [60]. The microcrystalline cellulose applied as gel electrolyte for the dye-sensitized solar cell. The photovoltaic efficiency of dye-sensitized solar cell based on cellulose gel can be optimized by regulating cellulose concentration and ionic liquid volume ratio. The photoconversion efficiency reaches 3.33% by using a gel composed of LiI (2 wt%), iodine (10 wt% of the whole weight of iodide), microcellulose (5 wt%), 4-*tert*-butylpyridine (10 wt%), and 1-methyl-3-propylimidazolium/1-ethyl-3-methylimidazolium thiocyanate (50/50 volume percentage) when stimulated more than 8 h under solar irradiation [61].

4.1 Lignin-Based Composites in Photocatalysis

4.1.1 Preparation Techniques of Lignin-Based Composite Photocatalyst

There are a number of methods employed for the synthesis of photocatalysts with lignin as a support (Table 2). Depositing or immobilizing a photoactive material on or within the pores of lignin-based supports might offer substantial gains in photocatalysis, especially if the photoactive component can be introduced in an oriented or assembled fashion. There are several methods used for the preparation of lignin-based photocatalyst such as solid-phase synthesis, solvent evaporation method,

Table 2 Potential use of different types of lignin for the synthesis of composite photocatalyst and their preparation methods

Entry no.	Lignin-based composite photocatalyst	Type of lignin	Method	References
1	Nano-ZnO lignin–amine composite	Aminated lignin	Solid-phase method	[81]
2	SLS–CuO/ZnO nanocomposites	Sodium lignosulphonates (SLS)	Solid-phase grinding method	[68]
3	Lignin@TiO_2 composites	Kraft lignin Organosolv lignin Low sulfonate content (LSC) Sodium lignin Sodium lignin without sugars Alkali lignin	Solvent evaporation method	[114]
4	Porous carbon–CeO_2 composites	Sodium lignin sulfonate	Cocalcination method	[20]
5	TiO_2–lignin composite	Alkali lignin	pH assisted precipitation	[63]
6	Lignin-based carbon/ZnO composite	Alkali lignin	One-pot carbonization method	[18]
7	Lignin-based carbon/ZnO nanocomposite	Alkali lignin	One-pot in situ method	[19]
8	Aminated lignin–CuO nanoparticles	Aminated lignin	Solid-phase method	[70]
9	Lignin–TiO_2 mixture	Commercial lignin from nonwoody biomasses like wheat straw and sarkanda grass by soda pulping process using aq. NaOH	Ball mill via dry milling and wet milling	[65]
10	Nano TiO_2–lignin composite	Alkali lignin	Hydrolysis precipitation method	[72]
11	LPQAS–ZnO crystallites	Alkali lignin	One-step precipitation method	[75]
12	SL–ZnO array	Sodium lignosulphonate	Precipitation method	[77]
13	Nano ZnO–AL	Alkali lignin	Solid-state reaction	[113]
14	SLS-functionalized MWNTs/SnO_2 hybrids SLS-functionalized MWNTs/CdS hybrids	Sodium lignosulfonate	Grinding-in situ formation method	[79]

cocalcination, one-pot in situ method, and pH-assisted precipitation (see Table 2 for details).

Precipitation is one of the most frequently applied and cost-efficient methods for the synthesis of photocatalyst composites. More than one layer of photoactive material can be deposited in a homogeneous distribution on the lignin-based support without using costly solvents [62].

Morsella et al. reported the synthesis of TiO_2–lignin composite by coating the TiO_2 nanoparticles with lignin as shell via pH-assisted precipitation (entry 5, Table 2) [63]. The method involves the solubilization of alkali lignin in an alkaline (B) or organic solvent (A) followed by the addition of TiO_2 particles. The resulting solution is sonicated to ensure homogenization and finally TiO_2–lignin clusters precipitated out by decreasing the pH of the solution with the help of acids [63]. The detailed scheme for the synthesis of TiO_2–lignin composite is given in Fig. 4.

Also, facile mechano-chemical processes like ball milling can be employed for the preparation of photocatalyst composites based on lignin [64]. For instance, lignin–TiO_2 composites were prepared (entry 9, Table 2) through dry and wet milling techniques in a ball mill [65]. In this example, lignin and TiO_2 with a mass ratio of 1:1 were milled for 6 h at 120 rpm. The obtained dry milled samples undergo wet milling by the addition of different solvents like water, hexane, or acetone with a solvent mass ratio of 1:2. The obtained composites were filtered and dried around 40 °C before application [65].

Furthermore, the one-pot in situ method has also been used for the preparation of composite photocatalysts. In this method, firstly the core photoactive particles are synthesized, followed by the addition of the coating agent or the template precursor. The reaction mixture is stirred and the resulting composite is washed, dried, and

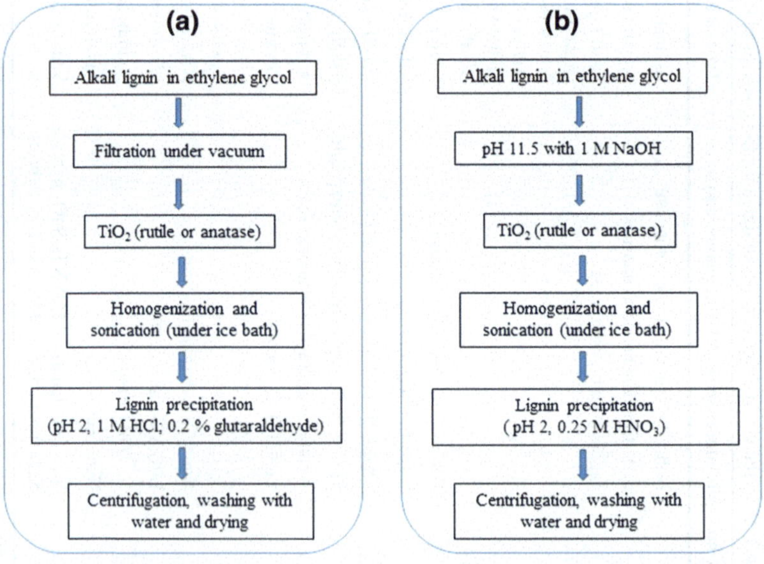

Fig. 4 Detailed protocols adopted for the preparation of the lignin/TiO_2 composite materials [63]

calcinated [66]. A recent study reports the synthesis of lignin-based carbon–ZnO nanocomposites using alkali lignin from pulping liquor and zinc nitrate as a precursor for support and ZnO nanoparticles, respectively [19]. The in situ method is considered to be a low-cost and environmentally friendly technique that can use industrial alkali lignin, while one of the drawbacks of the process is the possibility of the entrapment of impurities between the photoactive core and template [19, 66].

Another method used for the preparation of photocatalytic composites is solid-phase grinding. This synthesis approach involves the deposition or attachment of a substrate on a polymer support by grinding and mixing. After the completion of the reaction, the precipitates obtained is repeatedly washed with solvents to remove excess reagents [67]. Among the materials prepared by this method, nanocomposites of CuO–ZnO were synthesized (entry 2, Table 2) using sodium lignosulphonate as the support and zinc carbonate the precursor for the semiconductor component [68]. One of the advantages of solid-phase synthesis is the easy separation of the reactants from final products by washing and filtration [69]. Similarly, Wang et al. reported the synthesis of a CuO nano-photocatalyst based on aminated lignin by the solid-phase technique [70]. CuO particles were obtained through direct reaction of sodium hydroxide and copper nitrate with aminated lignin.

4.1.2 Chemical Interaction Between Lignin and Semiconductor

The hydroxyl groups along with other functional groups in lignin like carboxyl and carbonyl group can engage in specific interactions with the precursors of functional components such as polymers and photocatalysts during the formation of composites [71, 72]. Recently, lignin has been applied as a template to prepare mesoporous TiO_2 nanoparticles using $TiCl_4$ as precursor [72]. The highly electronegative hydroxyl moieties on the surface of the lignin develop a strong affinity towards electropositive metal ions, as shown in Fig. 5. The positively charged $Ti(OH)_n^{(4-n)+}$ formed during the partial hydrolysis of $TiCl_4$ has affinity for the nucleophilic ligand, resulting in adsorption on the surface of lignin via electrostatic forces of attraction. The adsorbed $Ti(OH)_n^{(4-n)+}$ further hydrolyzes and converts to Ti–(O–lignin)$_4$ over the surface of the lignin. After complete hydrolysis, well-dispersed TiO_2 nanoparticles are formed on the surface of the lignin (Fig. 5) [72, 73].

Amine groups can modify the surface of lignin consequently, increasing the molecular weight as well as the number of active groups on the surface. Synthesis

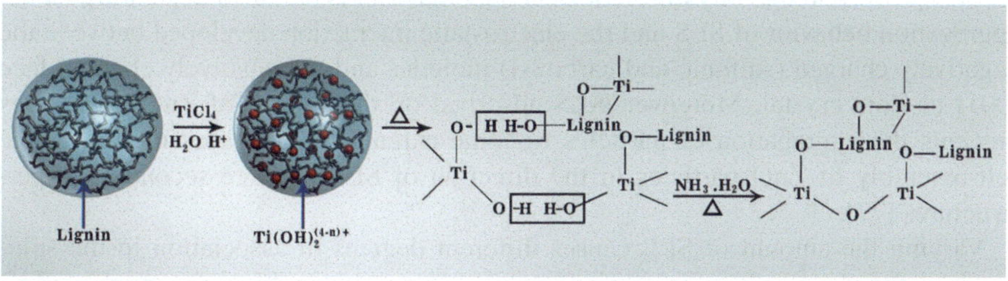

Fig. 5 Formation mechanism of mesoporous TiO_2 with lignin as a template [72]

of porous ZnO nanoparticles using zinc nitrate [$Zn(NO_3)_2$] and sodium oxalate ($Na_2C_2O_4$) and alkali lignin–amine as a template revealed that the ions dissociated from the lignin amine template during the reaction and, attractive electrostatic forces between Zn^{2+} and $C_2O_4^{2-}$ develop. The hydrogen bonding and van der Waal forces of attraction responsible for the direct contact between the ZnO particles has been reduced through steric hindrance caused by lignin amine molecule as template, and thus resulted in lower aggregation and generation of smaller-size particles (average of 15–44 nm) [18]. According to the findings, the native lignin is deficient in positively charged functional groups, leading to poor interaction with ZnO particles [18, 19]. Introduction of positively charged moieties to the alkali lignin through the process of quaternization developed strong affinity with the negatively charged ZnO nanoparticles, which was favorable for the fabrication of lignin–ZnO composite [18].

In another work, lignin–phosphate quaternary ammonium salt (LPQAS) was formed from modifying alkali lignin through a Mannich reaction. LQPAS has been applied as a surfactant for the assembly of ZnO crystallites via precipitation with NaOH as the precipitating agent. During the synthesis, LQPAS dissociates into phosphate ions (negatively charged) and quaternary ammonium ions (positively charged). During a chemical reaction, the positively charged quaternary ammonium ions develop an interaction with the negatively charged OH^- ions [74]. Subsequently, LQPAS acts as a slightly negatively charged surfactant, which results in an interaction with the positively charged face [001] of ZnO particles. This interaction represents the driving force for the assembly of ZnO nanoparticles. Reducing the pH value to 7 results in the deposition of porous ZnO nanoparticles. The size of the mesoporous nanoparticles mainly depends on the ratio of Zn^{2+} and OH^- ions, while the surface area of the ZnO particles is mainly determined by the amount of surfactant molecules rather than the molar ratio of Zn^{2+} and OH^- ions [75].

Lignosulphonates are derivatives of lignin, which can be obtained through the sulfite pulping process and subsequent sulfonation, degradation, and solubilization in water [76]. The functionalization of lignin with various hydrophilic (hydroxyl, sulfonic, and carboxyl groups) and hydrophobic (aliphatic and aromatic groups) moieties results in diverse surface characteristics. Miao et al. applied sodium lignosulphonates (SLS) as template for engineering ZnO nanomaterials [77]. The reaction proceeds with the interaction of positively charged zinc ions with negatively charged hydroxyl ions, resulting in the formation of $[Zn(OH)_4]^{2-}$, which is transformed to $Zn(OH)_2$ in an alkaline medium [74] and finally to crystalline ZnO particles [77]. Morphologically different ZnO nanomaterial is obtained depending on the aggregation behavior of SLS and the electrostatic interaction developed between the negatively charged (sulfonic and carboxyl) moieties and the positively charged face [001] of ZnO crystal. Moreover, SLS adsorbed on the surface of ZnO crystallites prevents the aggregation of particles to some extent and later contributing to the self-assembly of ZnO particles in the direction of SLS produce secondary superstructures [77].

Varying the amount of SLS causes different degrees of association in the solution and thereby resulting in fabrication of different hierarchical structure of ZnO nanoparticles. Lowering the concentration of sodium lignosulphonates causes aggregation attributed to steric repulsions that result in the formation of bars of ZnO clusters. Increasing the concentration of sodium lignosulphonates forms a bi-layer structure called mesh-lamina ZnO. Further increase in concentration changes the structure from bilayer to spherical bilayer, ultimately resulting in quasi-spherical particles of ZnO [77].

SLS have also been used for the surface functionalization of multi-walled carbon nanotubes (MWNTs). Sodium lignosulphonates act as a dispersing agent and are adsorbed on the surface of the MWNTs. Π–Π non-covalent stacking is mainly responsible for interaction between SLS and MWNTs. SLS is amphiphilic in nature, but dominated by hydrophobic groups with ether and C–C bonds. Other than hydrophobic linkages, Π–Π stacking interactions are also responsible for the adsorption of sodium lignosulphonate on the MWNTs. The steric repulsion caused by sodium lignosulphonate helps in minimizing the van der Waals forces of attraction at the surface contact. The anionic groups on the surface of the sodium lignosulphonate extrude outward to reduce the electrostatic forces of repulsion. The similar charges on the sodium lignosulphonates make MWNTs extremely hydrophilic and, consequently, responsible for their solubility and stability in aqueous medium [78]. SLS-functionalized MWNTs serves as a potential support for the fabrication of quantum dot hybrids. The uniform deposition of SnO_2 and CdS nanoparticles on the SLS-functionalized MWNTs template comprises the interaction between the positively charged ions of the nanoparticles (Sn^{4+} and Cd^{2+}) and the negatively charged groups of sodium lignosulphonates, as shown in Fig. 6. The SLS-functionalized MWNTs is an outstanding support for the fabrication of quantum dot hybrids ensuring the stability of over 6 months at ambient temperature [79].

4.1.3 Applications of Lignin-Based Composites in Photocatalysis

The interest in applying lignin in material engineering is increasing [9, 80] specifically in the case of preparing composite materials for photocatalytic applications [72]. The application of lignin as photocatalyst support allows controlling the size of the particles and to obtain a uniform distribution of the particles of the photocatalyst [70]. Additionally, lignin is derived from renewable biomass resource as a byproduct from the pulp and paper industry [81]; consequently, the application of lignin in material engineering will reduce the cost as well as help to fabricate environmentally

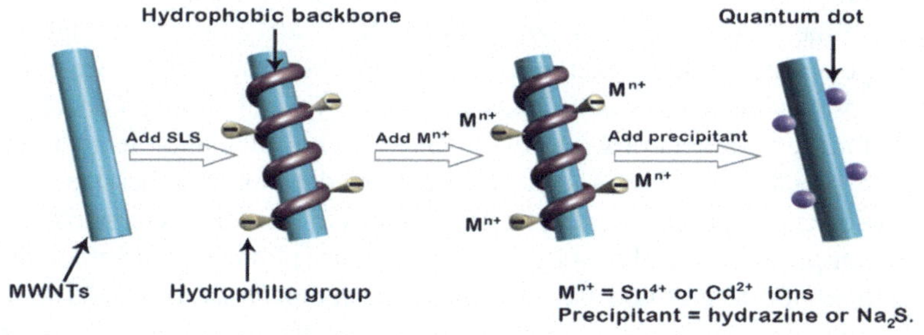

Fig. 6 Mechanism for the Functionalization of MWNTs and quantum dot decoration [79]

compatible composites. Hence, it is considered to be a promising support for the synthesis of composite photocatalyst. Recently, some attempts have been made to develop lignin-based photoactive composites to improve the efficiency of the photocatalytic process [18, 19]. In order to overcome the limitations of the photocatalytic reaction, there is a need to understand the basic photophysical and photochemical mechanism of the process. The interaction of photoactive material and light is an important factor in determining the efficiency of the photocatalytic reaction. Nevertheless, the redox reaction can be impeded by the large band gap between valence band and conduction band in the semiconductors [82, 83]. Therefore, high-energy UV radiation is required for carrying out the photochemical reaction, which is the principal constraint associated in upscaling of a photocatalytic conversion [82, 84].

Lignins form a dark-colored solution in most of the solvents, which makes it photocatalytically inactive, which may reduce the efficiency of the photocatalyst [85]. Nevertheless, it has been observed that lignin can be applied as template for the assembly of stable hybrid photocatalysts. The interaction between the components in the case of composites plays a significant role in determining the photocatalytic performance [86]. The properties of the composites are diverse due to the reinforcing synergy between the components of the hybrid photocatalyst. Consequently, this enhances the transfer and utilization efficacy of photogenerated electrons and intensify the separation of charges that synergistically boost photocatalysis by overcoming the chances of recombination of electrons and holes [82, 83, 86, 87]. Hence, the photocatalytic efficiency could be improved by fabricating a composite of photoactive material and appropriate support. Additionally, the chemical and physical stability of the composite photocatalyst mainly depends on the nature of the support [86]. The use of support stabilizes the textural properties upon thermal treatment, which enhances the disseminations of active sites and usually enhances the catalytic activity of pure oxides [88, 89].

Photocatalysts based on lignin semiconductor composites have great potential for the remediation of contaminated water and have received much attention in recent years [19, 72]. The use of lignin as a photocatalyst support would endorse distinctive functionality with excellent physicochemical properties for specific applications accredited to the interaction between semiconductor and support [90]. The following section provides an overview of recent applications of lignin-based composites for contaminant degradation. As reported in the recent literature, the assimilation between lignin and metal oxides, such as ZnO, TiO_2, or CuO, increases the degradation of pollutant compared to pristine metal oxide alone [19, 72, 77].

Lignin-amine (LA) mesoporous zinc oxide hybrid catalyst depicted high sunlight photocatalytic activity. Introduction of amine groups to the lignin via amination reaction improves the surface activity as well as the flocculation and decolorization efficiency for the treatment of wastewater [81, 91]. The calcination temperature during catalyst preparation plays a significant role in determining the size, morphology, microstructure, and photocatalytic performance of the ZnO nanophotocatalysts. ZnO–LA composite calcined at 400 °C exhibits higher photocatalytic efficiency than those calcined at 500 and 600 °C. Increasing the temperature from 400 to 600 °C resulted in an increase in size and decrease in specific surface area of the photocatalyst. In addition, doping of LA with ZnO precursor also contributes to acquire smaller size and high specific surface area of ZnO nanoparticles by preventing the agglomeration of ZnO particles. ZnO–LA annealed at 400 °C exhibited photocatalytic degradation efficiency of 99.2 and 96.4% for methyl orange (20 mg l^{-1}) under UV light irradiation (1 h) and under solar radiation (6 h), respectively. The solar photocatalytic performance of ZnO–LA is almost the same as that of TiO_2 (P25) [81]. During a photochemical reaction, the water adsorbed on the surface is oxidized by holes to $^{\cdot}OH$ radicals, while O_2 adsorbed on the surface of ZnO is reduced by the electrons. The hydroxyl radical and superoxide ion formed result in mineralization of methyl orange [90].

In addition to aminated lignin, sodium lingosulphonates have also been used for the fabrication of nano ZnO photocatalyst. Different ZnO morphologies such as nanoparticle-bar, nanomesh-lamina, and quasi-nanosphere were acquired [77], depending on the concentration and aggregation of sulphonated lignin and electrostatic interaction between the sulphonated lignin and ZnO crystals. The lignosulphonates–nanomesh ZnO composite displayed 100% degradation efficiency for methylene blue (5 ppm) under UV irradiation power of 12 W within 90 min (see entry 7, Table 3 for effect of irradiation power on degradation efficiency) [77]. In the course of photocatalytic degradation, holes generated react with either hydroxyl ions or adsorbed water molecules and form hydroxyl radicals. Simultaneously, the interaction of O_2 molecules and electrons produce superoxide anion radicals. Finally, the interaction of superoxide anion radicals and hydroxyl radicals with methylene blue results in its degradation [77, 92]. Compared to pristine ZnO particles, lignosulphonate-doped ZnO particles exhibited high photocatalytic efficiency due to improve surface state, high specific surface area with more hydroxyl groups, and smaller band gap of ZnO particles [93].

Moreover, quaternized alkali lignin–ZnO hybrid have been applied for the degradation of Rhodamine B (Rh B) and methyl orange (MO). Yet again, efficiency of composite is far better than the pure ZnO. After the light irradiation (see entry 3, Table 3 for light source details) for 30 min, methyl orange (15 mg l^{-1}), was completely degraded by the composite while pure ZnO showed 75.3% degradation efficiency even after 50 min. Quaternized alkali lignin–ZnO exhibited lower degradation efficiency for 15 mg l^{-1} Rhodamine B (79.2%) compared to methyl orange but was higher than that of pure ZnO (31.1%). Methyl orange mainly degraded by holes, accredited to the negative charge of the dye pushed towards ZnO as shown in Fig. 7. Whereas, the degradation of Rhodamine B is driven by $^{\cdot}O_2^-$ and $^{\cdot}OH$ radicals due to their strong oxidative abilities [18]. The applied composites were rather stable and no obvious decrease in photodegradation efficiency was observed in three successive recycling tests [18]. Thus, doping with carbon materials improved the efficiency of ZnO by inhibiting the photocorrosion of ZnO [19] and overcoming the limitation associated with the pure ZnO as photocatalyst such as charge separation and low quantum efficiency [94].

Similarly, porous carbon-based CeO_2 composites were fabricated applying lignin as support. Lignin decomposition contribute to the porosity of the template and ensure the uniform growth of CeO_2 nanorods to carbon–CeO_2 composite (Fig. 8).

The photocatalytic activity of carbon–CeO_2 hybrid was determined for the desulfurization of SO_2, which is extremely injurious to human health as well as for the

Table 3 Applications of lignin-based composites photocatalyst in photocatalytic degradation of organic substances

Entry no.	Type of composite	Contaminant	Light source	Temperature (°C)	Photocatalytic activity (% degradation/conversion)	Reaction time (min)	References
1	Nano-ZnO–LA	Methyl orange	UV light (300 W)	25	99.20	60	[81]
			Solar radiation	28–36	96.40	360	
2	SLS-CuO/ZnO nanocomposites	Rhodamine B	Visible light	25	91.50	240	[68]
		Congo Red	500 W Xe lamp		74.30		
3	Lignin-based carbon/ZnO composite	Methyl orange	Solar light (500-W Xe lamp)	15	99.90	30	[18]
		Rhodamine B			79.20	50	
4	Lignin-based carbon/ZnO nanocomposite	Methyl orange	Solar light (500-W Xe lamp)	15	98.90	30	[19]
5	AL–CuO nanoparticles	Methylene blue	UV light	25	97.80	90	[70]
		Methyl orange			66.70		
6	Nano TiO_2–lignin composite	Phenol	UV light (8-W mercury lamp)	25	97.90	120	[72]
7	SL–ZnO nanomesh lamina	Methylene blue	UV irradiation (WFH-203)				[77]
			12 W	25	100	90	
			8 W		96	120	
			5 W		88	120	
			2 W		80	120	
8	Nano ZnO–AL	Methyl orange	UV light	25	98	90	[113]

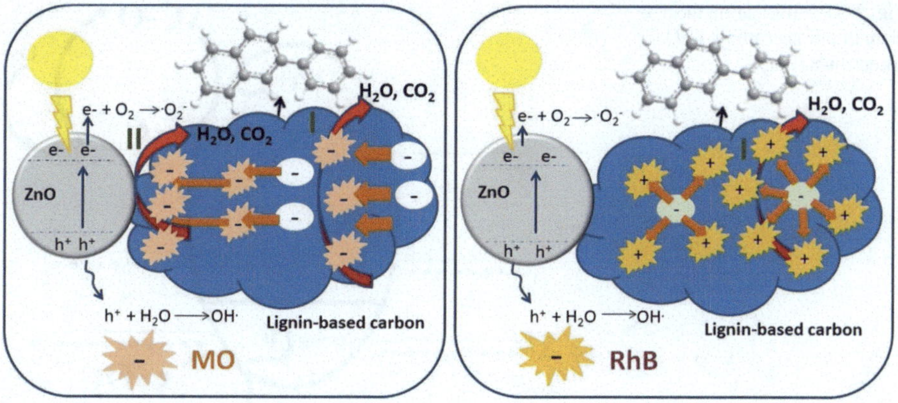

Fig. 7 Photocatalytic mechanism for the degradation of MO (**a**) and RhB (**b**) over the LC–ZnO composite [18]

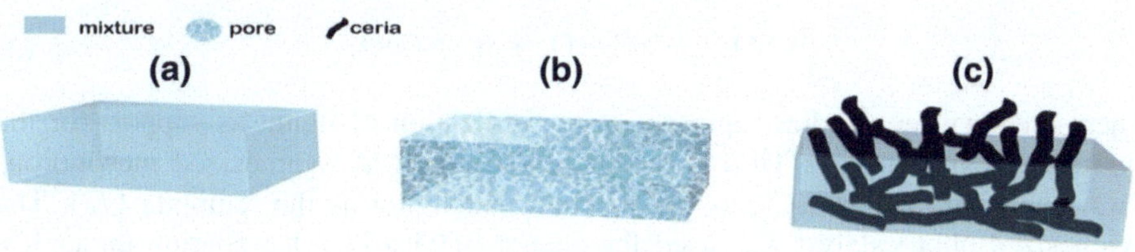

Fig. 8 Formation mechanism of porous carbon–CeO_2 composite. **a** Blend of lignin and cerium nitrate. **b** Partially decomposed porous lignin. **c** Porous carbon CeO_2 composite [20]

environment. CeO_2 plays a significant role in trapping SO_2 molecules, oxygen storage, and oxidizing ability of CeO_2 results in chemisorption of SO_2. Lignin support not only takes part in physisorption but is also involved in photocatalytic conversion of SO_2 [20, 95]. The reaction initiated with physical adsorption of SO_2, and then chemisorption took place simultaneously due to the oxidizing and oxygen storage properties of CeO_2. Porous carbon CeO_2 hybrid exhibited high desulfurization efficiency and significantly oxidize the adsorbed SO_2 with conversion ratio of 51.8%. Carbon–CeO_2 showed improved photocatalytic performance compared to pristine CeO_2, which is attributed to the possible involvement of carbon during a reaction.

The reaction mechanism for the desulfurization of SO_2 is as follows (Reactions 1–3) [21]

$$(1/2)xO_2 + xC \rightarrow xC(O) \tag{1}$$

The reaction proceeds with the formation of active site C(O) that later is converted to carbon–oxygen complex, which provides oxygen for the photocatalytic reaction. SO_2 reacts with the holes and O_2 from the CeO_2, and is thereby converted to SO_3 (Fig. 9). The electrons on the surface of the catalyst react with the carbon–oxygen complex, simultaneously CeO_2 retains the O_2 and generates carbon [20].

$$CeO_2 + xSO_2 \rightarrow xSO_3 + CeO_{2-x} \tag{2}$$

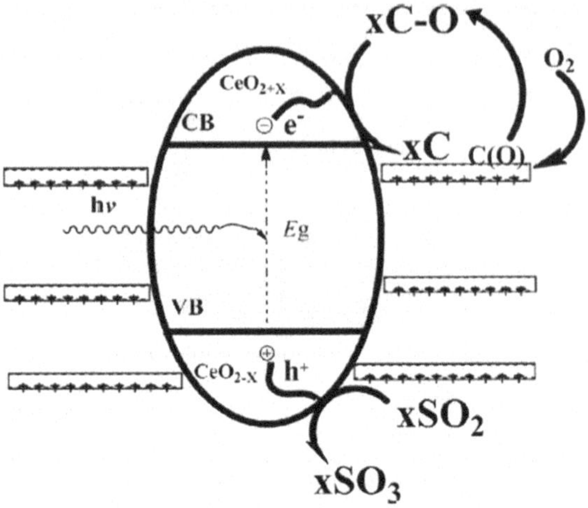

Fig. 9 Desulfurization mechanism of porous carbon–CeO_2 composites [20]

$$CeO_{2-x} + xC - O \rightarrow xC + CeO_2 \quad (3)$$

There are very few studies reported on the utilization of lignin as support for the synthesis of mesoporous TiO_2 photocatalyst. Chen et al. synthesized mesoporous TiO_2 composite using $TiCl_4$ as the reactant and lignin as the template [72]. The synthesized photocatalyst was used for phenol (0.05 g l^{-1}) degradation under UV light, resulting in degradation of 97.9% of phenol in 120 min (entry 6, Table 3). The photocatalytic performance of TiO_2–lignin composite was reported to be higher than the TiO_2 synthesized without template and commercial TiO_2 P25 that showed phenol degradation efficiency of 76.3 and 86.3%, respectively. The high photocatalytic activity of TiO_2–lignin composite is due to the high electronegativity difference between lignin and TiO_2 precursor that contributes to its uniform distribution for the formation of mesoporous TiO_2 particles. The lower surface hydroxyl group on TiO_2–lignin is another factor for the superior photocatalytic efficiency of composite that is ascribed to the stronger interaction between surface hydroxyl groups of TiO_2 precursor and lignin hydroxyl groups during hydrolysis. Moreover, lignin also contributes to the smaller crystal size and high specific surface area that ultimately improved the photocatalytic performance of the composite [72].

Recently, cupric oxide (CuO), a *p*-type semiconductor, has gained much attention as a photocatalyst due to its narrow band gap (1.2 eV). It has been widely used as a photocatalyst [96, 97] for the degradation of pollutants [97] and for the production of H_2 gas [98]. The size, morphology, microstructure, and photocatalytic performance of CuO nanoparticles was found to be enhanced by using aminated lignin (AL) as template. Calcination temperature and AL amount play a significant role in determining the photocatalytic activity of the CuO–AL composite. Under UV light irradiation for 90 min, CuO doped with AL (0.5 g) exhibited considerably higher photodegradation of 10 mg l^{-1} methylene blue (97.8%) and 10 mg l^{-1} methyl orange (66.7%) compared to undoped CuO. The optimum calcination temperature for CuO catalyst was 400 °C, which yields smaller crystallite size and high surface area, thus improving the photodegradation rate of organic dyes. Low temperature (300 °C) leads to incomplete decomposition of aminated lignin while high temperature (500 °C) resulted in aggregation of particles. The optimum dosage of aminated lignin for CuO doping was 0.5 g; further increase in dosage resulted in larger crystal size due to aggregation [70]. Moreover, optimizing the dosage and calcination temperature, the photocatalytic performance of the lignin-based composites can be maximized, in agreement with other studies [81].

4.2 Applications of Lignin-Based Materials in Photovoltaics

Appropriate engineering of photovoltaic cells is required for optimum light harvesting capacity and improved photo-inductive charge transfer [99]. The interface engineering of the anode is of utmost significance to improve the efficiency of the cell [100]. A power conversion efficiency of 10% has been achieved through cathode modification. However, in case of the anode, water-soluble conductive polymer poly(3,4-ethylene dioxythiophene):poly(styrene sulfonic acid) (PEDOT:PSS) has been widely used. The efficiency of the PEDOT:PSS principally driven by the conductivity and highest occupied molecular orbital (HOMO) energy level. In order to control HOMO energy level, perfluorinated ionomer (PFI) was applied as dopant for PEDOT, which exhibited better efficiency than PEDOT:PSS [101]. Similarly, the use of PSS as dopant also improved the conductivity of PEDOT [102]. Nevertheless, the microstructural and electrical inhomogeneities caused by PSS due to its non-conjugated structure configure PEDOT:PSS unsuitable for hole injection layer (HIL) [103].

The development of hole transport materials based on biomaterials are of great interest in the field of organic electronic devices. Lignin and its derivatives contain several aromatic rings that strongly absorb in the ultraviolet range of the electromagnetic spectrum [104]. In organic electronic devices, hole transport process is associated with the oxidation of electron-rich compounds such as thiophene in PEDOT:PSS and carbazole in poly(vinylcarbazole). Lignosulfonate poses exceptional hole-transfer characteristics owing to oxidation of phenols and j aggregation phenomenon. The aggregation behavior is responsible for the semiconductive nature of lignosulfonates [105].

Recently, various attempts have been made to improve the conductivity of PEDOT:PSS through different additives such as ionic liquids, surfactants, and organic solvents. Post-treatment of PEDOT:PSS with inorganic acid, polar solvents, salts, and zwitterions significantly enhanced the conductivity of PEDOT [106]. To significantly improve the conductivity and reduce the production cost, there is a need for renewable dopant to boost this technology. In view of green chemistry and green economics, lignin derivatives, as an alternative semiconductive material, have been employed as dopant for PEDOT. Li et al. investigated the potential of lignosulfonate (SL) and alkyl chain cross-linked lignosulfonate polymer (ASL) obtained by the introduction of alkyl chain to sulfomethylated lignin, as hole transport material for solar cells [105]. The mass ratio of PEDOT:SL has not shown considerable change in PCEs of the cell (see entry 2, Table 4 for detailed photovoltaic performance),

Table 4 Photovoltaic performances of PSCs with different proportions of different lignin-derived dopants for PEDOT

Entry no.	Anode	Device architecture	V_{OC} (V)	J_{SC} (mA cm^{-2})	FF (%)	PCE (%)	References
1	PEDOT:PSS	HTL/$CH_3NH_3PbI_3$/$PC_{61}BM$/Al	1.02	18.12	68.5	12.62	[103]
	PEDOT:GSL		0.98	19.21	74.7	14.10	
	PEDOT:GSLa reverse		1.031	20.1	72.08	14.94	
	PEDOT:GSLa forward		1.026	19.93	72.45	14.82	
2	PEDOT:PSS	ITO/HTM/PTB7:$PC_{71}BM$/Al	0.73	12.11	65.57	5.80	[106]
	PEDOT:ASL-1: 1		0.44	10.34	47.01	2.14	
	PEDOT:ASL-1: 2		0.59	12.47	53.39	3.93	
	PEDOT:SL-1: 1		0.70	13.22	62.60	5.79	
	PEDOT:SL-1: 2		0.69	13.27	62.96	5.76	
	PEDOT:SL-1: 6		0.68	12.64	62.03	5.33	
3	PEDOT:PSS	ITO/HTL/PTB7-Th:$PC_{71}BM$/PFN/Al	0.77	15.82	68.71	8.39	[22]
	PEDOT:GSL-1: 1		0.73	16.29	63.85	7.57	
	PEDOT:GSL-1: 2		0.73	16.27	64.81	7.74	
	PEDOT: GSL-1: 4		0.77	16.25	68.17	8.47	
	PEDOT: GSL-1: 6		0.77	16.29	67.04	8.37	
4	PEDOT:SL-1: 1	ITO/HEL/PTB7:$PC_{71}BM$/Al	0.58	15.06	58.87	5.19	[109]
	PEDOT:SL-1: 2		0.65	13.43	57.55	5.02	
	PEDOT:ASL1-1: 1		0.57	14.69	56.40	4.75	
	PEDOT:ASL1-1: 2		0.56	12.20	54.55	3.73	
	PEDOT:ASL2-1: 1		0.68	11.17	45.88	3.49	
	PEDOT:ASL3-1: 1		0.68	10.96	41.74	3.11	

with PEDOT:SL mass ratio of 1:1, 1:2, and 1:3, PCEs exhibited by the device (ITO/HTM/PTB7:$PC_{71}BM$/Al) were 5.79, 5.76, and 5.33%, respectively [105]. The PCE of PEDOT:SL is quite close to PEDOT:PSS (4.28 and 4.50%) as reported in the literature [107, 108]. Moreover, the results of PCEs of PEDOT:ASL with mass ratios of 1:1 and 1:2 were 2.14 and 3.93%, respectively. The results obtained depicted the positive influence of phenolic groups in lignosulfonates on the hole transport features of the materials. The hole mobility of lignosulfonate polymer is comparatively lower than lignosulfonate due to the reduced content of phenolic hydroxyl groups. The surface and nanoaggregate size of the film also contribute to the hole transport properties of the materials. The unique surface and variable size of nanoaggregates among PEDOT:SL and PEDOT:ASL films leads to the different hole transport properties of both materials [105].

Wu et al. reported the application of grafted sulfonated acetone–formaldehyde lignin (GSL), as a *p*-type semiconductive dopant for hole extracting layer [103]. GSL is a polymeric semiconductor derived by grafting the sulfonated acetone–formaldehyde (SAF) to alkali lignin (AL). The long aliphatic chain and large number of sulfonic groups on GSL make it a fine dispersant for being used as dopant for PEDOT. The conjugated structure of GSL makes it a good candidate for electron–hole mobility similar to other conjugated polymers used in organic electronics. GSL as hole transporting layer has shown promising results with the hole mobility of 2.27×10^{-6} $cm^2\ V^{-1}\ s^{-1}$ attributed to large number of hydroxyl moieties. Furthermore, GSL:PEDOT exhibited better conductivity and power conversion efficiency up to 14.94% than PEDOT:PSS (12.6%) with the device structure of HTL/$CH_3NH_3PbI_3$/$PC_{61}BM$/Al. The high efficiency of PEDOT:GSL is credited to the homogeneity and uniformity of the film surface, which is instigated by highly disperse GSL. Altogether, it will improve performance of the device by increasing charge transfer properties [103]. Furthermore, larger grain size of the PEDOT:GSL film results in higher current density [49]. Indium tin oxide (ITO) modified by PEDOT:GSL exhibited larger grain size (67 nm) than ITO transformed by PEDOT:PSS (61 nm). Consequently, PEDOT:GSL modified ITO as hole-extraction layer has better transport characteristics for hole collection due to its conjugated structure than PEDOT:PSS that lacks a conjugated structure [103].

SL and ASL have exceptional properties of forming Block-like self-assembly without any external interface in particular solvents. During the oxidation of SL, characteristic aggregation behavior is acquired by SL and ASL through the electron transport mechanism and their self-assembly. SL acquires distinctive assembly, attributable to its amphiphilic nature and presence of benzene rings that leads to its aggregation in particular solvents through Π–Π interactions and CH–Π interaction. With 1:3 H_2O: ethanol solution, the aggregates acquired for SL were of nano size, while micro-sized aggregates were obtained for ASL in the same set of conditions. Block-like aggregation behavior was more dominant in ASL compared to SL due to cross-linked alkyl chain polymerization in SL. Based on the aggregation behavior and electron transport characteristics of the SL and ASL, the materials have been applied as dopants to improve the conductivity of PEDOT [109]. The power conversion efficiency of polymer solar cell also depends on the aggregation behavior of the dopants that is ultimately affected by the hydroxyl group content [105, 109]. Moreover, the oxidative capacity of SL is much better than ASL due to the high phenolic hydroxyl group content. The reaction proceeds with the formation of radical cations and phenol radicals, formed by the oxidation of SL and phenolic hydroxyl groups, respectively. With ITO/HEL/PTB7:$PC_{71}BM$/Al device structure, the maximum PCE showed by PEDOT:SL with mass ratio of 1:1 was 5.19% that shows the potential of SL as effective dopant for PEDOT in organic electronic devices. SL exhibited the hole mobility of 2.95×10^{-6} cm^2 V^{-1} s^{-1}, which is higher in comparison to ASL that showed the hole mobility of 3.18×10^{-7} cm^2 V^{-1} s^{-1}. The results of the study also showed that hydroxyl group content is directly related to the hole mobility and PCE, whereas increase in hydroxyl group increased hole transport ability and PCEs and vice versa. Furthermore, the high pH of SL and ASL is an advantage of the conductive polymers over conventional dopant PSS that will prevent corrosion of ITO layers [109].

Hong et al. also investigated the GSL as potential dopant and stabilizer for PEDOT to enhance the performance of light-emitting and photovoltaic devices [22]. PEDOT:GSL films and aqueous dispersions with adjustable conductivities and work functions have been used for fabricating high-performance organic light-emitting diodes and polymer solar cells [22]. GSL has a number of advantages over other lignin-derived polymers such as lignosulfonates applied as dopant for PEDOT. GSL has high phenolic content that results in better oxidative capability of the polymer. The high degree of sulfonic group in GSL compared to lignosulfonate makes it a more suitable dispersant for excellent PEDOT dispersion. The addition of GSL in PEDOT also results in better film characteristics in comparison to PEDOT modified by lignosulfonate attributed to the superior dispersing characteristics of GSL. Altogether, the superior GSL contribute in improving the hole transport properties of PEDOT as a dopant. The oxidation peak of GSL-doped electrode obtained at 1.1 V that indicates that GSL HOMO energy level is − 5.5 eV and its oxidation can take place at comparatively low potential. GSL as hole transporting material exhibited a good hole mobility (2.27×10^{-6} cm^2 V^{-1} s^{-1}) credited to its phenolic structure. The power efficiency of PEDOT:GSL with the mass ratios of 1:1, 1:2, 1:4, and 1:6 were 1.51 , 6.04 , 12.91, and 14.67 lmW^{-1}, respectively. It is evident by the power efficiencies that by increasing the GSL content the hole injection and transport properties were improved. Similarly, PEDOT:GSL 1:4 and 1:6 displayed far better power efficiency than PEDOT:PSS (8.25l mW^{-1}). Using the device structure of ITO/HTL/PTB7-Th:$PC_{71}BM$/PFN/Al, the PCEs of PEDOT:GSL with mass ratios of 1:4 and 1:6 was 8.47 and 8.37%, respectively, which is quite analogous to PEDOT:PSS (8.39%) (entry 3, Table 4). Moreover, the homogenous film surface of PEDOT:GSL results in enhancement of hole injection features of the material due to the excellent dispersion property of GSL. Another important function for organic electronic devices is work function that also increases with increasing GSL content. The work function of PEDOT:GSL with mass ratios of 1:2, 1:4, and 1:6 were 4.92, 5.05, and 5.10 eV, respectively, which is comparable to PEDOT:PSS (5.02 eV) [22].

5 Conclusions, Future Perspectives, and Challenges

In this review, we discussed the recent progress in the field of photocatalysis and photovoltaics with a focus on lignin-based composite materials. The aims were to review the recent studies on the application of lignin-based materials for photocatalysis and photovoltaics-related environmental remediation and energy conversion, respectively, which will provide some useful implications for future research. Lignin as a biopolymer support showed promising potential in the field of heterogeneous photocatalysis explicitly in the context of the degradation of unwanted contaminants in the environment. A number of studies have reported the use of lignin as a commendable template for photocatalyst synthesis [18] attributed to its high specific surface area that improves the physical adsorption of the substrate [20] and superior photoelectron transfer characteristics owing to unique surface contact [18]. In addition, lignin-based composite materials have great potential to replace cost-intensive materials like graphene in the field of photocatalysis. The photodegradation efficiency of graphene oxide-based composites for methyl orange, methylene blue, and Rhodamine B were 87.2, 85.1, and 73.9%, after 60 min of UV photodegradation, respectively [110]. Lignin composite possesses excellent photocatalytic activity (see Table 3 for details), which is superior to those of graphene.

Many methods are suitable for the preparation of lignin-based composites. However, pH-assisted precipitation and solid-phase grinding have attracted increasing attention and some promising results have been reported. In future research on the photocatalysis via lignin-based composites, detailed investigations on the interaction of composite and substrate should be performed. Moreover, prospective applications may be expected in the field of photoelectric conversion [19], and for electrochemical storage systems [42], such as supercapacitors [15].

There are a number of studies discussed in this review demonstrating that lignin-based materials have wide applications in diverse fields ranging from photocatalysis to electrochemical energy devices and biomedicine. Developing lignin into functional materials, specifically its application as a support for solid composite photocatalysts, would present a great success. Moreover, lignin-based composites show improved photocatalytic efficiency, e.g., for the degradation of pollutants in aqueous media. Nevertheless, a deeper understanding of the underlying mechanism of interaction of lignin and the photoactive material are in demand.

Although considerable advancement has been made in materials development and understanding the structure–property relationships of organic photovoltaics materials and devices, there are still numerous open questions that need to be answered to achieve an increase in the photoconversion efficiency in order to reach at an economically feasible utilization. Particularly, the charge-carrier separation and mobility within the materials has to be improved. There are very few studies that have explored the potential of lignin as dopant for anode in photovoltaic devices. However, recent success in development of lignosulfonate-based dopants for semiconductive polymers with PCE analogous to PSS (see Table 4 for comparison) will pave the way for future research in lignin-based high-performance organic devices. Further research is required to understand the origin of such substantial electron transport properties of lignin as well as the mechanism of the lignin-based photovoltaic cells with improved performance.

The valorization of lignin not only encompasses technical and scientific developments but also economic aspects. There are striking opportunities for an economic gain from lignin valorization attributed to its low cost and profuse availability as a byproduct of the pulping industry and bioethanol refineries. Hence, the potential to apply the underutilized lignin sources stimulates the aspiration not only for the development of efficient isolation methods but also for the fabrication of new lignin-based products, which have high economic value in the coming years.

Acknowledgements This publication is part of a project that has received funding from the European Union's Horizon 2020 research and innovation programme under the Marie Skłodowska-Curie grant agreement No. 711859 and from the financial resources for science in the years 2017–2021 awarded for the implementation of an international co-financed project. Prof. Dr. J.C. Colmenares and Dr. V. Nair are very grateful for the partial support from the National Science Centre in Poland within Sonata Bis Project No. 2015/18/E/ST5/00306. Roger Gläser gratefully acknowledges support from the Leipzig Graduate School of Natural Sciences: Building with Molecules and Nano-objects as well as from the Research Academy Leipzig.

References

1. Tuck CO, Perez E, Horvath IT, Sheldon RA, Poliakoff M (2012) Science 337:695–699
2. Zeng X, Yao G, Wang Y, Jin F (2014) In: Jin F (ed) Application of hydrothermal reactions to biomass conversion. Springer, Berlin
3. Ren X, Wang P, Han X, Zhang G, Gu J, Ding C, Zheng X, Cao F (2017) ACS Sustain Chem Eng 5:6548–6556
4. Guan Q, Mao T, Zhang Q, Miao R, Ning P, Gu J, Tian S, Chen Q, Chai XS (2014) J Supercrit Fluids 95:413–421
5. Agrawal A, Kaushik N, Biswas S (2014) Scitech J 01:32–36
6. Hu J, Zhang Q, Lee DJ (2018) Bioresour Technol 247:1181–1183
7. Chatel G, De Oliveira Vigier K, Jérôme F (2014) ChemSusChem 7:2774–2787
8. Pu Y, Zhang D, Sigh PM, Ragauskas AJ (2008) Biofuels Bioprod Biorefin 2:58–73
9. Upton BM, Kasko AM (2016) Chem Rev 116:2275–2306
10. Zakzeski J, Bruijnincx PCA, Jongerius AL, Weckhuysen BM (2010) Chem Rev 110:3552–3599
11. Li SH, Liu S, Colmenares JC, Xu YJ (2016) Green Chem 18:594–607
12. Lee HV, Hamid SBA, Zain SK (2014) Sci World J 2014:1–20
13. Dorrestijn E, Laarhoven LJJ, Arends IWCE, Mulder P (2000) J Anal Appl Pyrolysis 54:153–192
14. Lu H, Zhao XS (2017) Sustain Energy Fuels 1:1265–1281
15. Richter AP, Bharti B, Armstrong HB, Brown JS, Plemmons D, Paunov VN, Stoyanov SD, Velev OD (2016) Langmuir 32:6468–6477
16. Lignin Market (Lignosulfonates, Kraft Lignin and others) for concrete additive, animal feed, dye stuff, and other applications: global industry perspective, comprehensive analysis and forecast 2014-2020. Market Research Store, Deerfield Beach. https://www.globe.com.ph/
17. Kun D, Pukánszky B (2017) Eur Polym J 93:618–641

18. Wang H, Qiua X, Liua W, Yang D (2017) Appl Surf Sci 426:206–216
19. Wang H, Qiu X, Zhong R, Fu F, Qian Y, Yang D (2017) Mater Chem Phys 199:193–202
20. Wang N, Fan H, Ai S (2015) Chem Eng J 260:785–790
21. Wang W, Tade MO, Shao Z (2015) Chem Soc Rev 44:5371–5408
22. Hong N, Xiao J, Li Y, Li Y, Wu Y, Yu W, Qiu X, Chen R, Yip HL, Huangd W, Cao Y (2016) J Mater Chem C 4:5297–5306
23. Virkutyte J, Jegatheesan V, Varma RS (2012) Bioresour Technol 113:288–293
24. Naseem A, Tabasum S, Zia KM, Zuber M, Ali M, Noreen A (2016) Int J Biol Macromol 93:296–313
25. Doherty WOS, Mousavioun P, Fellows CM (2011) Ind Crops Prod 33:259–276
26. Sjöström E (1993) Wood chemistry. Fundamentals and applications. Academic Press, San Diego
27. Smook GA (2002) Handbook for pulp and paper technologies. Angus Wilde Publications Inc., Vancouver
28. Watt C, Burgess H (1854) Improvement in the manufacture of paper from wood. US 11343
29. Koljonen K, Österberg M, Kleen M, Fuhrmann A, Stenius P (2004) Cellulose 11:209–224
30. Wunderlich B, Bodily DM, Kaplan MH (1964) J Appl Phys 35:95–102
31. Faix O, Meier D, Grobe I (1987) J Anal Appl Pyrolysis 11:403–416
32. Chauvette G, Heitz M, Rubio M, Khorami J, Chornet E, Menard H (1985) Thermochim Acta 84:1–5
33. Yang W, Owczarek JS, Fortunati E, Kozanecki M, Mazzaglia A, Balestra GM, Kenny JM, Torre L, Puglia D (2016) Ind Crops Prod 94:800–811
34. Barana D, Ali SD, Salanti A, Orlandi M, Castellani L, Hanel T, Zoia L (2016) ACS Sustain Chem Eng 4:5258–5267
35. Sarkanen KV, Ludwig CH (1971) Lignins occurrence formation structure and reactions. John Wiley and Sons, New York
36. Gupta VG, Tuohy M, Kubicek CP, Saddler J, Xu F (2014) Bioenergy research: advances and applications. Elsevier, Amsterdam
37. Thakur VK, Thakur MK, Raghavan P, Kessler MR (2014) ACS Sustain Chem Eng 2:1072–1092
38. Morandim-Giannetti AA, Agnelli JAM, Lancas BZ, Magnabosco R, Casarin SA, Bettini SHP (2012) Carbohydr Polym 87:2563–2568
39. Evans A, Strezov V, Evans T (2010) Energy Rev 14:1419–1427
40. Lima RB, Raza R, Qin H, Li J, Lindstromac ME, Zhu B (2013) RSC Adv 3:5083–5089
41. Jeon JW, Zhang LB, Lutkenhaus J, Laskar D, Lemmon J, Choi D, Nandasiri MI, Hashmi A, Xu J, Motkuri RK, Fernandez CA, Liu J, Tucker MP, McGrail PB, Yang B, Nune SK (2015) ChemSusChem 8:428–432
42. Geng X, Zhang Y, Jiao L, Yang L, Hamel J, Giummarella N, Henriksson G, Zhang L, Zhu H (2017) ACS Sustain Chem Eng 5:3553–3561
43. Shi M, Wei W, Jiang Z, Han H, Gaoa J, Xie J (2016) RSC Adv 6:25255–25266
44. Colmenares JC, Varma RS, Lisowski P (2016) Green Chem 18:5736–5750
45. Dai K, Zhang XH, Fan K, Zeng P, Peng TY (2014) J Nanomater 2014:694073–694080
46. Leary R, Westwood A (2011) Carbon 49:741–772
47. Khan G, Kim YK, Choi SK, Han DS, Wahab AA, Park H (2013) Bull Korean Chem Soc 34:1137–1144
48. Bagheri S, Julkapli NM, Hamid SBA (2015) Int J Photoenergy 1:1–30
49. Dong SS, Zhang JB, Gao LL, Wang YL, Zhou DD (2012) Trans Nonferrous Met Soc China 22:2477–2483
50. Lisowski P, Colmenares JC, Mašek O, Lisowski W, Lisovytskiy D, Kamińska A, Łomot D (2017) ACS Sustain Chem Eng 5:6274–6287
51. Ke D, Liu S, Dai K, Zhou J, Zhang L, Peng T (2009) J Phys Chem C 113:16021–16026
52. Zeng J, Liu S, Cai J, Zhang L (2010) J Phys Chem C 114:7806–7811
53. Paridaa B, Iniyanb S, Goic R (2011) Renew Sust Energ Rev 15:1625–1636
54. Yu J, Zheng Y, Huang J (2014) Polymers 6:2473–2509
55. Nazeeruddin MK, Baranoff E, Grätzel M (2011) Sol Energy 85:1172–1178
56. Gan Q, Bartoli FJ, Kafafi ZH (2013) Adv Mater 25:2385–2396
57. Spanggaard H, Krebs FC (2004) Sol Energy Mater Sol Cells 83:125–146
58. Zhou Y, Fuentes-Hernandez C, Khan TM, Liu JC, Hsu J, Shim JW, Dindar A, Youngblood JP, Moon RJ, Kippelen B (2013) Sci Rep 3:1536–1540

59. Hu L, Zheng G, Yao J, Liu N, Weil B, Eskilsson M, Karabulut E, Ruan Z, Fan S, Bloking JT, McGehee MD, Wagberg L, Cui Y (2013) Energy Environ Sci 6:513–518
60. Chiba Y, Islam A, Watanabe Y, Komiya RN, Koide LH (2006) Jpn J Appl Phys 45:638–640
61. Salvador GP, Pugliese D, Bella F, Chiappone A, Sacco A, Bianco S, Quaglio M (2014) Electrochim Acta 146:44–51
62. Hendrix Y, Lazaro A, Yu Q, Brouwers L (2015) World J Nanosci Eng 5:161–177
63. Morsella M, Giammatteo M, Arrizza L, Tonucci L, Bressana M, d'Alessandro N (2015) RSC Adv 5(57453):57461
64. Zhou J, Zhang M, Zhu Y (2015) Phys Chem Chem Phys 17:3647–3652
65. Nair V, Dhar P, Vinu R (2016) RSC Adv 6:18204–18216
66. Altavilla C (2016) Upconverting nanomaterials: perspectives, synthesis, and applications. CRC Press, New York
67. Neckers DC (1975) J Chem Educ 52:695–702
68. Wu F, Wang X, Hu S, Hao C, Gao H, Zhou S (2017) Int J Hydrog Energy 42:30098–30108
69. Hodge P, Sherrington DC (1980) Polymer-supported reactions in organic synthesis. John Wiley & Sons, Chichester
70. Wang X, Wu F, Duan Y, Wang Y, Hao C, Ge C (2016) RSC Adv 6:65644–65653
71. Jin W, Zhang Z, Wu G, Tolba R, Chen A (2014) RSC Adv 4:27843–27849
72. Chen X, Kuo DH, Lu D, Hou Y, Kuo YR (2016) Microporous Mesoporous Mater 223:145–151
73. Zhou Y, Ding EY, Li WD (2007) Mater Lett 61:5050–5052
74. Xingfu Z, Zhaolin H, Yiqun F, Su C, Weiping D, Nanping X (2008) J Phys Chem C 112:11722–11728
75. Guo YR, Yu FD, Fang GZ, Pan QJ (2013) J Alloys Compd 552:70–75
76. Nunez-Flores R, Gimenez B, Fernandez-Martın F, Lopez-Caballero M, Montero M, Gomez-Guillen M (2011) Food Hydrocoll 27:60–71
77. Miao TT, Guo YR, Pan QJ (2013) J Nanopart Res 15:1725
78. Liu A, Honma I, Ichihara M, Zhou H (2006) Nanotechnology 17:2845–2849
79. Liu Y, Gao L, Sun J (2007) J Phys Chem C 111:1223–1229
80. Kai D, Tan MJ, Chee PL, Chua YK, Yap YL, Loh XJ (2016) Green Chem 18:1175–1200
81. Wang X, Zhang Y, Hao C, Feng F, Yin H, Si N (2014) Ind Eng Chem Res 53:6585–6592
82. Zhang X, Chen YL, Liu RS, Tsai DP (2013) Rep Prog Phys 76:046401
83. Chaturvedi S, Dave PN (2012) Mater Sci Forum 734:273–294
84. Jiang L, Wang Y, Feng C (2012) Procedia Eng 45:993–997
85. Colmenares JC, Varma RS, Nair V (2017) Chem Soc Rev 46:6675–6686
86. Colmenares JC, Kuna E (2017) Molecules 22:790
87. Song L, Guo C, Li T, Zhang S (2017) Ceram Int 43:7901–7907
88. Ghosh T, Oh WC (2012) Asian J Chem 24:5419–5423
89. Schneider M, Widberger M, Duff DG, Mallant T, Maciejewsks M, Baiker A (1995) Scientific bases for the preparation of heterogeneous catalyst, Preparation of Catalyst, vol 6. Elsevier, Amsterdam
90. Singh AK, Nakate UT (2013) J Nanopart 2013:1–7
91. Fang R, Cheng XS, Xu XR (2010) Bioresour Technol 101:7323–7329
92. Uddin MT, Nicolas Y, Olivier C, Toupance T, Servant L, Muller MM, Kleebe HJ, Ziegler J, Jaegermann W (2012) Inorg Chem 51:7764–7773
93. Xiaohong W, Wenjing X, Chen H, Pengfei Z, Xiaoqi F, Naicha S (2013) Acta Metall Sin 49:1098–1104
94. Akir A, Hamdi A, Addad Y, Coffinier R, Boukherroub ADO (2017) Appl Surf Sci 400:461–470
95. Ferrizz RM, Gorte RJ, Vohs JM (2002) Catal Lett 82:123–129
96. Li X, Guo W, Huang H, Chen T, Zhang M, Wang Y (2014) J Nanosci Nanotechnol 5:3428–3432
97. Umadevi M, Christy AJ (2013) Spectrochim Acta A 109:133–137
98. Barreca D, Fornasiero P, Gasparotto A, Gombac V, Maccato C, Montini T, Tondello E (2009) ChemSusChem 2:230–233
99. Editorial (2012) Nat Photon 6:129
100. Mei A, Li X, Liu L, Ku Z, Liu T, Rong Y, Xu M, Hu M, Chen J, Yang Y, Han H, Graetzel M (2014) Science 345:295–298
101. Lim KG, Kim HB, Jeong J, Kim H, Kim JY, Lee TW (2014) Adv Mater 26:6461–6466
102. Palumbiny CM, Liu F, Russell TP, Hexemer A, Wang C, Muller-Buschbaum P (2015) Adv Mater 27:3391–3397

103. Wu Y, Wang J, Qiu X, Yang R, Lou H, Bao X, Li Y (2016) ACS Appl Mater Interfaces 8:12377–12383
104. Qian Y, Qiu X, Zhu S (2015) Green Chem 17:320–324
105. Li Y, Hong N (2015) J Mater Chem A 3:21537–21544
106. Ouyang J (2013) Displays 34:423–436
107. Srinivasan MV, Ito M, Kumar P, Abhirami K, Tsuda N, Yamada J, Shin PK, Ochiai S (2015) Ind Eng Chem Res 54:181–187
108. Guo S, Cao BY, Wang WJ, Moulin JF, Muller- Buschbaum P (2015) ACS Appl Mater Interfaces 7:4641–4649
109. Hong N, Qiu X, Deng W, He Z, Li Y (2015) RSC Adv 5:90913–90921
110. Wang J, Jin C, Sun Q, Zhang Q (2017) J Alloys Compd. 702:418–426
111. Diebold VB, Cowan WF, Walsh JK (1978) Solvent pulping process, US patent 4,100,016
112. Strassberger Z, Tanase S, Rothenberg G (2014) RSC Adv 4:25310–25318
113. Feng F, Hao C, Zhang H, Xie W, Wang X, Zhao Y (2015) J Mater Sci Mater Electron 26:6704–6711
114. Morsella M, Alessandro N, Lanterna AE, Scaiano JC (2016) ACS Omega 1:464–469

Topics in Current Chemistry (2018) 376:29
https://doi.org/10.1007/s41061-018-0208-1

REVIEW

Degradation of Vanillin During Lignin Valorization Under Alkaline Oxidation

Yuting Zhu[1,2,3,4] · Jing Liu[1,2,3,4] · Yuhe Liao[5] · Wei Lv[1,2,3] · Longlong Ma[1,2,3] · Chenguang Wang[1,2,3]

Received: 8 March 2018 / Accepted: 19 June 2018 / Published online: 2 July 2018

Abstract

The preparation of vanillin from lignin is one of the lignin valorization strategies. However, obtaining high vanillin yield is still a challenge. Therefore, the process of vanillin production and factors that affect yield of vanillin has attracted much attention. Here, oxidation of vanillin was performed to study its degradation behavior under lignin alkaline oxidation conditions. High-performance liquid chromatography, liquid chromatography–electrospray mass spectrometry, gas chromatography–mass spectrometer and gel permeation chromatography were employed to analyze the products including monomers and dimers. Results demonstrated that reaction temperature and time greatly affected vanillin degradation; vanillin can be completely converted in 5 h at 160 °C. At 160 °C, the main products of vanillin oxidation were small molecule acids and alcohols, other monophenols, and even condensed dimers. A possible vanillin degradation pathway was proposed. The results indicate that vanillin degradation and condensation are the main reasons for decreasing vanillin yield during lignin valorization under alkaline oxidation circumstances.

Keywords Lignin · Alkaline oxidation · Vanillin · Monophenol · Degradation · Condensation

---

Chapter 2 was originally published as Zhu, Y., Liu, J., Liao, Y., Lv, W., Ma, L. & Wang, C. Topics in Current Chemistry (2018) 376: 29. https://doi.org/10.1007/s41061-018-02208-1.

Electronic supplementary material The online version of this article (https://doi.org/10.1007/s41061-018-0208-1) contains supplementary material, which is available to authorized users.

✉ Chenguang Wang
wangcg@ms.giec.ac.cn

Extended author information available on the last page of the article

1 Introduction

Lignocellulose, as a renewable carbon resource for fuel, chemicals, and materials, is gaining more attention considering environmental concern and shorten of fossil resources. Chemo- and bio-catalysis technologies for (hemi-)cellulose conversion were well developed. However, lignin utilization is still a big challenge. Though some lignin conversion methods (such as hydrogenolysis [1–3], oxidation [4–7], bioengineering) have been developed, these methods are still in their early stages. Consequently, the recent focus of biomass conversion is lignin valorization, which not only broadens the raw material resource but also increases economic efficiency of current biorefinery industry [8–12]. Lignin is an aromatics polymer, which is an idea source for producing value-added chemicals like aromatics [8, 13, 14].

Vanillin (3-methoxy-4-hydroxy benzaldehyde) is an important aroma chemical produced worldwide and widely used as flavoring agents for chocolate and ice cream, intermediate of drugs like Aldomet, L-Dopa Trimethoprim; ripeners and sunscreens, even catalysts [15–17]. Total annual vanillin production is about 20,000 tons, 15% came from lignin, 85% came from petroleum-based raw material guaiacol, and less than 1% was extracted from vanilla beans [15, 18, 19]. Given its sustainability, lignin has a great potential to become the primary source for vanillin in the future, compared to the exhausted petroleum resource.

Although vanillin has high economic value, the yield of vanillin from lignin is still much lower than its theoretical level. To find out the reasons for the limited yield of vanillin, effects of lignin sources [20], structures [21], pretreatment [22], catalyst [4, 23, 24], reaction conditions on lignin conversion to vanillin have been extensively studied [13]. For instance, Rodrigues et al. investigated the effect of some lignin features (including wood species, pulping process, and lignin isolation process) on its oxidative conversion to high-added-value phenolic aldehydes. The frequency of phenolic hydroxyl groups in noncondensed structures was declaimed to be one possible limiting factor of phenol yield [25]. Wu et al. used steam-explosion hardwood lignin to produce aldehydes (mainly vanillin, syringaldehyde, and hydroxybenzaldehyde) under alkaline conditions. Monophenol yields significantly increased when $CuSO_4$ and $FeCl_3$ were employed as catalysts. Consequently, 14.6 wt% of lignin was converted to three aldehydes with vanillin yield of 4.6 wt% [26]. Recently, the oxidative degradation of monophenols has become new research hotspot. Rodrigues and coworkers systematically investigated the kinetics of vanillin oxidation and found that vanillin oxidation is a first-order reaction under high-alkalinity conditions ($pH > 12$) [27]. Sultanov et al. studied the degradation of a variety of guaiacyl and syringyl lignin model compounds under alkaline oxidative conditions (1 M KOH, 0.1 MPa O_2, 70 °C), and compared the activity difference between guaiacyl and syringyl model compounds so as reactivity among various 4-substituted syringols with different substituents. They found that vanillin and vanillic acid were stable under alkali oxidation conditions. Although oxidization of vanillin under more severe conditions (1.25 M NaOH, 150 °C, 1 MPa O_2) was conducted, and a mixture of carboxylic acids and hydroxyl acids were obtained, nevertheless specific structural information of degradation products was not given [28]. TiO_2 photocatalytic oxidation of vanillin can produce ring open intermediates like formic acid, acetic acid, and oxalic acid. *Trans*-ferulic acid can also open rings to produce these intermediates in photocatalytic oxidation [29]. Vanillin can also be selectively oxidized into vanillic acid under the promotion of gold nanoparticles supported on alumina and titania in alkaline aqueous media at 80 °C in the presence of pressurized oxygen, which resulted in selectivity up to 99% at conversions over 90% [16]. So, degradation of vanillin under aerobic alkaline conditions may be a key factors for the low yield of vanillin.

Here, we prepared vanillin derived from pine wood oxidation in alkaline media and investigated the degradation of vanillin under the same alkaline oxidative conditions. HSQC, GC–MS, LC–MS, and LC analysis were employed to analyze the products including dimers. We found that vanillin was not simply converted to small molecules of acids and alcohols by further oxidation but also to dimers via condensation. On the basis of our findings, a possible degradation mechanism of vanillin under alkaline oxidation conditions was proposed. This work led us to clarify limited vanillin yield during lignin alkaline oxidation from a new perspective, i.e., degradation of vanillin.

2 Methods

2.1 Materials

Pine wood was produced in South China (lignin, 23%). High purity O_2 (99.99%) was obtained from Yi gas Gases Co. ltd (Guangdong, China). NaOH, chloroform, and tetrahydrofuran (THF) were purchased from Macklin Biochemical Technology Co., Ltd. (Shanghai, China). Standards for vanillin (99%), *p*-hydroxybenzaldehyde (99%), vanillic acid (98%), acetic acid (98%), formic acid (98%), oxalic acid (99.9%), methanol (99.5%), and ethanol (99.5%) were also obtained from Macklin Biochemical Technology Co., Ltd. (Shanghai, China). All chemicals were analytical reagents and used without purification.

2.2 Alkaline Oxidation

Batch experiments were performed in a 100-ml stainless-steel autoclave reactor equipped with a thermocouple and a magnetic coupling mechanical stirring rod. In a typical experiment, the reactor was loaded with 50 ml NaOH aqueous solution and 1 g pine or 0.2 g vanillin, and was sealed. Then, the reactor was purged three times and pressurized with O_2 to 1 MPa. During the reaction, the reactor was stirred at a predefined speed. A heating procedure preceded as follows: first, heating from room temperature (about 30 °C) to target temperature with heating rate of 2.5 °C/min; then, heating at the target temperature for a fixed time. After the reaction, the reactor was cooled in air for about 3 h, and then was opened to collect the reaction solution.

Ten milliliters of the reaction solution was acidified to pH 2–3 with 0.5 ml concentrated hydrochloric acid (35%). Acidified liquid was diluted with 1.5 ml of deionized water, then was filtered (over a 0.45-μm Teflon filtrate pad) for quantitative analysis of formic acid, acetic acid, methanol, and ethanol by HPLC with a SH1011 column.

The acidified liquid was distilled to remove water, small acids, and HCl under vacuum at 40 °C, and then 5 ml of acetonitrile was added to dissolve the solid products; 2.5 ml acetonitrile phase was mixed with 5 ml of water (containing 0.5 g/l acetophenone) and filtered (over a 0.45-μm Teflon filtrate pad) for quantitative analysis of aromatics by HPLC with a C18 column. The rest 2.5 ml acetonitrile phase was distilled to remove solvent, and then dissolved in 2.5 ml of deionized water. This solution was applied to quantitative analysis of heavy acid (oxalic acid) by HPLC with a SH1011 column.

Ten milliliters of acidified liquid was extracted with THF until the THF layer appeared colorless. All the THF extractions were collected together. THF phase was added with a small amount of $NaHCO_3$ to neutralize the residual acid from acidification, and then added with anhydrous Na_2SO_4 to absorb the residual water in the solution. The resulting anhydrous THF phase was concentrated to 2 ml and filtrated over a 0.45-μm Teflon filtrate pad.

2.3 Analysis Methods

2.3.1 High-Performance Liquid Chromatography (HPLC)

The HPLC test was performed on an HPLC instrument (WATERs e2695) equipped with a 2489 UV–Vis detector and 2414 RI Detector. Agilent ZORBAX Eclipse XDB-C18 column (4.6 mm × 150 mm, 5 μm) was used to quantitative analysis of aromatics. Eluent, 20% acetonitrile (containing 0.1% trifluoroacetic acid) and 80% water (containing 0.1% trifluoroacetic acid); flow rate, 1 ml/min; injection volume, 10 μl; detection temperature, 30 °C; wave length, 260 nm; retention time, 20 min. To determine small acids and alcohols, the HPLC test was performed by using a Shodex SUGAR SH1011 (8 mm × 300 mm, 6 μm) column at 30 °C. The eluent was 5 mmol/l H_2SO_4 aqueous solutions with flow rate of 0.5 ml/min; injection volume was 10 μl, and retention time was 35 min.

2.3.2 Liquid Chromatography–Electrospray Mass Spectrometry (LC–MS)

LC–MS test was preceded at Agilent 1290–6540, 1260–6420 LC–MS system using a C18 column at 30 °C. Eluent A was acetonitrile, and eluent B was water (0.1% ammonium acetate). Separation was achieved with 20% A/80% B at 0–3 min, 60% A/40% B at 3–10 min, 100% A at 10–15 min. Flow rate was 0.1 ml/min and injection volume was 10 μl.

2.3.3 Gel Permeation Chromatography (GPC)

GPC analysis was performed on an Agilent 1260 LC system equipped with a refractive index detector (RID) by using PL gel mixed-C column (7.5 mm × 300 mm, 5 μm) at 35 °C. THF was used as eluent at a flow rate of 1 ml•min^{-1}. The sample injection volume was 20 μl.

2.3.4 Gas Chromatography–Mass Spectrometer (GC–MS)

Before GC–MS analysis, a derivatization step was added to increase volatility of vanillin oxidation products. The derivatization experiment was conducted according to Ref. [1]. Briefly, 5-ml THF extractions were dried by vacuum distillation at 35 °C, then mixed with 0.5 ml of pyridine and 0.5 ml of *N*-methyl-*N*-(trimethylsilyl)trifluoroacetamide and sealed. The mixture was put in an oven preheated to 80 °C and heated for 30 min. For qualitative analysis, derivatives were analyzed by TRACE 1300ISQ GC/MS (Thermo Fisher Scientific, Waltham, MA, USA) equipped with a TG-5MS column. The following operating conditions were used: injector and detector temperatures were 250 and 280 °C, respectively. The temperature program was set as follows: 60 °C, 2 min; 60–250 °C, heating rate 10 °C•min^{-1}; 250 °C, 15 min.

2.3.5 Heteronuclear Single Quantum Coherence (2D HSQC NMR)

To investigate the alkaline oxidation performance of vanillin, 2D HSQC NMR spectrum of THF extracts was used to record structural information of heavy products. 2D HSQC NMR analysis was run on a BRUKER AVANCE III 400-MHz spectrometer. Around 100 mg of sample was dissolved in 1.5 ml deuterated dimethyl sulfoxide (DMSO-d6). The parameters of collecting and processing are listed as follows: spectral frequency, 400.15 MHz for f1 (^{13}C), 100.61 MHz for f2 (^{1}H); spectral width, 22,137.686 Hz for f1 dimension, 4807.692 Hz for f2 dimension; number of the collected complex points for f2 dimension was 2048 with a recycle delay of 2 s; number of scan for f1 dimension was 72 with 256 time increments. The $^{1}J_{CH}$ used was 145 Hz. The solvent DMSO-d6 peak (δ_C/δ_H 39.5/2.49 ppm) was used as an internal reference. MestReNova software (9.0.1 version) was used for the spectra analysis.

2.3.6 Calculation Formulas

For the alkaline oxidation of lignin, yields of monophenol ($Y_{\text{monophenol}}$) and small acids and alcohols ($Y_{\text{small molecule}}$), selectivity of vanillin [30] and monophenol theoretical yield ($Y_{\text{theoretical}}$) were calculated by Eqs. (1)–(4), respectively. Yields of small molecules were calculated on basis of pine wood weight (Eq. 2), since acids and alcohols can both from oxidation of lignin and (hemi-)cellulose.

$$Y_{\text{monophenol}}\,(\%) = W_{\text{monophenol}} \div (W_{\text{pine}} \times 23\%) \times 100\% \tag{1}$$

$$Y_{\text{small molecule}}\,(\%) = W_{\text{small molecule}} \div W_{\text{pine}} \times 100\% \tag{2}$$

$$S_{\text{v}}\,(\%) = W_{\text{v}} \div W_{(\text{total monophenols})} \times 100\% \tag{3}$$

$$Y_{theoretical}\,(\%) = P\,(\text{C–O–C}) \times \text{P}\,(\text{C–O–C}) \tag{4}$$

$W_{monophenol}$: the weight of monophenol; W_{Pine}: the weight of pine wood; $Y_{small\ molecule}$: the weight of small molecule; W_v: the weight of vanillin; $W_{(total\ monophenols)}$: the weight of total monophenols. P (C–O–C): C–O–O linkages (including β-O-4, α-O-4, 4-O-5) occurrence probability in lignin structure. According to Refs. [13, 25], C–O–O linkages occurrence probability softwood lignin are: 43–50% (β-O-4), 6–8% (α-O-4), 4% (4-O-5). Thus, highest P (C–O–C) is up to 62% in total, so $Y_{theoretical}$ of softwood can reach 38%.

For the alkaline oxidation of vanillin, the vanillin conversion (C_v), mole yields of products ($Y_{product}$) and carbon balance (CB) were calculated by Eqs. (5)–(7), respectively.

$$C_v\,(\%) = W_{v1} \div W_v \times 100\% \tag{5}$$

$$Y_{product}\,(\text{mol}\%) = n_{product} \times X \div 8n_v \times 100\ \text{mol}\% \tag{6}$$

$$CB\,(\text{mol}\%) = \sum Y_{product} + (W_{v1} \div W_v) \times 100\ \text{mol}\% \tag{7}$$

$Y_{monophenol}$: the weight of monophenol; W_v: the weight of initial vanillin. W_{v1}: the weight of remained vanillin; $n_{product}$: the moles of product; X: carbon numbers of product formula; n_v: the moles of initial vanillin.

3 Results and Discussion

3.1 Alkaline Oxidation of Pine Lignin

In order to study the evolution of real vanillin under lignin alkaline oxidation conditions, a set of experiments of oxidative depolymerization of pine wood under different reaction temperatures and times were conducted. According to the GC–MS spectrum (Figure S1), oxidation products involve monophenols, carboxylic acids, alcohols, and hydroxyl acid, while small molecules such as formic acid, acetic acid, methanol, and ethanol were removed by the vacuum distillation and can be detected by HPLC. Monophenols include vanillin, *p*-hydroxybenzaldehyde, *p*-hydroxyacetophenone, vanillin acid, acetovanillone, ferulic acid, benzyl alcohol, and 4-ethylphenol. Vanillin is the major monophenol (selectivity ≥ 70%) owing to lignin in softwood (pine) mainly consisting of guaiacyl unit (type-G lignin), which can be converted into vanillin by a well-recognized mechanism of retro-aldol condensation [21, 31, 32].

As shown in Table 1, when the reaction time is 1 h, yields of vanillin and other monophenols increased from 120 to 160 °C and then decreased over 160 °C. The highest yield of vanillin (21%) was obtained at 160 °C, which was twice the yield of 120 °C; however, yields of small molecule compounds increased from 120 to 160 °C (Table 1, entries 1–5). At 160 °C, vanillin yield reached 12.9% in 10 min and increased to 21.1% when reaction time prolonged to 1 h, then

Table 1 The yield of product from pine alkaline oxidation

Entry	Condition	Yield (%)							Vanillin selectivity (%)
		Vanillin	Other monophenols	Formic acid	Acetic acid	Oxalic acid	MeOH	EtOH	
1	120 °C	8.6	3.3	1.4	3.3	0.6	1.2	0.0	72.6
2	150 °C	15.1	4.9	1.6	3.8	3.4	2.3	0.0	75.6
3	160 °C	21.1	8.3	3.5	6.1	4.5	3.2	0.2	71.8
4	170 °C	16.0	6.0	3.8	6.5	5.3	3.5	0.5	73.1
5	200 °C	15.1	6.5	4.5	6.6	5.8	3.6	0.8	70.0
6	10 min	12.9	4.2	2.1	3.4	1.2	0.6	0.0	75.5
7	30 min	15.2	5.8	2.6	4.1	2.0	1.2	0.1	72.6
8	90 min	17.9	5.6	4.3	6.3	4.5	3.6	1.2	76.1
9	180 min	15.5	5.9	4.9	7.5	5.5	3.7	1.5	72.5

Conditions: (entries 1–5) 1 g pine, 7.5 wt% NaOH, 1 MPa O_2, 400 rpm, 1 h; (entries 6–9) 1 g pine, 7.5 wt% NaOH, 1 MPa O_2, 400 rpm, 160 °C

reduced to 15.5% after reaction for 180 min. Meanwhile, the increase of reaction time led to an increase of yields of small molecule products (Table 1, entries 3, 7–9).

Type-G lignin-rich softwood (pine) was used as a raw material in alkaline oxidation experiments, but vanillin yield obtained (21%) was much lower than the theoretical value (38%, calculated on basis of ester linkages content [3]). Clarify side reactions and byproducts are of great importance for answering the question of why vanillin yield is low. Here, in addition to monophenols, by-products are small molecules, long-chain aliphatic alcohol, as well as large molecular weight lignin fragments (not listed in the table). Given that oxidation of cellulose and hemicellulose also give carboxylic acid and alcohols, analysis is very difficult in the real biomass system. Therefore, degradation of pure vanillin under alkaline oxidation conditions will be discussed below.

3.2 Alkaline Oxidation of Vanillin

According to a previous study, reaction temperature and time greatly affect vanillin yield. In order to figure out the effect of reaction temperature and time, what happened to vanillin during lignin valorization need to be clear. GC–MS analysis showed that oxidation products of vanillin are similar to the by-products of real lignin system, i.e., monophenols, carboxylic acids, alcohols, and hydroxyl acids (Figure S2). Vanillin is unstable under alkaline aerobic conditions and it will be oxidative depolymerized and re-polymerized. The conversion of vanillin and distribution of products is greatly influenced by reaction conditions. Here, degradation of vanillin under different reaction conditions will be discussed in detail.

3.2.1 Temperature

Under aerobic alkaline condition (7.5 wt% NaOH, 1 MPa O_2, 400 rpm, 30 min), an increase of temperature from 50 to 160 °C resulted in vanillin conversion increased from 11.0 to 73.9% (Table 2, entries 1–3). The molar yield of acid products was higher than that of alcohols because alcohols can be further oxidized to carboxylic acids. A small amount of formic acids (1.1 mol%) and acetic acid (3.6 mol%) can be generated at 50 °C (Table 2, entry 1). Oxidative cleavage of the methoxyl group on vanillin resulted in the formation of formic acid and *p*-hydroxybenzaldehyde, while the benzene ring-opening reaction was attributed to the generation of both formic acid and acetic acid [29]. As the reaction temperature increased to 100 °C, trace vanillic acid (0.1 mol%) was detected (Table 2, entry 2). The formation of vanillic acid resulted from the selectivity oxidation of vanillin under high pH conditions [16]. A small amount of oxalic acid (0.7 mol%) and methanol (1.9 mol%) were formed, except formic and acetic acid. Oxalic acid may be derivatives from ring-opening of vanillin or vanillic acid [27, 29]. Vanillin decomposed significantly when the reaction temperature increased to 160 °C, with high vanillin conversion (73.9%) and total yields of identified products increased to 38.2 mol% (Table 2, entry 3). In addition, vanillic acid yield was only slightly increased because a portion of vanillic acid is prone to degrade into small molecule acids and alcohols and other substances at high temperatures (Table 2, entry 3). In Figure S2, a large number of methyl 2-hydroxy-2-(4-hydroxy-3-methoxyphenyl)acetate (16.96 min) accompanied by some trace monophenols with carbon number over 8 (14.46, 14.79, 18.05, 19.60, 23.37, 24.52 min) and hydroxybenzaldehyde (11.67 min) were generated. *p*-Hydroxybenzaldehyde probably resulted from methoxylation of vanillin, while other monophenols might be the products of reaction between vanillin and small acids or alcohols. Especially, acetovanillone, *p*-hydroxybenzaldehyde, and ferulic acid generated in the vanillin oxidation system can also be observed in the alkaline oxidation of lignin (Figure S1), which means the production of these chemicals in lignin oxidation may come from both lignin itself and vanillin further oxidation. The formation of these monophenols implies that vanillin in the lignin oxidation system might convert to other monophenols and thus leaded to a decrease of vanillin selectivity. In addition, further degradation of monophenols by ring-opening cause the appearance of trace long-chain acids and esters.

Moreover, oxidation of vanillic acid was carried out to confirm whether small molecule products were generated from vanillic acid. All small molecule products were produced and acetic acid was the main product, indicating that partial degradation of vanillin resulted in the formation of vanillic acid and followed by decomposition of vanillic acid. It is noteworthy that oxidation of vanillic acid gave similar long-chain acids and esters with oxidation of vanillin (Figure S3). This finding leads us to conclude that vanillic acid is a crucial intermediate.

Obviously, most of the vanillin decomposed at 160 °C in only 30 min; nevertheless, alkaline oxidation of lignin to vanillin gave highest yield after reaction for 1 h, the loss of vanillin could be considerable.

Table 2 Yield of product from vanillin alkaline oxidation

Entry	Condition	Conversion (%)	Yield (mol %)								Carbon balance (mol%)
			Vanillin	Vanillic acid	Formic acid	Acetic acid	Oxalic acid	MeOH	EtOH	Others[a]	
1	50 °C	11.0	89.0	0.0	1.1	3.6	0.0	0.0	0.0	6.3	93.7
2	100 °C	26.0	74.0	0.1	2.9	5.6	0.7	1.9	0.0	14.8	85.2
3	160 °C	73.9	26.1	0.5	7.9	13.8	11.7	4.3	0.0	36.1	63.9
4	10 min	63.6	36.3	0.1	4.8	8.2	2.9	2.7	1.8	43.2	56.8
5	1 h	82.5	17.5	0.1	15.1	18.6	9.0	3.6	0.0	42.1	57.9
6	5 h	97.9	0.0	0.0	21.5	29.9	8.8	3.9	0.0	35.9	64.1
7	1100 rpm	42.4	57.6	0.5	7.8	13.2	2.5	1.6	0.0	16.8	83.2
8	2 MPa O_2	78.9	21.1	0.0	13.9	19.2	4.8	3.8	0.0	37.2	62.8
9	15 wt% NaOH	38.3	61.8	1.4	4.6	10.0	3.7	0.5	0.0	18.0	82.0
10	0.1 g CuO, 30 min	87.5	12.5	2.3	11.6	14.4	16.5	24.2	1.1	17.4	82.6
11	0.1 g CuO, 10 h	98.7	1.3	1.3	14.0	10.6	10.0	13.7	0.0	49.1	50.9
12	Vanillic acid	84.7	0.0	25.3	6.4	28.3	5.5	9.8	3.5	11.2	88.8

[a]Others is 100 minus carbon balance. Conditions: (entries 1–3) 0.2 g vanillin, 7.5% NaOH, 1 MPa O_2, 400 rpm, 30 min. (entries 4–6) 0.2 g vanillin, 7.5 wt% NaOH, 1 MPa O_2, 160 °C, 400 rpm. (Entry 7) 0.2 g vanillin, 7.5 wt% NaOH, 160 °C, 400 rpm, 30 min. (Entry 8) 0.2 g vanillin, 50 ml 7.5 wt% NaOH, 1 MPa O_2, 160 °C, 30 min. (Entry 9) 0.2 g vanillin, 1 MPa O_2, 160 °C, 400 rpm, 30 min. (Entries 10 and 11) 0.2 g vanillin, 50 ml 7.5 wt% NaOH, 1 MPa O_2, 160 °C, 400 rpm. (Entry 12) 0.2 g vanillic acid, 50 ml 7.5 wt% NaOH, 1 MPa O_2, 160 ° C, 400 rpm, 30 min

3.2.2 Time

In order to study the effects of reaction time on vanillin degradation, several experiments were conducted under temperature of 160 °C with reaction time differed from 10 min to 5 h (7.5 wt% NaOH, 1 MPa O_2, 400 rpm) (Table 2, entries 3–6). Vanillin conversion was 63.6 and 20.3% of vanillin was converted to small molecules of acids and alcohols in only 10 min (Table 2, entry 4), indicating that vanillin was very unstable and ring-opening reaction of vanillin could occur under high temperature [29]. After 30 min, yields of small molecule products increased to 37.7 mol%; oxalic acid yield was tripled compared to that of initial amount (2.9 mol%) (Table 2, entry 3). As the reaction time prolonged to 5 h, vanillin was almost completely degraded, long-chain acids and esters remained in the system (Figure S4); total yield of small molecule products significantly increased to 63.1 mol%, especially for formic acid and acetic acid, while yields of oxalic acid and methanol barely changed (Table 2, entries 5–6). Above results implying that vanillin is very unstable at relatively harsh condition, which could be totally converted with extend of reaction time.

3.2.3 Stirring Speed, O_2 Pressure, and Alkali Concentration

Except for temperature and time, stirring speed, O_2 pressure, and alkali concentration can also affect the conversion of vanillin. Reaction conditions of 7.5 wt% NaOH, 1 MPa O_2, 400 rpm, 160 °C, and 30 min was chosen as contrast reaction (Table 2, entry 3). When increasing the stirring speed to 1100 rpm, conversion of vanillin was decreased from 73.9 to 42.4%; however, yields of vanillic acid and small molecule products had no noteworthy change (Table 2, entries 3 and 7). Higher O_2 pressure (2 MPa) resulted in slightly increased vanillin conversion but achieved different product distribution, i.e., more formic acid and acetic acid while less oxalic acid (Table 2, entries 3 and 8). The sharp decrease of oxalic acid yield may be caused by the excessive oxidative degradation of oxalic acid by excess O_2. Meanwhile, vanillic acid yield dropped to zero, probably due to the fact that vanillic acid is more unstable under harsh reaction conditions, and degraded into more stable small molecule acids and alcohols. When changing the alkali concentration from 7.5 to 15 wt%, vanillin conversion significantly decreased to 38.3% and was accompanied by an increase in vanillic acid yield (from 0.5 to 1.4 mol%), mainly due to the fact that high alkali concentrations can protect the vanillin from decomposed to small molecules but accelerate the oxidation of vanillin to vanillic acid [16, 29, 33–37].

3.2.4 Catalyst

To obtain higher vanillin, metal oxide catalysts are widely used to selective oxidation of lignin to vanillin [5, 15, 24, 38]. Thus, stability of vanillin under aerobic alkaline conditions in the presence of catalyst is also investigated. The employment of CuO led to a significant increase of oxalic acid and methanol yields. It is likely that the presence of CuO accelerated the ring-opening reaction of vanillin and inhibited the degradation of oxalic acid (Table 2, entry 10). Extending the reaction time to 10 h, yields of formic acid and methanol increased while the yields of acetic acid and oxalic acid decreased; ultimately yields of the four products were considerable (Table 2, entry 11). More long-chain acids and esters were generated (Figure S5).

3.3 Analysis of Unknown Products

Under severe reaction conditions, vanillin degradation not only included oxidation and ring-opening reactions in the alkali oxidation system, but also involved a complex polymerization reaction that generated large molecular weight products. Structural information of these unknown products needs to be confirmed. According to Table 2, vanillin mainly degraded to small acids and alcohols along with a small amount of vanillic acid. Vanillin was almost converted after 5 h, but mole carbon balance was only 64.1 mol%, the other 35.9 mol% was still unclear (Table 2, entry 6). Similarly, more or less unknown products existed in other reactions (Table 2). Yield of the known product was less than the conversion of the starting material, probably due to that the small molecule products in the system degraded to CO_2 and water or formed macromolecular products that are difficult to detect. CO_2 was not detected in the waste gas because CO_2 molecules generated under alkaline conditions were fixed in the solution in the form of CO_3^{2-}, which cannot enter the gas phase. To verify the formation of macromolecules, GPC analysis of heavy products was performed and the results are presented in Fig. 1.

No large molecular weight products were generated at a low temperature of 50 °C, while the remaining conditions gave large molecular weight products (Fig. 1, Table 3). In a short reaction time of 30 min, condensation aggravated with the increase of reaction temperature was not obvious (Fig. 1a). Condensation was accelerated as stirring speed, alkali concentration, or O_2 pressure increased, accompanied by a widening of distribution of large molecular weight products (Fig. 1c; Table 3, entries 7–9). On the contrary, condensation became more obvious with the increase of reaction time at 160 °C. The M_w of macromolecules increased from 194 to 364 when the reaction time was extended from 30 min to 5 h (Fig. 1b). Furthermore, the addition of CuO resulted in M_w of macromolecules dramatically increased from 200 to 260 at 30 min, and reached 469 after 10-h reaction (Fig. 1d; Table 3, entries 10 and 11). In addition, vanillic acid is less stable than vanillin under the same alkaline oxidation conditions and suffered faster degradation (Table 2, entry 12) and severe condensation to give high M_w of 452 (Fig. 1d; Table 3, entry 12), which can help explain the low vanillic acid yield during vanillin oxidation (Table 2).

To shed light on the evolution of vanillin degradation in the alkali oxidation system and obtain more accurate condensation products information, LC–MS and HSQC analysis on the heavy products of reaction 3 was performed, and the results are shown in Figs. 2 and 3.

As can be seen in Fig. 2, all condensation products contained structure blocks of vanillin ($m/z = 151.0405$) or vanillic acid ($m/z = 167.0350$). The molecular weights of the products were $m/z = 191.0714$, 235.1368, 248.9611, and 264.9662, respectively. Correlation with structural information from HSQC NMR spectrum, seven monomers of I–VII were obtained, all of them were monophenols. Among

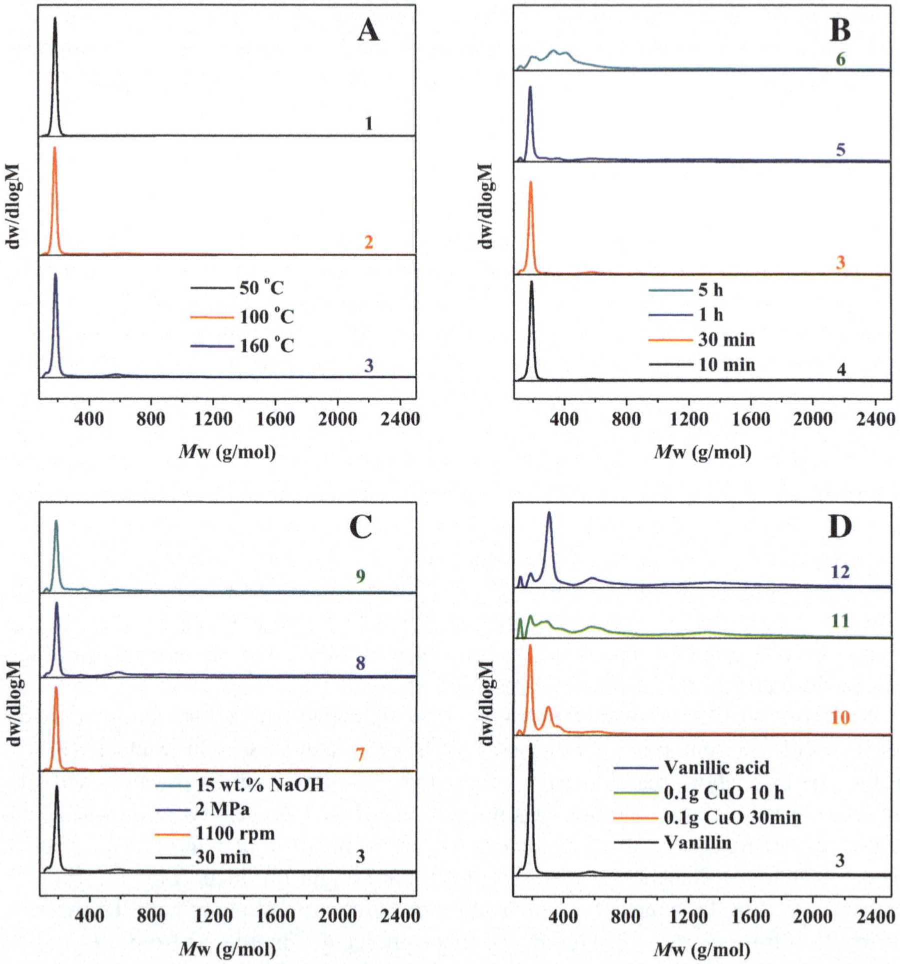

Fig. 1 Molecular weight distribution of heavy products. Conditions: (entries 1–3) 0.2 g vanillin, 7.5 wt% NaOH, 1 MPa O_2, 400 rpm, 30 min. (Entries 4–6) 0.2 g vanillin, 7.5 wt% NaOH, 1 MPa O_2, 160 °C, 400 rpm. (Entries 7) 0.2 g vanillin, 7.5 wt% NaOH, 1 MPa O_2, 160 °C, 30 min. (Entry 8) 0.2 g vanillin, 50 ml 7.5 wt% NaOH, 400 rpm, 160 °C, 30 min. (Entries 9) 0.2 g vanillin, 1 MPa O_2, 160 °C, 400 rpm, 30 min. (Entries 10 and 11) 0.2 g vanillin, 7.5 wt% NaOH, 1 MPa O_2, 160 °C, 400 rpm. (Entry 12) 0.2 g vanillic acid, 50 ml 7.5 wt% NaOH, 1 MPa O_2, 160 ° C, 400 rpm, 30 min

them, III was vanillin and IV was vanillic acid, the rest of the products could be the products of ring-opening addition of vanillin.

From the HSQC NMR spectrum of heavy products, ^{13}C–^{1}H correlations exhibited three main regions, including aliphatic (δ_C/δ_H = 0–50/0–2.5 ppm), side-chain (δ_C/δ_H = 50–95/2.5–6.0 ppm), and aromatic region (δ_C/δ_H = 5.0–8.0/100–135 ppm). In addition, strong signals at δ_C/δ_H = 191.27/9.27–10.24 ppm beyond these three regions were caused by aldehyde groups of aromatic aldehydes.

Table 3 GPC analysis of heavy products

Entry	Condition	M_n	M_w	*PD*	Entry	Condition	M_n	M_w	*PD*
1	50 °C	186	188	1.0	7	1100 rpm	199	239	1.2
2	120 °C	191	212	1.1	8	2 MPa O_2	213	338	1.6
3	160 °C	189	200	1.1	9	15 wt% NaOH	204	255	1.3
4	10 min	189	194	1.0	10	0.1 g CuO, 30 min	220	260	1.2
5	1 h	216	331	1.5	11	0.1 g CuO, 10 h	276	469	1.7
6	5 h	289	364	1.3	12	Vanillic acid	300	452	1.5

PD $=M_w/M_n$. Conditions: (entries 1–3) 0.2 g vanillin, 7.5 wt% NaOH, 1 MPa O_2, 400 rpm, 30 min. (Entries 4–6) 0.2 g vanillin, 7.5 wt% NaOH, 1 MPa O_2, 160 °C, 400 rpm. (Entries 7) 0.2 g vanillin, 7.5 wt% NaOH, 1 MPa O_2, 160 °C, 30 min. (Entry 8) 0.2 g vanillin, 50 ml 7.5 wt% NaOH, 400 rpm, 160 °C, 30 min. (Entries 9) 0.2 g vanillin, 1 MPa O_2, 160 °C, 400 rpm, 30 min. (Entries 10 and 11) 0.2 g vanillin, 7.5 wt% NaOH, 1 MPa O_2, 160 °C, 400 rpm. (Entry 12) 0.2 g vanillic acid, 50 ml 7.5 wt% NaOH, 1 MPa O_2, 160 ° C, 400 rpm, 30 min

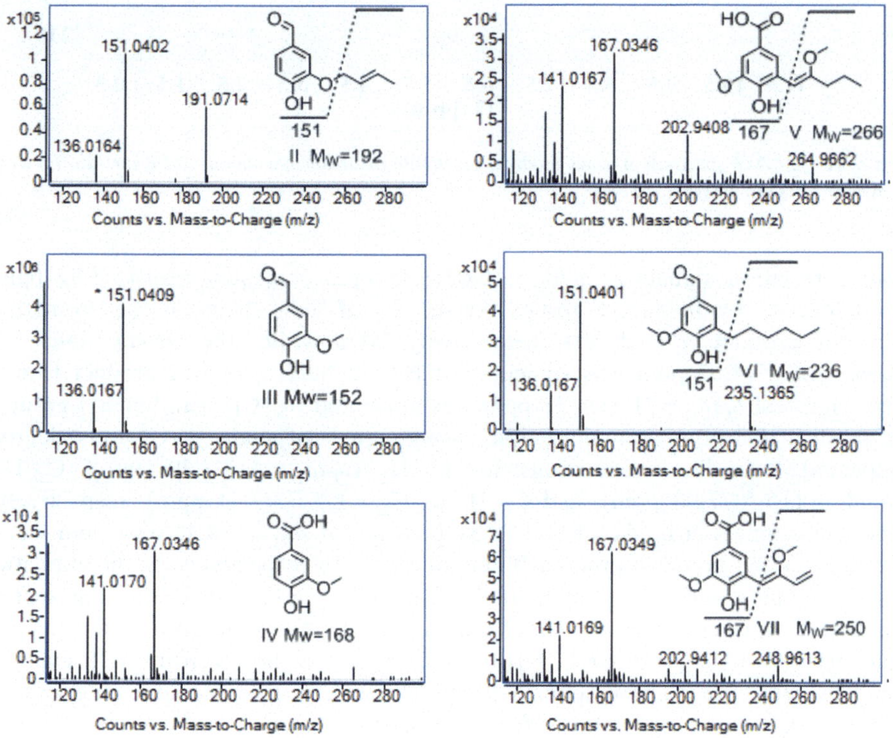

Fig. 2 LC–MS analysis of heavy products from vanillin oxidation. Conditions: 0.2 g vanillin, 7.5 wt% NaOH, 1 MPa O_2, 400 rpm, 30 min

In the side-chain region, three strong signals all resulted from methoxyl groups, the signal at $\delta_C/\delta_H=56.02/3.85$ ppm attributed to methoxyl groups linked to aromatic ring is clearly shown in the spectrum (Fig. 3) [39, 40], while the other

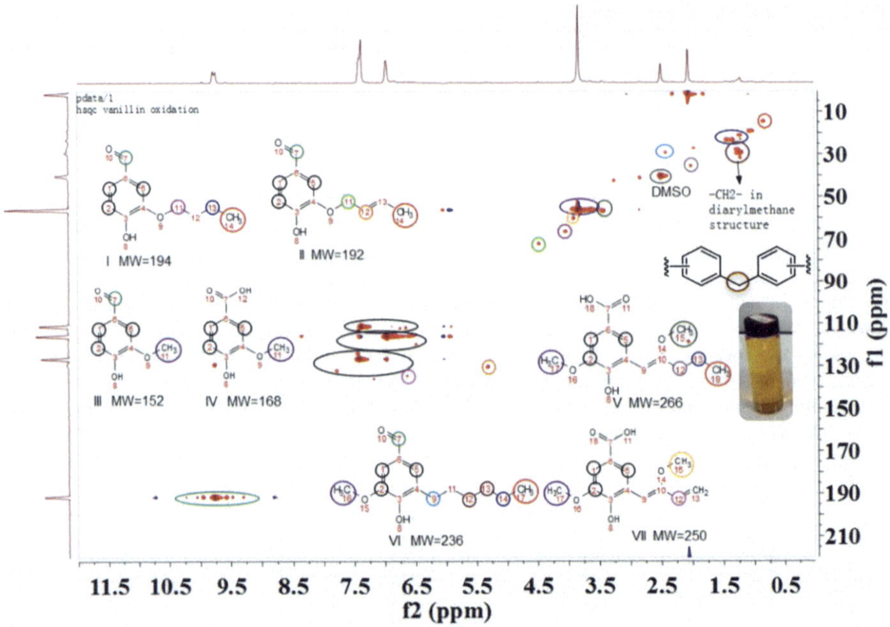

Fig. 3 HSQC NMR spectrum of heavy products of vanillin oxidation. Conditions: 0.2 g vanillin, 7.5 wt% NaOH, 1 MPa O_2, 400 rpm, 30 min

two correlation signals at $\delta_C/\delta_H=56.02/3.45$ ppm and $\delta_C/\delta_H=59.75/3.92$ ppm were caused by methoxyl groups on the C_β of side chain on the aromatic-ring of product V and VII, respectively. Meanwhile, the strong signal at $\delta_C/\delta_H=66.09/4.05$ ppm was signed to –CH_2– in butyl group of product I, and the signal at $\delta_C/\delta_H=71.81/4.48$ ppm was attributed to –CH_2– in butenyl groups of product II. In the aromatic region, strong signals of guaiacyl (G) units clearly appeared, including correlations for C_2–H_2 ($\delta_C/\delta_H=112/7.39$ ppm), C_5–H_5 ($\delta_C/\delta_H=115.54/7.00$ ppm), and C_6–H_6 ($\delta_C/\delta_H=126.75/7.44$ ppm), respectively [39–41]. Signals at $\delta_C/\delta_H=129.98/5.31$ ppm and $\delta_C/\delta_H=134.46/6.64$ ppm were signed to vinyl group of product II and product VII, respectively. In the aliphatic region, signals were caused by methyl or methylene in the side chain of aromatic ring. It is worth mentioning that strong signals of methylene in the diarylmethane structure ($\delta_C/\delta_H=29.67/1.25$ ppm) were found in the spectrum [42]. Diarylmethane structure is chromophoric group, which can well explain the colorless of solution become orange-yellow after reaction (Fig. 3).

During the alkaline oxidation of vanillin, the formation of monophenols I, II, V, VI, and VII might be attributable to substitution of C_5 and methoxyl groups in benzene of vanillin by another vanillin subsequently ring-opening reaction, or by fragments (aliphatic alcohols and acids, Figure S2) from degradation of vanillin. Moreover, the presence of diarylmethane structure in heavy products leads us to conclude that more complex condensation happened between two vanillin molecules or other monophenols and formed dimer.

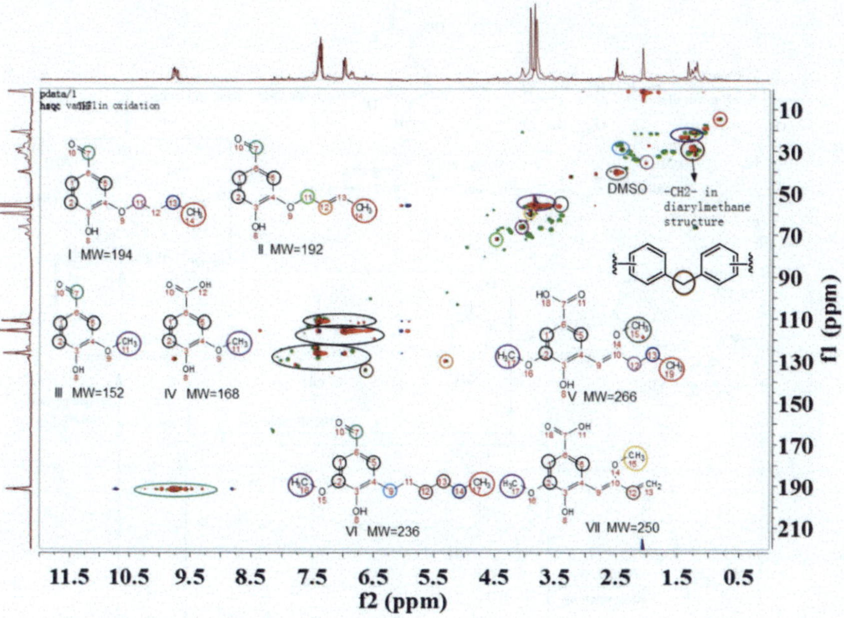

Fig. 4 HSQC NMR spectrum of heavy products of pine (green) and vanillin (red). Condition: (green) 1 g pine, 7.5 wt% NaOH, 1 MPa O_2, 400 rpm, 1 h; (red) 0.2 g vanillin, 7.5 wt% NaOH, 1 MPa O_2, 400 rpm, 30 min

To further confirm product evolution during alkaline oxidation of lignin, HSQC analysis of heavy product of lignin (from pine) alkaline oxidation under the same condition was performed. All products (I–VII) can be clearly found in the HSQC spectrum, which means that vanillin underwent the same degradation and condensation process (Fig. 4). From the results, we have obtained that instability of monophenols is a key factor of limited monophenols yield from lignin.

3.4 Presumable Mechanism of Vanillin Degradation

Correlation the structural information of the small molecule products and condensed products, a possible mechanism for vanillin degradation was deduced, as shown in Scheme 1. Vanillin degradation may occur through three routes. One route is form vanillic acid after direct oxidation and vanillic acid further degraded via ring-opening to form small molecules such as oxalic acid, acetic acid, formic acid, ethanol, methanol, and other fragments undetected (Route 1). In addition, the other two routes both start from condensation and end by ring opening. In Route 2, fragments re-condensed with vanillin to produce monophenols, such as I, II, V, VI, VII, and X (Figure S2, 16.96 min). For Route 3, vanillin molecules and/or other monophenols condense into dimers like VIII and IX. Subsequently, dimers decomposed to complex monophenols of II, V, VI, and VI. Then, all monophenols obtained by three routes will further degrade to long-chain acids

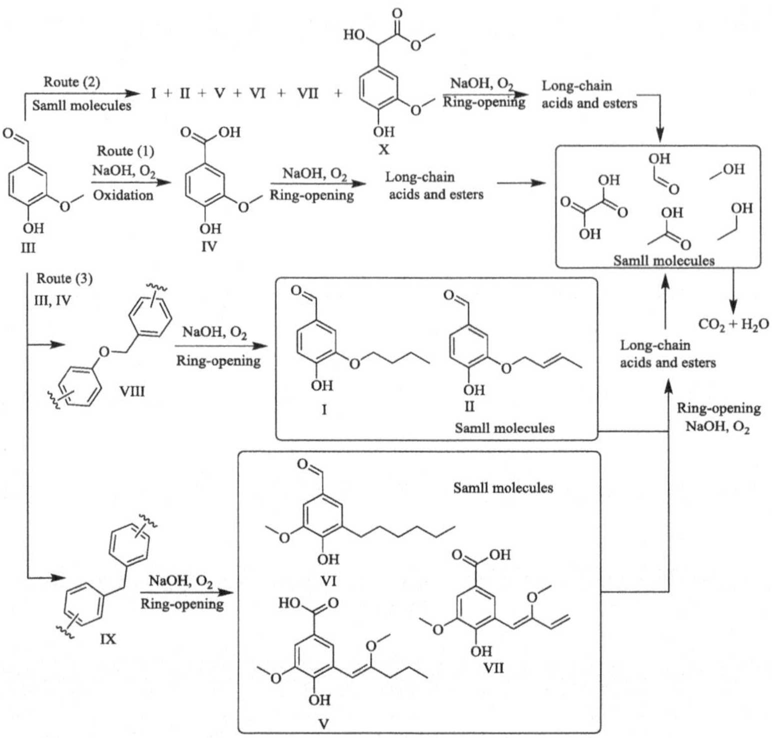

Scheme 1 Presumable mechanism of vanillin degradation

and esters via ring opening, and subsequently decomposed to small molecules. These small molecules will be completely oxidized to CO_2 and water.

4 Conclusions

In summary, vanillin is unstable under alkaline aerobic conditions, especially under high temperature. Under a temperature of 160 °C for 30 min, 73.9% vanillin was transformed into small acids, alcohols, and other monophenols, even dimers. LC–MS and HSQC were used to study the structural features of heavy products, which confirmed that vanillin condensed in the presence of NaOH and O_2. Moreover, the degradation of pure vanillin was similar to those vanillin generated during the alkaline oxidation of lignin. In addition, a presumable mechanism for the vanillin degradation was provided. These results revealed that degradation of vanillin is a crucial limiting factor of low vanillin yield by lignin alkaline oxidation. Therefore, in practical application, mild reaction conditions and in situ extraction of monophenols can be a useful strategy to avoid unnecessary condensation and decomposing, and thus increase vanillin yield.

Acknowledgements This work was supported by NSFC (National Natural Science Foundation of China) project (nos. 51476175, 51606205), the National Natural Science Foundation of China (no. 51536009) and Chinese Academy of Sciences "one hundred talented plan" (no. y507y51001).

References

1. Van den Bosch S, Schutyser W, Vanholme R, Driessen T, Koelewijn SF, Renders T, De Meester B, Huijgen WJJ, Dehaen W, Courtin CM, Lagrain B, Boerjan W, Sels BF (2015) Reductive lignocellulose fractionation into soluble lignin-derived phenolic monomers and dimers and processable carbohydrate pulps. Energy Environ Sci 8(6):1748–1763. https://doi.org/10.1039/c5ee00204d
2. Renders T, Van den Bosch S, Koelewijn S-F, Schutyser W, Sels BF (2017) Lignin-first biomass fractionation, the advent of active stabilisation strategies. Energy Environ Sci 10(7):1551–1557. https://doi.org/10.1039/C7EE01298E
3. Song Q, Wang F, Cai JY, Wang YH, Zhang JJ, Yu WQ, Xu J (2013) Lignin depolymerization (LDP) in alcohol over nickel-based catalysts via a fragmentation–hydrogenolysis process. Energy Environ Sci 6(3):994–1028. https://doi.org/10.1039/c2ee23741e
4. Deng HB, Lin L, Sun Y, Pang CS, Zhuang JP, Ouyang PK, Li JJ, Liu SJ (2009) Activity and stability of perovskite-type oxide $LaCoO_3$ catalyst in lignin catalytic wet oxidation to aromatic aldehydes process. Energy Fuels 23:19–24. https://doi.org/10.1021/ef8005349
5. Zakzeski J, Dębczak A, Bruijnincx PCA, Weckhuysen BM (2011) Catalytic oxidation of aromatic oxygenates by the heterogeneous catalyst Co-ZIF-9. Appl Catal A 394(1–2):79–85. https://doi.org/10.1016/j.apcata.2010.12.026
6. Sales FG, Abreu CAM, Pereira JAFR (2004) Catalytic wet-air oxidation of lignin in a three-phase reactor with aromatic aldehyde production. Braz J Chem Eng 21:211–218. https://doi.org/10.1590/S0104-66322004000200010
7. Werhan H, Assmann N, von Rohr PR (2013) Lignin oxidation studies in a continuous two-phase flow microreactor. Chem Eng Process 73:29–37. https://doi.org/10.1016/j.cep.2013.06.015
8. Galkin MV, Smit AT, Subbotina E, Artemenko KA, Bergquist J, Huijgen WJ, Samec JS (2016) Hydrogen-free catalytic fractionation of woody biomass. Chemsuschem 9(23):3280–3287. https://doi.org/10.1002/cssc.201600648
9. Zakzeski J, Jongerius AL, Weckhuysen BM (2010) Transition metal catalyzed oxidation of Alcell lignin, soda lignin, and lignin model compounds in ionic liquids. Green Chem 12(7):1225–1236. https://doi.org/10.1039/c001389g
10. Lv W, Si Z, Tian Z, Wang C, Zhang Q, Xu Y, Wang T, Ma L (2017) Synergistic effect of EtOAc/H_2O biphasic solvent and Ru/C catalyst for cornstalk hydrolysis residue depolymerization. Acs Sustain Chem Eng 5(4):2981–2993. https://doi.org/10.1021/acssuschemeng.6b02535
11. Zhang XH, Zhang Q, Wang TJ, Ma LL, Yu YX, Chen LG (2013) Hydrodeoxygenation of lignin-derived phenolic compounds to hydrocarbons over Ni/SiO_2–ZrO_2 catalysts. Bioresour Technol 134:73–80. https://doi.org/10.1016/j.biortech.2013.02.039
12. Liu Y, Chen LG, Wang TJ, Zhang Q, Wang CG, Yan JY, Ma LL (2015) One-pot catalytic conversion of raw lignocellulosic biomass into gasoline alkanes and chemicals over $LiTaMoO_6$ and Ru/C in aqueous phosphoric acid. Acs Sustain Chem Eng 3(8):1745–1755. https://doi.org/10.1021/acssuschemeng.5b00256
13. Li C, Zhao X, Wang A, Huber GW, Zhang T (2015) Catalytic transformation of lignin for the production of chemicals and fuels. Chem Rev 115(21):11559–11624. https://doi.org/10.1021/acs.chemrev.5b00155
14. Lange H, Decina S, Crestini C (2013) Oxidative upgrade of lignin—recent routes reviewed. Eur Polym J 49(6):1151–1173. https://doi.org/10.1016/j.eurpolymj.2013.03.002
15. Fache M, Boutevin B, Caillol S (2016) Vanillin production from lignin and its use as a renewable chemical. Acs Sustain Chem Eng 4(1):35–46. https://doi.org/10.1021/acssuschemeng.5b01344
16. Rautiainen S, Chen J, Vehkamäki M, Repo T (2016) Oxidation of vanillin with supported gold nanoparticles. Top Catal 59(13–14):1138–1142. https://doi.org/10.1007/s11244-016-0633-8
17. Patil DG, Magdum PA, Nandibewoor ST (2015) Mechanistic investigations of uncatalyzed and ruthenium(III) catalyzed oxidation of vanillin by periodate in aqueous alkaline medium. J Solution Chem 44(6):1205–1223. https://doi.org/10.1007/s10953-015-0341-1

18. Fache M, Boutevin B, Caillol S (2015) Vanillin, a key-intermediate of biobased polymers. Eur Polym J 68(SI):488–502. https://doi.org/10.1016/j.eurpolymj.2015.03.050
19. Bomgardner MM (2014) Following many routes to naturally derived vanillin. Chem Eng News 92(6):14–15. https://doi.org/10.1021/cen-09232-bus2
20. Tarabanko VE, Koropatchinskaya NV, Kudryashev AV, Kuznetsov BN (1995) Influence of lignin origin on the efficiency of the catalytic oxidation of lignin into vanillin and syringaldehyde. Russ Chem Bull 44:367–371. https://doi.org/10.1007/bf00702154
21. Tomlinson GH, Hibbert H (1936) Studies on lignin and related compounds. XXV. Mechanism of vanillin formation from spruce lignin sulfonic acids in relation to lignin structure. J Am Chem Soc 58:348–353. https://doi.org/10.1021/ja01293a047
22. Kim JS, Lee YY, Kim TH (2016) A review on alkaline pretreatment technology for bioconversion of lignocellulosic biomass. Bioresour Technol 199:42–48. https://doi.org/10.1016/j.biortech.2015.08.085
23. Behling R, Valange S, Chatel G (2016) Heterogeneous catalytic oxidation for lignin valorization into valuable chemicals: what results? What limitations? What trends? Green Chem 18(7):1839–1854. https://doi.org/10.1039/c5gc03061g
24. Deng WP, Zhang HX, Wu XJ, Li RS, Zhang QH, Wang Y (2015) Oxidative conversion of lignin and lignin model compounds catalyzed by CeO_2-supported Pd nanoparticles. Green Chem 17(11):5009–5018. https://doi.org/10.1039/C5GC01473E
25. Rodrigues Pinto PC, Borges da Silva EA, Rodrigues AE (2011) Insights into oxidative conversion of lignin to high-added-value phenolic aldehyde. Ind Eng Chem Res 50:741–748. https://doi.org/10.1021/ie102132a
26. Wu GX, Heitz M, Chornet E (1994) Improved alkaline oxidation process for the production of aldehydes (vanillin and syringaldehyde) from steam-explosion hardwood lignin. Ind Eng Chern Res 33:718–723. https://doi.org/10.1021/ie00027a034
27. Fargues C, Mathias Á, Silva J, Rodrígues A (1996) Kinetics of vanillin oxidation. Chem Eng Technol 19(2):127–136. https://doi.org/10.1002/ceat.270190206
28. Sultanov VS, Wallis AFA (1991) Reactivities of guaiacyl and syringyl lignin model phenols towards oxidation with oxygen-alkali. J Wood Chem Technol 11(3):291–305. https://doi.org/10.1080/02773819108050276
29. Augugliaro V, Camera-Roda G, Loddo V, Palmisano G, Palmisano L, Parrino F, Puma MA (2012) Synthesis of vanillin in water by TiO_2 photocatalysis. Appl Catal B 111–112:555–561. https://doi.org/10.1016/j.apcatb.2011.11.007
30. Bjørsvik H-R (1999) Fine chemicals from lignosulfonates. 1. Synthesis of vanillin by oxidation of lignosulfonates. Org Process Res Dev 3:330–340. https://doi.org/10.1021/op9900028
31. Tarabanko VE, Petukhov DV, Selyutin GE (2004) New mechanism for the catalytic oxidation of lignin to vanillin. Kinet Catal 45(4):569–577. https://doi.org/10.1023/B:KICA.0000038087.95130.a5
32. Tarabanko VE, Fomova NA, Kuznetsov BN, Ivanchenko NM, Kudryashev AV (1995) On the mechanism of vanillin formation in the catalytic oxidation of lignin with oxygen. React Kinet Catal Lett 55:161–170. https://doi.org/10.1007/BF02075847
33. Mathias AL, Rodrigues AB (1995) Production of vanillin by oxidation of pine kraft lignins with oxygen. Holzforschung 49(3):273–278. https://doi.org/10.1515/hfsg.1995.49.3.273
34. Borges da Silva EA, Zabkova M, Araújo JD, Cateto CA, Barreiro MF, Belgacem MN, Rodrigues AE (2009) An integrated process to produce vanillin and lignin-based polyurethanes from Kraft lignin. Chem Eng Res Des 87(9):1276–1292. https://doi.org/10.1016/j.cherd.2009.05.008
35. Araújo JDP, Grande CA, Rodrigues AE (2010) Vanillin production from lignin oxidation in a batch reactor. Chem Eng Res Des 88(8):1024–1032. https://doi.org/10.1016/j.cherd.2010.01.021
36. Klinke HB, Ahring BK, Schmidt S, Thomsen AB (2002) Characterization of degradation products from alkaline wet oxidation of wheat straw. Bioresour Technol 82(1):15–26. https://doi.org/10.1016/S0960-8524(01)00152-3
37. Fargues C, Mathias Á, Rodrigues A (1996) Kinetics of vanillin production from kraft lignin oxidation. Ind Eng Chem Res 35(1):28–36. https://doi.org/10.1021/ie950267k
38. Shilpy M, Ehsan MA, Ali TH, Abd Hamid SB, Ali ME (2015) Performance of cobalt titanate towards H_2O_2 based catalytic oxidation of lignin model compound. Rsc Adv 5(97):79644–79653. https://doi.org/10.1039/c5ra14227j
39. Lahive CW, Deuss PJ, Lancefield CS, Sun Z, Cordes DB, Young CM, Tran F, Slawin AM, de Vries JG, Kamer PC, Westwood NJ, Barta K (2016) Advanced model compounds for understanding

acid-catalyzed lignin depolymerization: identification of renewable aromatics and a lignin-derived solvent. J Am Chem Soc 138(28):8900–8911. https://doi.org/10.1021/jacs.6b04144
40. Jiang ZC, Zhang H, He T, Lv XY, Yi J, Li JM, Hu CW (2016) Understanding the cleavage of inter- and intramolecular linkages in corncob residue for utilization of lignin to produce monophenols. Green Chem 18(14):4109–4115. https://doi.org/10.1039/c6gc00798h
41. Yang HT, Xie YM, Zheng X, Pu YQ, Huang F, Meng XZ, Wu WB, Ragauskas A, Yao L (2016) Comparative study of lignin characteristics from wheat straw obtained by soda-AQ and kraft pretreatment and effect on the following enzymatic hydrolysis process. Bioresour Technol 207:361–369. https://doi.org/10.1016/j.biortech.2016.01.123
42. Froass PM, Ragauskas AJ, J-e Jiang (1996) Chemical structure of residual lignin from kraft pulp. J Wood Chem Technol 16(4):347–365. https://doi.org/10.1080/02773819608545820

Affiliations

Yuting Zhu[1,2,3,4] · Jing Liu[1,2,3,4] · Yuhe Liao[5] · Wei Lv[1,2,3] · Longlong Ma[1,2,3] · Chenguang Wang[1,2,3]

[1] Guangzhou Institute of Energy Conversion, Chinese Academy of Sciences, Guangzhou 510640, People's Republic of China

[2] Key Laboratory of Renewable Energy, Chinese Academy of Sciences, Guangzhou 510640, People's Republic of China

[3] Guangdong Provincial Key Laboratory of New and Renewable Energy Research and Development, Guangzhou 510640, People's Republic of China

[4] University of Chinese Academy of Sciences, Beijing 100049, People's Republic of China

[5] Center for Surface Chemistry and Catalysis, Katholieke Universiteit Leuven, Celestijnenlaan 200F, 3001 Heverlee, Belgium

Topics in Current Chemistry (2018) 376:30
https://doi.org/10.1007/s41061-018-0207-2

REVIEW

Perspective on Lignin Oxidation: Advances, Challenges, and Future Directions

Thijs Vangeel[1] · Wouter Schutyser[1] · Tom Renders[1] · Bert F. Sels[1]

Received: 28 February 2018 / Accepted: 19 June 2018 / Published online: 4 July 2018

Abstract
Lignin valorization has gained increasing attention over the past decade. Being the world's largest source of renewable aromatics, its valorization could pave the way towards more profitable and more sustainable lignocellulose biorefineries. Many lignin valorization strategies focus on the disassembly of lignin into aromatic monomers, which can serve as platform molecules for the chemical industry. Within this framework, the oxidative conversion of lignin is of great interest because it enables the formation of highly functionalized, valuable compounds. This work provides a brief overview and critical discussion of lignin oxidation research. In the first part, oxidative conversion of lignin models and isolated lignin streams is reviewed. The second part highlights a number of challenges with respect to the substrate, catalyst, and operating conditions, and proposes some future directions regarding the oxidative conversion of lignin.

Keywords Lignin · Oxidation · Catalysis · Biorefinery · Model compounds

1 Introduction

Lignocellulosic biomass is a promising feedstock for renewable materials, chemicals, and fuels. It is primarily composed of the carbohydrates cellulose and hemicellulose on the one hand, and lignin, a complex phenolic polymer, on the other hand [1]. Lignocellulose processing takes place in a so-called lignocellulose biorefinery, which usually involves biomass fractionation into a carbohydrate- and lignin-derived product stream. While the carbohydrates are used for the production of

---

Chapter 3 was originally published as Vangeel, T., Schutyser, W., Renders, T. & Sels, B. F. Topics in Current Chemistry (2018) 376: 30. https://doi.org/10.1007/s41061-018-0207-2.

Thijs Vangeel and Wouter Schutyser contributed equally to this work.

✉ Bert F. Sels
bert.sels@kuleuven.be

1 Center for Surface Chemistry and Catalysis, KU Leuven, Celestijnenlaan 200F, 3001 Leuven, Belgium

high-value products [2–4], the lignin fraction is primarily burned for energy production. However, lignin constitutes the largest source of bio-aromatics and its conversion into high-value products could significantly augment the viability of lignocellulose biorefineries [1, 5–7]. Therefore, lignin valorization is one of the foremost challenges in current biorefinery research [1, 5–8].

One promising valorization route is the conversion of lignin into chemicals. This involves disassembling of the lignin structure into its phenolic building blocks, which can further be transformed into targeted end products [1]. Lignin is mainly composed of three phenyl propane units, with either a *p*-hydroxyphenyl—(H), guaiacyl—(G), or syringyl—(S) core. These units are connected through various ether and carbon–carbon bonds [1, 6, 8–11]. The β-*O*-4 ether bond is the most abundant linkage, at least in native lignin (see Fig. 1). Lignin can also contain other building blocks, such as *p*-hydroxybenzoic acid (for instance in poplar), and *p*-coumaric and ferulic acid (in herbaceous crops), which are mainly connected to lignin through ester linkages [1]. While native lignin is highly reactive towards depolymerization, lignin streams isolated from biorefinery processes such kraft, sulphite, or organosolv pulping are much more recalcitrant, which is due to severe structural degradation taking place during the biorefinery process [1, 9]. Structural degradation involves cleavage of labile ether and ester linkages (mainly the β-*O*-4 ether bond), and formation of stable carbon–carbon linkages through recondensation [1, 8, 12]. Breaking these stable carbon–carbon bonds in a following depolymerization step is very difficult, explaining the low reactivity of degraded lignins. In order to facilitate lignin depolymerization, recently developed biorefinery methods, such as reductive catalytic fractionation (RCF) [13], formaldehyde-assisted fractionation [14], butanosolv pulping [15], or ammonia pretreatment [16, 17], make it possible to minimize lignin condensation during the fractionation stage [12].

Irrespective of the isolation method, various lignin conversion approaches exist, each with their own advantages and limitations. Major strategies include alkaline, acidic, thermal, reductive and oxidative lignin conversion [1, 6, 8, 9]. Oxidative conversion is of great interest, as it can generate highly functionalized, valuable structures such as vanillin [1, 8, 9, 18, 19]. A lot of recent research is devoted to lignin oxidation, which has mainly been performed on either lignin model compounds or real, isolated lignin streams. In this contribution, a brief overview of the available oxidative systems for model compounds and isolated lignins is provided. In a second part, critical remarks, challenges and future directions concerning the oxidation of lignin are addressed.

2 Overview of Lignin Oxidation Research

2.1 Oxidation of Dimeric Model Compounds

This section gives a brief overview of the main catalytic systems available for the oxidation of lignin model compounds. In order to keep it concise, only catalytic systems targeting β-*O*-4 and β-1 dimeric models (Fig. 1) are included. These models can either be phenolic (with a free hydroxyl group at the 4-position of the A ring) or non-phenolic (unsubstituted or with an alkoxy group at the 4-position). The catalytic methods are divided into three groups: (1) methods that enable the cleavage of non-oxidized lignin dimeric structures, (2) methods that oxidize side-chain aliphatic alcohol groups without cleaving the inter-unit linkages, thus generating oxidized dimeric structures, and (3) methods that catalyze the cleavage of oxidized dimeric structures. In the latter group, also non-oxidative methods are included. Schematic overviews of the most common catalytic systems are depicted in Figs. 1, 3, and 4.

2.1.1 Oxidative Cleavage of Non-oxidized Dimeric Structures

Oxovanadium complexes are among the most studied catalysts for the oxidation of dimeric lignin model compounds (Fig. 1) [20–28]. Using air or oxygen as oxidant (i.e., aerobic oxidation), a wide range of oxidation products can be obtained and the selectivity can be tuned by changing the substrate or the ligand structure [23]. The most studied oxovanadium complexes are depicted in Fig. 2. Complex 1a catalyzes the redox-neutral cleavage of β-*O*-4 models (phenolic and non-phenolic), resulting in aryl enones (**2**) [23, 25]. Although no oxidant is required for the reaction, optimal activity was achieved under aerobic conditions [25]. Complex 2a exhibits a remarkably different selectivity. Non-phenolic β-*O*-4 and β-1 models were oxidized primarily into C_α ketones (**1**), while the phenolic analogues were converted mainly into *p*-quinones (**4**), next to C_α ketones [21, 23]. Complex 3a catalyzed the oxidation of non-phenolic β-*O*-4 compounds into a mixture of C_α ketones, aryl enones and aromatic acids (**3**) [22], while phenolic β-1 models were converted primarily into C_γ aldehydes (**5**) [21]. In addition to model compounds, also organosolv lignin was subjected to oxidation and depolymerization with complexes 1a, 2a, and 3a. All three catalysts induced a decrease in molecular weight and an increase in side-chain oxidation, but the effects were most pronounced for complex 2a [28].

The catalytic system CuX/L/TEMPO [X: e.g., Cl, OTf; L: e.g., (bi)pyridine, 2,6-lutidine; TEMPO: (2,2,6,6-tetramethylpiperidin-1-yl)oxyl, structure in Fig. 2] is also commonly used for aerobic oxidation of lignin models, and shows a different product selectivity in the conversion of phenolic and non-phenolic lignin dimeric

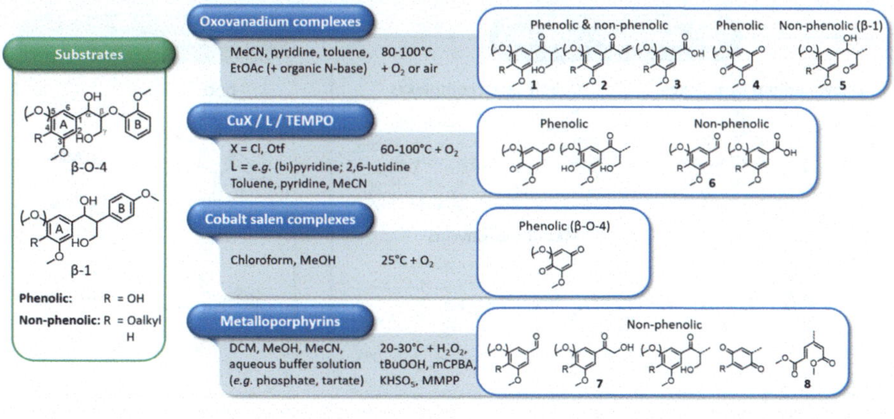

Fig. 1 Overview of catalytic systems used for the oxidation of non-oxidized dimeric lignin models

models. Non-phenolic models were oxidized primarily into aldehydes (**6**), while phenolic compounds were converted into *p*-quinones and C_α ketones [22, 29]. The depolymerization of organosolv lignin was inefficient, which was ascribed to radical coupling reactions [28].

Cobalt salen complexes enable aerobic lignin oxidation at room temperature (representative structure in Fig. 2). Non-phenolic β-*O*-4 models showed low reactivity and generated polymerization products. Phenolic β-*O*-4 models on the other hand were converted into *p*-quinones [30]. Cobalt salen complexes bearing bulky *N*-bases enabled, in addition to S-models, the selective conversion of less reactive G-models into the corresponding *p*-quinones [31]. Direct oxidation of organosolv lignin yielded aldehydes and *p*-quinones in roughly equal proportions, but the absolute monomer yield was low (3.5 wt%) [31].

Synthetic *metalloporphyrins* represent biomimetic systems for lignin peroxidase (LiP) and manganese dependent peroxidase enzymes (MnP) [9, 10, 18, 32], and enable the oxidation of lignin models at room temperature. These catalysts form highly oxidized metallo–oxo complexes upon reaction with the oxidant species. The basic structure of metalloporphyrins is shown in Fig. 2 [18, 32]. Catalyst properties like activity, selectivity, solubility, and stability can be modified by changing the R_1 and R_2 substituents [18, 32]. Also, the reaction medium, which can comprise organic solvents or aqueous buffer solutions, strongly impacts the reaction [33, 34]. Oxidation of dimeric lignin model compounds has mainly been investigated with Fe- and Mn-containing porphyrins, with various oxidants such as H_2O_2, tBuOOH, $KHSO_5$ and magnesium monoperoxyphthalate (MMPP) [10, 33–36]. In general, Mn complexes performed better than the corresponding Fe complexes in the oxidation of lignin models [32]. Mostly non-phenolic (methylated or ethylated) β-*O*-4 and β-1 model compounds have been studied, which yielded products such as aromatic aldehydes, 2-hydroxyethanone-substituted aromatics (**7**), C_α ketones, quinones and muconic acids (**8**) [33–36]. To address problems such as catalyst loss and degradation, immobilized metalloporphyrins, for instance on silica gels or clays, were examined for lignin oxidation and showed promising results [9, 10, 32, 37].

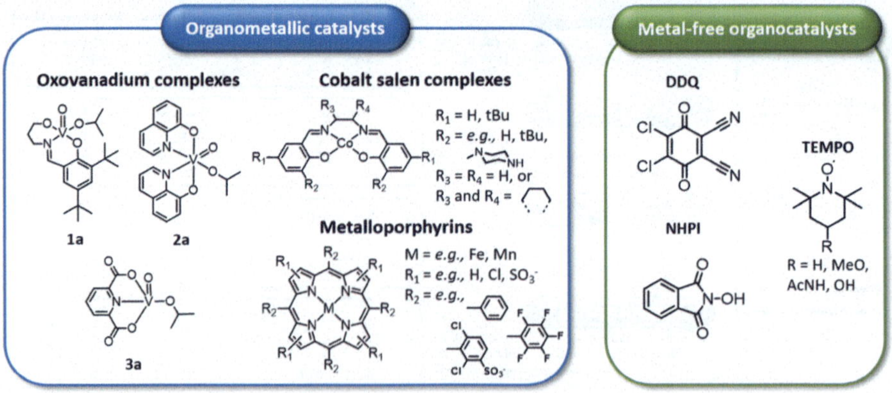

Fig. 2 Main catalysts for the oxidation of dimeric lignin models

In addition to the catalytic systems described above, several other systems have been studied for the oxidative cleavage of dimeric models. Non-phenolic β-*O*-4 models were converted into (1) aromatic aldehydes by persulfates (non-catalytic) [38] and an Fe complex with H_2O_2 or tBuOOH in DMSO [39], (2) aromatic acids by $CuCl_2$/polybenzoxazine composite catalysts in combination with H_2O_2 [40], (3) a mixture of aromatic aldehydes and acids by Cu- and V-containing hydrotalcites with oxygen [41] and (4) aceto-derivatives by Pd/CeO_2 with oxygen [42]. Non-phenolic β-1 models were transformed into aromatic aldehydes by persulfates [38], Cu–*N*-heterocyclic carbene with oxygen [43] and CuO_x/CeO_2/TiO_2 anatase nanotubes with oxygen under visible light irradiation [44]. While conversion of phenolic β-*O*-4 models was ineffective with persulfates [38], graphene oxide in presence of oxygen could transform these into products such as guaiacol, *p*-quinones, aromatic acids and aldehydes [45]. Methyltrioxorhenium ($MeReO_3$) in combination with H_2O_2 was able to convert both phenolic and non-phenolic β-*O*-4 models into various oxygenated products (e.g., 2-hydroxyethanone-, 1-hydroxyacetic acid-substituted aromatics and aromatic acids) [46].

2.1.2 Oxidation of Side-Chain Alcohol Groups in Dimeric Structures

Oxidation of the side-chain alcohol groups in β-*O*-4 models constitutes an intriguing strategy to either (1) weaken the β-*O*-4 bond and thus facilitate its cleavage, (2) promote other depolymerization routes (e.g., Baeyer–Villager oxidation), or (3) add functionality to the lignin polymer. In this section, catalytic systems that enable the selective oxidation of the benzylic alcohol (α-OH) or terminal alcohol (γ-OH), yielding the corresponding C_α ketone or C_γ aldehyde, respectively, will be discussed. An overview is provided in Fig. 3.

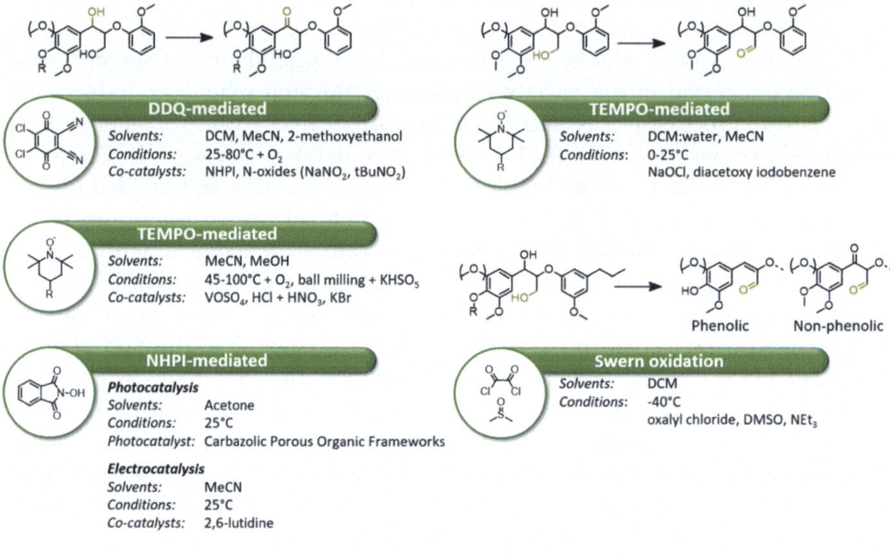

Fig. 3 Selective oxidation of α-OH or γ-OH groups in dimeric lignin models

A variety of catalytic systems has been presented for the selective benzylic oxidation with oxygen. The most common catalytic systems involve the (metal-free) organocatalysts DDQ (2,3-dichloro-5,6-dicyano-1,4-benzoquinone) [47, 48], TEMPO [49, 50] and NHPI (*N*-hydroxyphthalimide) [51] (Fig. 2), in combination with nitrogen oxides (e.g., HNO_3, tBuONO) or other co-catalysts (e.g., $VOSO_4$). NHPI has also been applied in electrocatalytic [52] and photocatalytic [53] oxidation of the benzylic OH group. In addition, metal catalysts, such as Pd/C [54] and certain oxovanadium complexes (vide supra), enable selective benzylic oxidation. Many of these catalytic systems have been successfully subjected to isolated lignin substrates, with HSQC NMR being used to track the formation of C_α ketones [47, 49, 51, 52, 55, 56]. Next to aerobic oxidation, selective benzylic oxidation is possible with other oxidants, such as DDQ and TEMPO [49, 57, 58]. Ball milling with $KHSO_5$ as oxidant and KBr/HO-TEMPO as catalysts was recently shown to enable a selective oxidation of the benzylic OH group [59]. Furthermore, dehydrogenation of the benzylic alcohol group, with concomitant H_2 production, was demonstrated with an Ir-complex [55].

Only a few studies have reported the selective oxidation of the terminal γ-OH group into the corresponding C_γ aldehyde. Different (exotic) systems were tested, often using oxidants other than oxygen. TEMPO was found to catalyze this reaction using NaOCl [49] or (diacetoxyiodo)benzene [60] as oxidant. Swern oxidation (i.e., oxidation with DMSO and oxalyl chloride) enabled the oxidation of the γ-OH group in phenolic β-*O*-4 models, and also dehydrated the benzylic OH group, generating enol–ether compounds [61]. In non-phenolic models on the other hand, both OH groups (α and γ) were oxidized into the corresponding carbonyl groups.

2.1.3 Cleavage of Oxidized Dimeric Structures

The C_α ketones and C_γ aldehydes generated through oxidation of dimeric models can be converted into a variety of products through oxidative, reductive or redox-neutral processing, as shown in Fig. 4. A first oxidative approach for cleaving C_α ketones (of both β-*O*-4 and β-1 models) is Baeyer–Villiger oxidation, which has been performed with H_2O_2 or a peroxy acid and yielded the corresponding esters [48, 62, 63]. These esters could be cleaved through subsequent solvolysis [62]. A second oxidative approach is the Cu-catalyzed aerobic cleavage of the C_α–C_β bond in C_α ketones (β-*O*-4 and β-1), which produced the corresponding acids or esters [50, 64–66]. Reductive approaches focus on cleaving the weakened β-*O*-4 bond in C_α ketones. This has been done through photocatalysis (**11**) [53, 67], (transfer) hydrogenolysis (**9, 10**) [51, 54], or using zinc as selective reductant (**11**) [47, 58]. Redox-neutral cleavage of the β-*O*-4 bond in C_α ketones was possible by reaction in aqueous formic acid/formate, which yielded mainly 1,2-propanedione-substituted phenolics (**12**) [68]. C_γ aldehydes of β-*O*-4 models on the other hand could be converted into aromatic aldehydes through a base-catalyzed retro-aldol reaction [60].

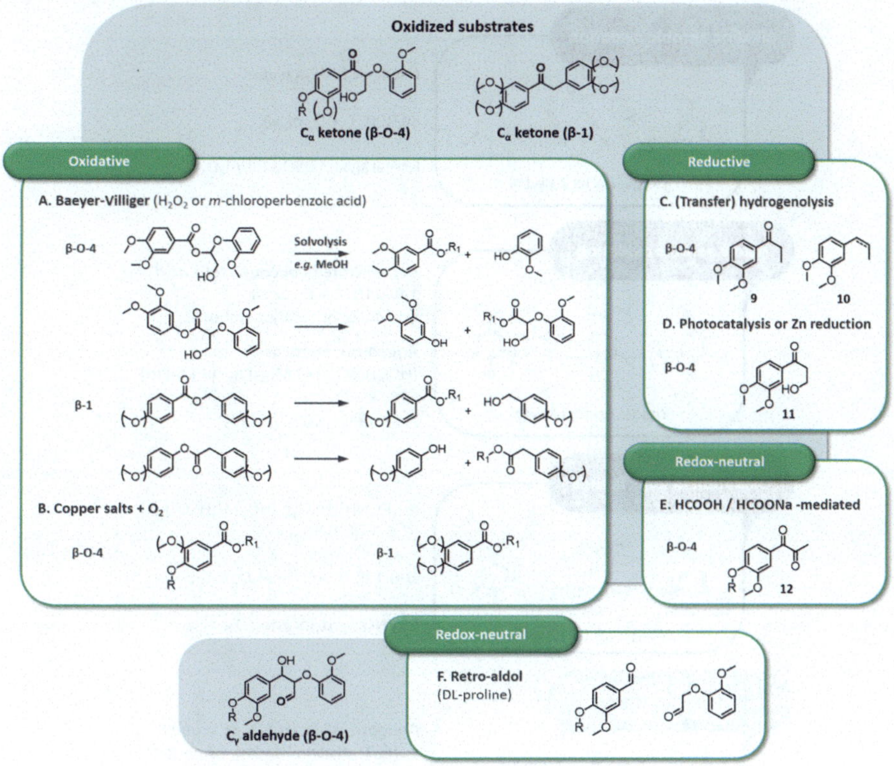

Fig. 4 Cleavage of oxidized dimeric lignin models

2.2 Oxidation of Isolated Lignin Substrates

Next to the conversion of model compounds, extensive research has been devoted to the oxidative depolymerization of isolated lignins. This topic was recently reviewed in detail by our group [1]. The main processes for lignin oxidation are outlined in Fig. 5.

The most common oxidants for lignin conversion are oxygen and hydrogen peroxide, but also nitrobenzene and CuO are frequently used for analytical purposes [69]. When aromatic products are targeted, the oxidation is mainly performed aerobically in aqueous alkaline media (usually NaOH), since this enables the selective production of aromatic aldehydes such as vanillin and syringaldehyde [1, 19, 70, 71]. Also some aromatic acids and acetophenone-like compounds are obtained. The reaction can be performed without a catalyst, although catalysts such as $CuSO_4$ have been shown to enhance the product yields [1, 19, 70, 71]. Aerobic lignin oxidation into aromatics has also been demonstrated in acidic media, either in concentrated acetic acid with a Co–Mn–Zr–Br catalyst system [72] or with diluted aqueous mineral acids such as HCl or $H_3PMo_{12}O$ [40, 73, 74], yielding aromatic aldehydes and acids (or esters). Other studies have examined the oxidation of lignin in ionic liquids (ILs) with oxygen or hydrogen

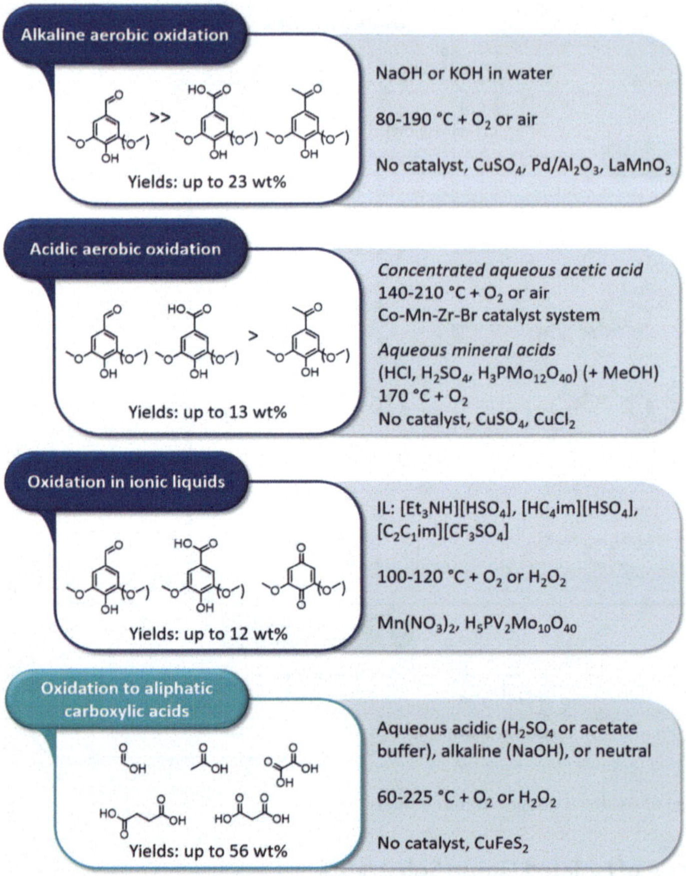

Fig. 5 Overview of main processes for oxidative conversion of real lignin substrates

peroxide [1, 75–77]. in an alcohol solvent (methanol) with oxygen [42], and with a peroxy acid (peracetic acid) [78]. In these processes, aromatic aldehydes and acids, quinones and hydroquinones were obtained. Next to aromatic products, lignin oxidation can also be steered towards the generation of small carboxylic acids, such as formic, acetic and oxalic acid [1, 79, 80]. This has been demonstrated under acidic, alkaline and neutral conditions, with oxygen or hydrogen peroxide as oxidant.

3 Challenges in Lignin Oxidation

While lignin oxidation holds great potential for production of chemicals, there are still a lot of difficulties to overcome. Some challenges related to lignin oxidation are discussed here. A distinction is made between (1) substrate- and (2) catalyst/process-related challenges.

3.1 Substrate-Related Challenges

Most studies on lignin oxidation utilize monomeric or dimeric lignin model compounds as substrate. Although many of these substrates closely mimic the structure of real lignin, considerable research is performed with oversimplified model compounds that do not represent the actual lignin structures. For instance, β-*O*-4 model compounds often lack the phenolic OH group, γ-CH_2OH group, and even the α-OH group, while these all significantly influence the reactivity towards oxidation. Especially the influence of the phenolic OH group is frequently overlooked, as most oxidation studies use non-phenolic β-*O*-4 model compounds. The presence or absence of a phenolic OH group usually implies a different reaction mechanism, resulting in a different product selectivity. For instance, phenolic moieties can form phenoxyl radicals and phenolate anions, which enable additional reaction mechanisms or stabilization through electron delocalization. Many oxidative systems can selectively convert non-phenolic β-*O*-4 models, while the conversion of phenolic models is either unselective, proceeds slower, or shows a completely different product selectivity [23, 27, 38, 49, 81].

In order to be applicable to real lignin, an oxidation system should be able to convert phenolic model compounds, because free, non-etherified phenolic units constitute a significant fraction of the aromatic units in lignin (7–13% in native wood lignin [82]). This amount increases upon processing as lignin ether-bonds are cleaved. Furthermore, the question needs to be asked if the non-phenolic model compounds that are used in lignin depolymerization studies adequately represent the reactivity of non-phenolic, etherified units in the lignin polymer. Non-phenolic units in lignin constitute moieties in which the phenolic OH group of the A unit (Fig. 1) is connected to another propylphenol unit, either in the α, β, or 5-position. In the main non-phenolic model compounds, i.e., methylated β-*O*-4 compounds, the phenolic OH group is connected to a methyl group, and this bond is much more persistent than ether bonds in lignin. This is evidenced by a large number of depolymerization studies on methylated β-*O*-4 models, in which the β-*O*-4 ether linkage is cleaved (since the B unit is obtained as a 'free' phenolic compound), while the ether linkage between the methyl group and the A unit is retained (since the A unit is obtained as a methylated phenolic compound) [25, 39, 68, 83, 84]. The presence of the methyl group might thus hinder or alter the conversion routes that take place in actual lignin.

Besides examining the conversion of relevant phenolic model compounds, it is important to evaluate the conversion of real lignins, because its complex three-dimensional structure might significantly impact the reaction. Several studies have investigated the conversion of isolated lignins, in addition to model compounds, but the analysis is often limited to assessing the decrease in molecular weight (by GPC) or the introduction of new functional groups (by NMR) [39–41, 46, 55, 61]. However, analyzing the structure (and yield) of the volatile compounds (by GC) is very informative, as this makes it possible to verify if similar products are obtained from real lignin as from the model compounds. Also here, a discrepancy is frequently observed in the product distribution from non-phenolic model compounds and lignin. For instance, Wang et al. obtained a mixture of acetophenone and an aromatic ester from the oxidation of a non-phenolic β-*O*-4 model compound, while the conversion of real lignin generated mainly aromatic aldehydes [42]. Lu et al. obtained vanillic acid as main product in the oxidation of a non-phenolic model, while real lignin primarily yielded vanillin. In the latter study, conversion of a phenolic model also generated vanillin as main product, which seems to indicate that the real lignin largely follows the same reaction course as the phenolic model compound [81]. Other studies however reported very similar products obtained from the conversion of non-phenolic models and real lignin substrates [47, 68].

3.2 Catalyst- and Process-Related Challenges

Lignin oxidation studies are mainly performed with homogeneous catalysts. These are usually difficult to separate and regenerate, while this is of utmost importance for large-scale processes. Furthermore, many studies use expensive catalyst systems comprising complex structures (e.g., oxovanadium complexes, salen complexes and metalloporphyrins) and multiple cocatalysts (e.g., CuCl/TEMPO/2,6-lutidine [29], $NaNO_2$/DDQ/NHPI [51], and HCl/HNO_3/AcNH-TEMPO [49]). Also, the complexity of real lignin is out of tune with the high specificity offered by these complex catalyst systems. Thus, applicability to real lignin and recyclability of the catalyst(s) should be primary objectives in lignin research. Catalyst recyclability is generally more practical for heterogeneous catalysts, although it might be difficult to attain a similar catalytic activity and selectivity. A large number of studies however demonstrate the effective oxidation of lignin with heterogeneous catalysts, which is discussed in dedicated reviews [85, 86].

Another challenge is related to the media used for lignin oxidation, which cause certain process and/or down-stream operational problems. Oxidation of lignin, and specifically of lignin model compounds, is frequently performed in organic solvents. These solvents are often not green, hazardous and/or flammable (e.g., pyridine, toluene, acetonitrile, methanol) [87]. The combination of a flammable solvent with oxygen is potentially explosive. Oxidation reactions should be performed outside the flammable region, which depends on the solvent (fuel) and oxygen concentration in the gas phase. Data regarding the flammable regions, however, are scarce and sometimes researchers seem to be operating within the flammable region, as they utilize a flammable solvent at elevated temperature under pressurized pure oxygen [88]. Working with very low oxygen concentrations is usually key to avoid explosions, although low oxygen concentrations also retard the oxidation reactions. Even if there is no explosion danger, solvent loss due to oxidation is usually inevitable. Next to flammable solvents, other reaction media that are often used for lignin oxidation are acidic or highly alkaline solutions, and ionic liquids. Especially alkaline solutions are commonly applied, as lignin oxidation in these media enables the selective production of vanillin (and syringaldehyde) [19]. However, the use of strongly alkaline and acidic media requires expensive reactor equipment, and, particularly for alkaline media, isolation of the reaction products is cumbersome. The conventional route to isolate the reaction products is to acidify the reaction medium and extract the product with an organic solvent, thus requiring large amounts of acid and organic solvent, and producing large volumes of saline waste water [19, 89]. Advances have been made to avoid the acidification step or the use of organic solvents (e.g., extraction with supercritical CO_2 or adsorption on zeolites or resins), but further research is necessary to enable a more sustainable isolation of the reaction products [19, 89]. Regarding the use of ILs, the IL cost, toxicity, oxidative stability, recuperation, and the ease of product isolation need to be considered.

4 Future Directions

Future research efforts should be directed towards the development of oxidative systems that enable a safe and sustainable conversion of lignin with high product yield and selectivity, straightforward recyclability of catalyst and solvent, minimal solvent loss, and easy isolation of products. In order to demonstrate the feasibility of such systems for lignin oxidation, studies should assess the conversion of relevant lignin model compounds (including phenolic models) and preferably also real lignin substrates. Besides studying the conversion of β-*O*-4 model compounds, it is important to develop oxidative systems for the conversion of carbon–carbon (C–C)-linked dimeric models (such as β-1 model compounds), as C–C linkages constitute a significant fraction of the inter-unit linkages in lignin. Exciting results on the oxidative cleavage of C–C-linked dimers have been presented in recent years [21, 43, 44]. Next to native C–C bonds in lignin (like β-1 bonds), C–C bonds are also formed during biomass fractionation, which are suggested to comprise linkages between the side-chain α-position and a phenolic ring (for instance in the 5-position). Research on the oxidative cleavage of such C–C-linked dimers might enable the effective depolymerization of condensed lignin substrates such as Kraft lignin.

Additionally, lignin oxidation research can complement new developments in biorefinery research. Recent research efforts have been directed to the creation of lignocellulose fractionation strategies that prevent lignin condensation. Two intriguing developments that have gained recent attention are reductive catalytic fractionation (RCF), in which lignin isolation and reductive depolymerization are combined [13], and formaldehyde-assisted fractionation, in which the β-*O*-4 bonds are stabilized by formaldehyde during fractionation [14]. Both methods generate a lignin stream with high potential for chemicals production, for instance through oxidative upgrading, but their chemical structure also poses new challenges for upgrading. For instance, the lignin products from RCF usually lack inter-unit ether bonds, α-OH groups and unsaturated side chains [13]. Therefore, oxidative upgrading of these structures requires the oxidation of saturated, non-functionalized benzylic carbon atoms. In formaldehyde-stabilized lignins, the β-*O*-4 bonds are almost completely preserved as acetal structures [14], which should enable a rather mild oxidative depolymerization, although recuperation of the formaldehyde might be challenging.

In line with RCF, lignin isolation might also be combined with oxidative depolymerization. This involves the direct oxidation of native lignin, and has been successfully performed by alkaline aerobic oxidation, which provided high yields of aromatic aldehydes from native wood lignin [90]. However, while RCF effectively retains the carbohydrates in a solid pulp (and thus effectively fractionates the lignin and carbohydrates), the applied conditions during alkaline aerobic oxidation inevitably solubilize and convert a substantial part of the carbohydrate fraction [91]. Other oxidative depolymerization strategies, as described above (e.g., acidic aerobic oxidation, oxidation in ionic liquids), are expected to better preserve the carbohydrates, and might possibly enable an efficient 'oxidative fractionation'.

Finally, the application potential of the oxidation products requires further exploration. Lignin oxidation can yield a variety of products, such as aromatic acids and aldehydes, *p*-quinones and aliphatic (di)carboxylic acids, but also novel structures are frequently obtained. Aromatic acids and aldehydes are regarded as promising polymer building blocks, and vanillin is an important flavor and fragrance component [70]. Lignin-derived quinones might be used in the production of hydrogen peroxide and could be applied as energy carrier in batteries [18, 92]. Aromatic aldehydes, acids and *p*-quinones might also be used as precursor for pharmaceuticals or other fine chemicals [6]. Aliphatic (di)carboxylic acids already have various applications in food, polymer and pharmaceutical industries [18]. New structures with highly functionalized side-chains, such as 1,2-propanedione-(**12**) [68] and 3-hydroxy-1-propanone-(**11**) substituted phenolics [47], might constitute platform compounds for a myriad of new chemicals. For the latter structures, various possible transformation routes were recently assessed by Westwood et al. [47]. Expanding the application window of both conventional and new oxidation products will strengthen the role of lignin oxidation in future biorefineries.

Acknowledgements This work was performed in the framework of Catalisti (formerly FISCH)-SBO project ARBOREF, FWO-SBO project BioWood, EU Interreg Vlaanderen-Nederland project BIO-HArT and EOS Excellence of Science project BioFact. T.V., W.S. and T.R. acknowledge the Research Foundation Flanders (FWO Vlaanderen) for (post-)doctoral fellowships.

References

1. Schutyser W, Renders T, Van den Bosch S, Koelewijn SF, Beckham GT, Sels BF (2018) Chemicals from lignin: an interplay of lignocellulose fractionation, depolymerisation, and upgrading. Chem Soc Rev 47(3):852–908
2. De Clercq R, Dusselier M, Sels BF (2017) Heterogeneous catalysis for bio-based polyester monomers from cellulosic biomass: advances, challenges and prospects. Green Chem 19(21):5012–5040
3. Isikgor FH, Becer CR (2015) Lignocellulosic biomass: a sustainable platform for the production of bio-based chemicals and polymers. Polym Chem 6(25):4497–4559
4. Delidovich I, Hausoul PJC, Deng L, Pfützenreuter R, Rose M, Palkovits R (2016) Alternative monomers based on lignocellulose and their use for polymer production. Chem Rev 116(3):1540–1599
5. Ragauskas AJ, Beckham GT, Biddy MJ, Chandra R, Chen F, Davis MF, Davison BH, Dixon RA, Gilna P, Keller M (2014) Lignin valorization: improving lignin processing in the biorefinery. Science 344(6185):1246843
6. Sun Z, Fridrich B, de Santi A, Elangovan S, Barta K (2018) Bright side of lignin depolymerization: toward new platform chemicals. Chem Rev 118(2):614–678
7. Key RE, Bozell JJ (2016) Progress toward lignin valorization via selective catalytic technologies and the tailoring of biosynthetic pathways. ACS Sustain Chem Eng 4(10):5123–5135
8. Rinaldi R, Jastrzebski R, Clough MT, Ralph J, Kennema M, Bruijnincx PC, Weckhuysen BM (2016) Paving the way for lignin valorisation: recent advances in bioengineering, biorefining and catalysis. Angew Chem Int Ed 55(29):8164–8215

9. Li C, Zhao X, Wang A, Huber GW, Zhang T (2015) Catalytic transformation of lignin for the production of chemicals and fuels. Chem Rev 115(21):11559–11624
10. Zakzeski J, Bruijnincx PC, Jongerius AL, Weckhuysen BM (2010) The catalytic valorization of lignin for the production of renewable chemicals. Chem Rev 110(6):3552–3599
11. Gillet S, Aguedo M, Petitjean L, Morais A, da Costa Lopes A, Łukasik R, Anastas P (2017) Lignin transformations for high value applications: towards targeted modifications using green chemistry. Green Chem 19(18):4200–4233
12. Renders T, Van den Bosch S, Koelewijn S-F, Schutyser W, Sels B (2017) Lignin-first biomass fractionation: the advent of active stabilisation strategies. Energy Environ Sci 10(7):1551–1557
13. Van den Bosch S, Schutyser W, Vanholme R, Driessen T, Koelewijn SF, Renders T, De Meester B, Huijgen WJJ, Dehaen W, Courtin CM, Lagrain B, Boerjan W, Sels BF (2015) Reductive lignocellulose fractionation into soluble lignin-derived phenolic monomers and dimers and processable carbohydrate pulps. Energy Environ Sci 8(6):1748–1763
14. Shuai L, Amiri MT, Questell-Santiago YM, Héroguel F, Li Y, Kim H, Meilan R, Chapple C, Ralph J, Luterbacher JS (2016) Formaldehyde stabilization facilitates lignin monomer production during biomass depolymerization. Science 354(6310):329–333
15. Deuss PJ, Lancefield CS, Narani A, de Vries JG, Westwood NJ, Barta K (2017) Phenolic acetals from lignins of varying compositions via iron(iii) triflate catalysed depolymerisation. Green Chem 19(12):2774–2782
16. da Costa Sousa L, Jin M, Chundawat SPS, Bokade V, Tang X, Azarpira A, Lu F, Avci U, Humpula J, Uppugundla N, Gunawan C, Pattathil S, Cheh AM, Kothari N, Kumar R, Ralph J, Hahn MG, Wyman CE, Singh S, Simmons BA, Dale BE, Balan V (2016) Next-generation ammonia pretreatment enhances cellulosic biofuel production. Energy Environ Sci 9(4):1215–1223
17. Mittal A, Katahira R, Donohoe BS, Pattathil S, Kandemkavil S, Reed ML, Biddy MJ, Beckham GT (2017) Ammonia pretreatment of corn stover enables facile lignin extraction. ACS Sustain Chem Eng 5(3):2544–2561
18. Ma R, Xu Y, Zhang X (2015) Catalytic oxidation of biorefinery lignin to value-added chemicals to support sustainable biofuel production. Chemsuschem 8(1):24–51
19. Rodrigues Pinto PC, Borges da Silva EA, Rodrigues AE (2012) Lignin as source of fine chemicals: vanillin and syringaldehyde. In: Baskar C, Baskar S, Dhillon RS (eds) Biomass conversion: the interface of biotechnology, chemistry and materials science. Springer, Berlin, pp 381–420
20. Hanson SK, Baker RT, Gordon JC, Scott BL, Thorn DL (2010) Aerobic oxidation of lignin models using a base metal vanadium catalyst. Inorg Chem 49(12):5611–5618
21. Sedai B, Díaz-Urrutia C, Baker RT, Wu R, Silks LP, Hanson SK (2013) Aerobic oxidation of β-1 lignin model compounds with copper and oxovanadium catalysts. ACS Catal 3(12):3111–3122
22. Sedai B, Díaz-Urrutia C, Baker RT, Wu R, Silks LP, Hanson SK (2011) Comparison of copper and vanadium homogeneous catalysts for aerobic oxidation of lignin models. ACS Cataly 1(7):794–804
23. Hanson SK, Wu R, Silks LA (2012) C–C or C–O bond cleavage in a phenolic lignin model compound: selectivity depends on vanadium catalyst. Angew Chem Int Ed 51(14):3410–3413
24. Jiang Y-Y, Yan L, Yu H-Z, Zhang Q, Fu Y (2016) Mechanism of vanadium-catalyzed selective C–O and C–C cleavage of lignin model compound. ACS Catal 6(7):4399–4410
25. Son S, Toste FD (2010) Non-oxidative vanadium-catalyzed C–O bond cleavage: application to degradation of lignin model compounds. Angew Chem Int Ed 49(22):3791–3794
26. Chan JM, Bauer S, Sorek H, Sreekumar S, Wang K, Toste FD (2013) Studies on the vanadium-catalyzed nonoxidative depolymerization of *Miscanthus giganteus*-derived lignin. ACS Catal 3(6):1369–1377
27. Hanson SK, Baker RT (2015) Knocking on wood: base metal complexes as catalysts for selective oxidation of lignin models and extracts. Acc Chem Res 48(7):2037–2048
28. Díaz-Urrutia C, Chen W-C, Crites C-O, Daccache J, Korobkov I, Baker RT (2015) Towards lignin valorisation: comparing homogeneous catalysts for the aerobic oxidation and depolymerisation of organosolv lignin. RSC Adv 5(86):70502–70511
29. Sedai B, Baker RT (2014) Copper catalysts for selective C–C bond cleavage of β-*O*-4 lignin model compounds. Adv Synth Catal 356(17):3563–3574
30. Canevali C, Orlandi M, Pardi L, Rindone B, Scotti R, Sipila J, Morazzoni F (2002) Oxidative degradation of monomeric and dimeric phenylpropanoids: reactivity and mechanistic investigation. J Chem Soc Dalton Trans 15:3007–3014
31. Biannic B, Bozell JJ (2013) Efficient cobalt-catalyzed oxidative conversion of lignin models to benzoquinones. Org Lett 15(11):2730–2733

32. Lange H, Decina S, Crestini C (2013) Oxidative upgrade of lignin—recent routes reviewed. Eur Polym J 49(6):1151–1173
33. Labat G, Meunier B (1989) Factors controlling the reactivity of a ligninase model based on the association of potassium monopersulfate to manganese and iron porphyrin complexes. J Org Chem 54(21):5008–5011
34. Artaud I, Ben-Aziza K, Mansuy D (1993) Iron porphyrin-catalyzed oxidation of 1, 2-dimethoxyarenes: a discussion of the different reactions involved and the competition between the formation of methoxyquinones or muconic dimethyl esters. J Org Chem 58(12):3373–3380
35. Shimada M, Habe T, Umezawa T, Higuchi T, Okamoto T (1984) The CC bond cleavage of a lignin model compound, 1, 2-diarylpropane-1, 3-diol, with a heme-enzyme model catalyst tetraphenylporphyrinatoiron (III) chloride in the presence of tert-butylhydroperoxide. Biochem Biophys Res Commun 122(3):1247–1252
36. Cui F, Wijesekera T, Dolphin D, Farrell R, Skerker P (1993) Biomimetic degradation of lignin. J Biotechnol 30(1):15–26
37. Zucca P, Rescigno A, Rinaldi AC, Sanjust E (2014) Biomimetic metalloporphines and metalloporphyrins as potential tools for delignification: molecular mechanisms and application perspectives. J Mol Catal A Chem 388–389:2–34
38. Luo F-X, Zhou T-G, Li X, Luo Y-L, Shi Z-J (2015) Fragmentation of structural units of lignin promoted by persulfate through selective C–C cleavage under mild conditions. Orga Chem Front 2(9):1066–1070
39. Mottweiler J, Rinesch T, Besson C, Buendia J, Bolm C (2015) Iron-catalysed oxidative cleavage of lignin and β-*O*-4 lignin model compounds with peroxides in DMSO. Green Chem 17(11):5001–5008
40. Ren X, Wang P, Han X, Zhang G, Gu J, Ding C, Zheng X, Cao F (2017) Depolymerization of lignin to aromatics by selectively oxidizing cleavage of C–C and C–O bonds using $CuCl_2$/polybenzoxazine catalysts at room temperature. ACS Sustain Chem Eng 5(8):6548–6556
41. Mottweiler J, Puche M, Räuber C, Schmidt T, Concepción P, Corma A, Bolm C (2015) Copper-and vanadium-catalyzed oxidative cleavage of lignin using dioxygen. Chemsuschem 8(12):2106–2113
42. Deng W, Zhang H, Wu X, Li R, Zhang Q, Wang Y (2015) Oxidative conversion of lignin and lignin model compounds catalyzed by CeO_2-supported Pd nanoparticles. Green Chem 17(11):5009–5018
43. Zhou Z-Z, Liu M, Li C-J (2017) Selective copper-*N*-heterocyclic carbene (copper–NHC)-catalyzed aerobic cleavage of β-1 lignin models to aldehydes. ACS Catal 7(5):3344–3348
44. Hou T, Luo N, Li H, Heggen M, Lu J, Wang Y, Wang F (2017) Yin and Yang dual characters of CuOx clusters for C–C bond oxidation driven by visible light. ACS Catal 7(6):3850–3859
45. Blandez JF, Navalón S, Alvaro M, Garcia H (2015) Graphenes as metal-free catalysts for the oxidative depolymerization of lignin models. ChemCatChem 7(18):3020–3026
46. Crestini C, Pro P, Neri V, Saladino R (2005) Methyltrioxorhenium: a new catalyst for the activation of hydrogen peroxide to the oxidation of lignin and lignin model compounds. Bioorg Med Chem 13(7):2569–2578
47. Lancefield CS, Ojo OS, Tran F, Westwood NJ (2015) Isolation of functionalized phenolic monomers through selective oxidation and C–O bond cleavage of the β-*O*-4 linkages in lignin. Angew Chem Int Ed 54(1):258–262
48. Patil ND, Yao SG, Meier MS, Mobley JK, Crocker M (2015) Selective cleavage of the C_{alpha}–C_{beta} linkage in lignin model compounds via Baeyer–Villiger oxidation. Org Biomol Chem 13(11):3243–3254
49. Rahimi A, Azarpira A, Kim H, Ralph J, Stahl SS (2013) Chemoselective metal-free aerobic alcohol oxidation in lignin. J Am Chem Soc 135(17):6415–6418
50. Wang M, Lu J, Zhang X, Li L, Li H, Luo N, Wang F (2016) Two-step, catalytic C–C bond oxidative cleavage process converts lignin models and extracts to aromatic acids. ACS Catal 6(9):6086–6090
51. Zhang C, Li H, Lu J, Zhang X, MacArthur KE, Heggen M, Wang F (2017) Promoting lignin depolymerization and restraining the condensation via an oxidation–hydrogenation strategy. ACS Catal 7(5):3419–3429
52. Bosque I, Magallanes G, Rigoulet M, Kärkäs MD, Stephenson CR (2017) Redox catalysis facilitates lignin depolymerization. ACS Centr Sci 3:621
53. Luo J, Zhang X, Lu J, Zhang J (2017) Fine tuning the redox potentials of carbazolic porous organic frameworks for visible-light photoredox catalytic degradation of lignin β-*O*-4 models. ACS Catal 7(8):5062–5070

54. Dawange M, Galkin MV, Samec JS (2015) Selective aerobic benzylic alcohol oxidation of lignin model compounds: route to aryl ketones. ChemCatChem 7(3):401–404
55. Zhu R, Wang B, Cui M, Deng J, Li X, Ma Y, Fu Y (2016) Chemoselective oxidant-free dehydrogenation of alcohols in lignin using Cp* Ir catalysts. Green Chem 18(7):2029–2036
56. Das A, Rahimi A, Ulbrich A, Alherech M, Motagamwala AH, Bhalla A, da Costa Sousa L, Balan V, Dumesic JA, Hegg EL, Dale BE, Ralph J, Coon JJ, Stahl SS (2018) Lignin conversion to low-molecular-weight aromatics via an aerobic oxidation-hydrolysis sequence: comparison of different lignin sources. ACS Sustain Chem Eng 6(3):3367–3374
57. Guo H, Miles-Barrett D, Neal AR, Zhang T, Li C, Westwood NJ (2018) Unravelling the enigma of ligninOX: can the oxidation of lignin be controlled? Chem Sci 9(3):702–711
58. Lancefield CS, Rashid GMM, Bouxin F, Wasak A, Tu W-C, Hallett J, Zein S, Rodríguez J, Jackson SD, Westwood NJ, Bugg TDH (2016) Investigation of the chemocatalytic and biocatalytic valorization of a range of different lignin preparations: the importance of β-*O*-4 content. ACS Sustain Chem Eng 4(12):6921–6930
59. Dabral S, Wotruba H, Hernández JG, Bolm C (2018) Mechanochemical oxidation and cleavage of lignin β-*O*-4 model compounds and lignin. ACS Sustain Chem Eng 6(3):3242–3254
60. Dabral S, Hernández J, Kamer P, Bolm C (2017) Organocatalytic chemoselective primary alcohol oxidation and subsequent cleavage of lignin model compounds and lignin. ChemSusChem 10(13):2707–2713
61. Mobley JK, Yao SG, Crocker M, Meier M (2015) Oxidation of lignin and lignin β-*O*-4 model compounds via activated dimethyl sulfoxide. RSC Adv 5(127):105136–105148
62. Wang Y, Wang Q, He J, Zhang Y (2017) Highly effective C-C bond cleavage of lignin model compounds. Green Chem 19(13):3135–3141
63. Jennings JA, Parkin S, Munson E, Delaney SP, Calahan JL, Isaacs M, Hong K, Crocker M (2017) Regioselective Baeyer–Villiger oxidation of lignin model compounds with tin beta zeolite catalyst and hydrogen peroxide. RSC Adv 7(42):25987–25997
64. Wang M, Li L, Lu J, Li H, Zhang X, Liu H, Luo N, Wang F (2017) Acid promoted C–C bond oxidative cleavage of β-*O*-4 and β-1 lignin models to esters over a copper catalyst. Green Chem 19(3):702–706
65. Liu H, Wang M, Li H, Luo N, Xu S, Wang F (2017) New protocol of copper-catalyzed oxidative C(CO)–C bond cleavage of aryl and aliphatic ketones to organic acids using O_2 as the terminal oxidant. J Catal 346:170–179
66. Rinesch T, Mottweiler J, Puche M, Concepción P, Corma A, Bolm C (2017) Mechanistic investigation of the catalyzed cleavage for the lignin β-*O*-4 linkage: implications for vanillin and vanillic acid formation. ACS Sustain Chem Eng 5(11):9818–9825
67. Nguyen JD, Matsuura BS, Stephenson CR (2014) A photochemical strategy for lignin degradation at room temperature. J Am Chem Soc 136(4):1218–1221
68. Rahimi A, Ulbrich A, Coon JJ, Stahl SS (2014) Formic-acid-induced depolymerization of oxidized lignin to aromatics. Nature 515(7526):249–252
69. Chen CL (1992) Nitrobenzene and cupric oxide oxidations. In: Lin SY, Dence CW (eds) Methods in lignin chemistry. Springer, Berlin, pp 301–321
70. Fache M, Boutevin B, Caillol S (2016) Vanillin production from lignin and its use as a renewable chemical. ACS Sustain Chem Eng 4(1):35–46
71. Santos SG, Marques AP, Lima DL, Evtuguin DV, Esteves VI (2010) Kinetics of eucalypt lignosulfonate oxidation to aromatic aldehydes by oxygen in alkaline medium. Ind Eng Chem Res 50(1):291–298
72. Partenheimer W (2009) The aerobic oxidative cleavage of lignin to produce hydroxyaromatic benzaldehydes and carboxylic acids via metal/bromide catalysts in acetic acid/water mixtures. Adv Synth Catal 351(3):456–466
73. Voitl T, Rudolf von Rohr P (2008) Oxidation of lignin using aqueous polyoxometalates in the presence of alcohols. Chemsuschem 1(8–9):763–769
74. Werhan H, Mir JM, Voitl T, Von Rohr PR (2011) Acidic oxidation of kraft lignin into aromatic monomers catalyzed by transition metal salts. Holzforschung 65(5):703–709
75. Chatel G, Rogers RD (2013) Oxidation of lignin using ionic liquids—an innovative strategy to produce renewable chemicals. ACS Sustain Chem Eng 2(3):322–339
76. De Gregorio GF, Prado R, Vriamont C, Erdocia X, Labidi J, Hallett JP, Welton T (2016) Oxidative depolymerization of lignin using a novel polyoxometalate-protic ionic liquid system. ACS Sustain Chem Eng 4(11):6031–6036

77. Stärk K, Taccardi N, Bösmann A, Wasserscheid P (2010) Oxidative depolymerization of lignin in ionic liquids. Chemsuschem 3(6):719–723
78. Ma R, Guo M, Lin K-T, Hebert VR, Zhang J, Wolcott MP, Quintero M, Ramasamy KK, Chen X, Zhang X (2016) Peracetic acid depolymerization of biorefinery lignin for production of selective monomeric phenolic compounds. Chem A Eur J 22(31):10884–10891
79. Xiang Q, Lee Y (2000) Oxidative cracking of precipitated hardwood lignin by hydrogen peroxide. Appl Biochem Biotechnol 84(1–9):153–162
80. Ma R, Guo M, Zhang X (2014) Selective conversion of biorefinery lignin into dicarboxylic acids. Chemsuschem 7(2):412–415
81. Azarpira A, Ralph J, Lu F (2014) Catalytic alkaline oxidation of lignin and its model compounds: a pathway to aromatic biochemicals. BioEnergy Res 7(1):78–86
82. Lai Y-Z, Guo X-P (1991) Variation of the phenolic hydroxyl group content in wood lignins. Wood Sci Technol 25(6):467–472
83. Galkin MV, Dahlstrand C, Samec JSM (2015) Mild and robust redox-neutral Pd/C-catalyzed lignol β-*O*-4′ bond cleavage through a low-energy-barrier pathway. Chemsuschem 8(13):2187–2192
84. Deuss PJ, Scott M, Tran F, Westwood NJ, de Vries JG, Barta K (2015) Aromatic monomers by in situ conversion of reactive intermediates in the acid-catalyzed depolymerization of lignin. J Am Chem Soc 137(23):7456–7467
85. Behling R, Valange S, Chatel G (2016) Heterogeneous catalytic oxidation for lignin valorization into valuable chemicals: what results? What limitations? What trends? Green Chem 18(7):1839–1854
86. Das L, Kolar P, Sharma-Shivappa R (2012) Heterogeneous catalytic oxidation of lignin into value-added chemicals. Biofuels 3(2):155–166
87. Prat D, Hayler J, Wells A (2014) A survey of solvent selection guides. Green Chem 16(10):4546–4551
88. Osterberg PM, Niemeier JK, Welch CJ, Hawkins JM, Martinelli JR, Johnson TE, Root TW, Stahl SS (2015) Experimental limiting oxygen concentrations for nine organic solvents at temperatures and pressures relevant to aerobic oxidations in the pharmaceutical industry. Org Process Res Dev 19(11):1537–1543
89. Mota MIF, Rodrigues Pinto PC, Loureiro JM, Rodrigues AE (2016) Recovery of vanillin and syringaldehyde from lignin oxidation: a review of separation and purification processes. Sep Purif Rev 45(3):227–259
90. Taraban'ko VE, Koropatchinskaya NV, Kudryashev AV, Kuznetsov BN (1995) Influence of lignin origin on the efficiency of the catalytic oxidation of lignin into vanillin and syringaldehyde. Russ Chem Bull 44(2):367–371
91. Tarabanko VE, Kaygorodov KL, Skiba EA, Tarabanko N, Chelbina YV, Baybakova OV, Kuznetsov BN, Djakovitch L (2017) Processing pine wood into vanillin and glucose by sequential catalytic oxidation and enzymatic hydrolysis. J Wood Chem Technol 37(1):43–51
92. Huskinson B, Marshak MP, Suh C, Er S, Gerhardt MR, Galvin CJ, Chen X, Aspuru-Guzik A, Gordon RG, Aziz MJ (2014) A metal-free organic–inorganic aqueous flow battery. Nature 505:195

Topics in Current Chemistry (2018) 376:32
https://doi.org/10.1007/s41061-018-0211-6

REVIEW

Thermosetting Polymers from Lignin Model Compounds and Depolymerized Lignins

Elias Feghali, et al. *[full author details at the end of the article]*

Received: 26 January 2018 / Accepted: 3 July 2018 / Published online: 10 July 2018

Abstract

Lignin is the most abundant source of renewable ready-made aromatic chemicals for making sustainable polymers. However, the structural heterogeneity, high polydispersity, limited chemical functionality and solubility of most technical lignins makes them challenging to use in developing new bio-based polymers. Recently, greater focus has been given to developing polymers from low molecular weight lignin-based building blocks such as lignin monomers or lignin-derived bio-oils that can be obtained by chemical depolymerization of lignins. Lignin monomers or bio-oils have additional hydroxyl functionality, are more homogeneous and can lead to higher levels of lignin substitution for non-renewables in polymer formulations. These potential polymer feed stocks, however, present their own challenges in terms of production (i.e., yields and separation), pre-polymerization reactions and processability. This review provides an overview of recent developments on polymeric materials produced from lignin-based model compounds and depolymerized lignin bio-oils with a focus on thermosetting materials. Particular emphasis is given to epoxy resins, polyurethanes and phenol-formaldehyde resins as this is where the research shows the greatest overlap between the model compounds and bio-oils. The common goal of the research is the development of new economically viable strategies for using lignin as a replacement for petroleum-derived chemicals in aromatic-based polymers.

Keywords Lignin · Depolymerization · Lignin model compounds · Polymers · Thermosets

Chapter 4 was originally published as Feghali, E., Torr, K. M., van de Pas, D. J., Ortiz, P., Vanbroekhoven, K., Eevers, W. & Vendamme, R. Topics in Current Chemistry (2018) 376: 32. https://doi.org/10.1007/s41061-018-0211-6

1 Introduction

Producing renewable chemicals from biomass has attracted increased research interest in recent years [1, 2], driven by factors such as depleting fossil resources, environmental concerns, growth of the bioeconomy, and demand for green chemicals and energy security. A key challenge for the modern biorefinery is deriving maximum value from the biomass resource. While many valuable fibers, fuels and chemical products are commercially produced from polysaccharides found in lignocellulosic biomass, limited value is currently extracted from the lignin. Lignosulfonate products with a global production of around 1.2 million tons/year have by far the largest market [3]. While lignin offers an abundant source of renewable aromatic compounds, valorizing lignin has proven both technically and economically difficult to achieve. Nonetheless, lignin offers many advantages over polysaccharides when targeting new sustainable polymers with novel functionality and performance. Aromatic rings within polymers offer structural rigidity and stability, and confer good thermal and mechanical properties to the material. Petrochemical-derived aromatic polymers, such as bisphenol A-derived polycarbonates and epoxy resins, polyurethanes, polystyrene and polyethylene terephthalate (PET) cannot be easily recycled, and are frequently derived from monomers with high toxicity or environmental concerns. Consequently, producing aromatic polymers with a high renewable carbon content is an important goal with lignin offering the only renewable source of ready-made aromatic chemicals available in larger quantities and at low cost.

Of the various strategies employed to incorporate lignins into a range of polymer materials, blending [3, 4] and direct functionalization (e.g., glycidylation) of the lignin [5] is generally limited by poor compatibility, leading to reduced mechanical performance at high levels of incorporated lignin [3, 5, 6]. Chemical modifications of lignin, such as phenolation and oxypropylation, have been used successfully to enhance reactivity for functionalization, and have led to higher degrees of lignin substitution. The extensive research in this area has been recently reviewed [3, 7–10].

One of the major challenges in using technical lignins in new polymer materials is their heterogeneity and high polydispersity [7]. Fractionating lignins by membrane separation techniques, solvent fractionation and selective precipitation can produce more homogeneous lignin fractions with reduced polydispersity [7]. Alternatively, lignin can be depolymerized to produce either monomer feedstock chemicals or low molecular weight lignins with additional hydroxyl functionality. The main chemical approaches to depolymerizing lignins, including acid/base catalyzed depolymerization, sub/supercritical fluid-based depolymerization, oxidative depolymerization and catalytic reductive depolymerization, have been recently reviewed [5, 10–13].

The ability to chemically depolymerize lignin depends on the chemical structure of the lignin and the method of processing. Native lignins have a complex chemical structure resulting from the radical polymerization of three aromatic alcohols; *p*-coumaryl, coniferyl, and sinapyl alcohol in proportions, which are dictated by the type of lignocellulosic species [11]. These monolignol units are linked together in a variety of ways with the most common linkages in lignins being β-*O*-4 ether bonds [12]. The chemical structures of technical lignins, e.g., kraft lignin, lignosulfonates, and organosolv lignins, can be highly-modified and are heavily dependent on the method of isolation [12]. Depolymerization strategies applied to technical lignins typically lead to low yields of monomers and complex mixtures of products due to (1) the modified structure of technical lignins, (2) the requirement for forcing reaction conditions, and (3) competing re-polymerization reactions that can occur alongside depolymerization. Vanillin is the only commercial monomer product currently produced from lignin. The high value of vanillin makes it economically viable to produce from lignosulfonates despite a low yield and a complex purification process. Higher yields of monomers can be achieved when depolymerization strategies are applied to native unmodified lignins [14–17]. However, the separation and purification of the monomer chemicals can be a major challenge. These operations typically account for 60–80% of the process cost of most mature chemical processes [18].

A potentially more economically viable, nearer-term strategy for using depolymerized lignins to make new polymer materials is to deconstruct the lignin to a mixture of monomers, dimers and oligomers and to use this mixture directly for polymer synthesis without purification. This strategy can be applied to technical lignins or can take a lignin-first approach [17] in which the native lignin is first separated from the biomass matrix by catalytic processing, producing a solvent-soluble depolymerized lignin bio-oil. Using this approach to produce aromatic feedstock chemicals for new polymers is an emerging area of research that is underdeveloped compared to using technical lignins directly. This review describes recent research on lignin-based polymers derived from low molecular weight aromatic building blocks with particular emphasis on thermosetting materials. The first section focuses on polymers from possible lignin-derived monomers and the second section on depolymerized lignins.

2 Lignin Model Compounds

Despite the great interest in the depolymerization of lignin shown by the chemical community, there is not yet a mature technology to separate the different components into monomers or families of compounds [19]. Vanillin **1** is currently the only monomer produced from lignin on an industrial scale [20]. Its two functional groups (phenolic and aldehyde) enable high chemical versatility, which is further enhanced by the chemical modifications (oxidation or reduction) that can be performed on the aldehyde moiety (Fig. 1) [20]. These features make using vanillin a nearer-term opportunity compared to other lignin-derived monomers [21]. Consequently, it has received the highest attention. The complex structure of lignin results in a range of compounds being produced during depolymerization. Moreover, the compounds differ depending on the depolymerization method that is used [22–25]. The potential of being able to access such a wide range of aromatic monomers has led many research groups to use these compounds as models in their polymerization studies.

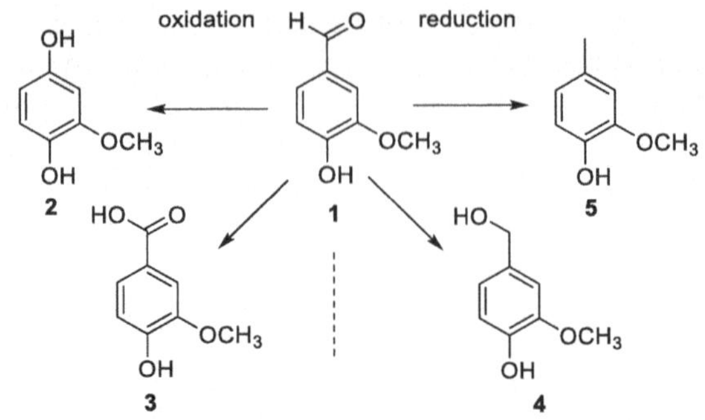

Fig. 1 Chemical modifications of the aldehyde group of vanillin

This expanding field has been recently reviewed [2, 13, 26, 27]. As mentioned in the Introduction, the use of depolymerized lignins and low molecular weight lignin fractions is an attractive option, and perhaps more economically and technically feasible in the nearer-term. Although uncommon, there are interesting reports on mixtures of model compounds intended to mimic these bio-oils. Part of this review pays special attention to these studies as they have not been previously reviewed. The use of vanillin in thermosets, as well as model compounds that can potentially be obtained from lignin, will also be discussed. Model compounds that are part of lignin biosynthesis rather than from depolymerized lignins are outside the scope of this review.

2.1 Epoxy Resins

Epoxy polymers are used across a wide range of industrial applications, and currently 75% of them are derived from the diglycidyl ether of bisphenol A (BADGE) (Fig. 2)—a reprotoxic substance [28]. The use of renewable resources could improve the sustainability of these types of polymers as well as addressing toxicity issues. Although there has been extensive research on the use of lignin for the development of epoxy resins [9], the increased solubility of the monomers over native and technical lignins is a key advantage that drives research in this area.

The synthesis of bisphenol A (BPA) is carried out by acid-catalyzed electrophilic aromatic substitution of phenol with acetone (Fig. 2). Stanzione et al. [29] reported a bio-based alternative by condensing vanillyl alcohol **4** with guaiacol

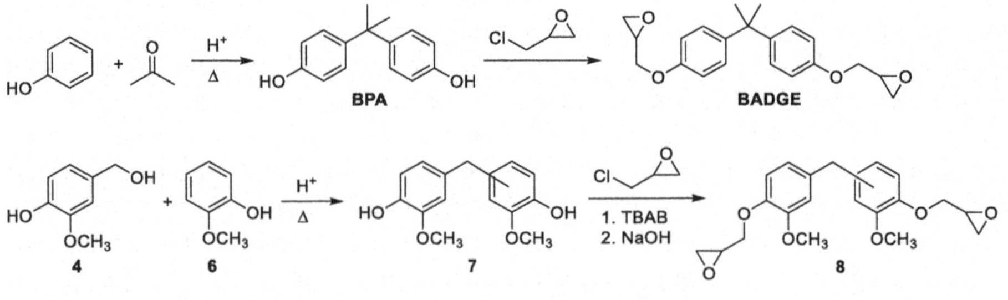

Fig. 2 Commercial synthesis of the diglycidyl ether of bisphenol A (BADGE), and the synthesis of the bio-based alternative bisguaiacol [29]

6 to yield bisguaiacol isomers **7** (Fig. 2, major isomer *p–p*). The latter were then reacted with epichlorohydrin to form the corresponding diglycidyl ether **8**. In addition, the diglycidyl ethers of certain monomers [vanillyl alcohol, gastrodigenin (or 4-hydroxymethylphenol) and hydroquinone] were prepared to study the influence of different structural parameters on the viscoelastic properties of the polymers. The epoxy resins were cured, either by themselves or together with a commercial BADGE resin, using a stoichiometric amount of commercial 4,4-methylenebiscyclohexanamine (Amicure PACM) [29]. The cured resins had a lower glass transition temperature (T_g) than the BADGE resin. Both the methoxy substituent and the methylene spacer between the aromatic ring and the epoxide were found to lower the T_g of the cured resins. Building on that knowledge, the group expanded the work by substituting the Amicure PACM with cellulose-derived furanyl diamine to further increase the bio-based content [30]. As before, the bio-based epoxies were blended with BADGE. This lead to thermosets with higher levels of bio-based components that had good thermal stability and thermomechanical properties, similar to commercial BADGE resins.

The group of Caillol has been very active in the field of renewable polymers, including epoxies [31]. Their interest has been focused on using vanillin **1** and vanillin derivatives (Fig. 1) [20]. Methoxyhydroquinone **2**, vanillic acid **3** and vanillyl alcohol **4** were glycidylated to obtain the corresponding epoxy monomers [32]. These monomers were cross-linked with isophorone diamine (IPDA) in an epoxy-amine ratio of 2:1 to give cured resins. The T_g of these resins derived from **2** and **3** were 97 °C and 152 °C, respectively. Interestingly, the latter had the same α relaxation temperature (T_α) as the BADGE-based resin. However, the diepoxy monomers were all solids at room temperature, which would hamper their industrial applicability. To address this problem, they applied the common industrial strategy of first making the oligomers. This was done by using **2** as chain extender of the glycidylated monomers derived from **2** [33]. The oligomers were then cured with IPDA. The researchers found that the thermomechanical properties could be tuned by changing the length of the starting oligomer. In doing so, they were able to formulate epoxy thermosets with T_g and T_α values between 80 and 110 °C.

Of special interest is the research that this group has carried out on the use of model mixtures of compounds that can be derived from the lignin to vanillin process [34]. Aware of the economic and environmental cost barriers associated with purifying the monomers, their goal was to demonstrate proof of concept on using these types of mixtures for making epoxy polymers. They prepared mixtures of products that resembled those obtained by alkaline depolymerization of both softwood (G) and hardwood (G and S) lignin (Table 1). The mixtures contained different proportions of phenolic acids (**3**, **9**), aldehydes (**1**, **10**, **11**) and ketones (**12**, **13**) (Fig. 3). Acids gave diepoxy monomers but aldehydes and ketones needed to be oxidized through Dakin oxidation (Fig. 3). The mixtures were then glycidylated to obtain epoxy monomers using a similar procedure to that used with the isolated monomers. They found that the glycidylated mixture derived from G-lignin was paste-like and partially crystallized, while the mixture derived from of G- and S-lignin did not crystallize, and was a highly viscous liquid. Curing with IPDA yielded brown

Table 1 Model mixtures of products from alkaline depolymerization of softwood and hardwood lignins [34]

Compound	Softwood (%)	Hardwood (%)
Vanillic acid (**3**)	6.6	2.2
Syringic acid (**9**)	–	7.1
Vanillin (**1**)	72.6	24.8
Syringaldehyde (**10**)	3.0	51.7
p-Hydroxybenzaldehyde (**11**)	4.5	2.3
Acetovanillone(**12**)	13.3	2.3
Acetosyringone (**13**)	–	9.6

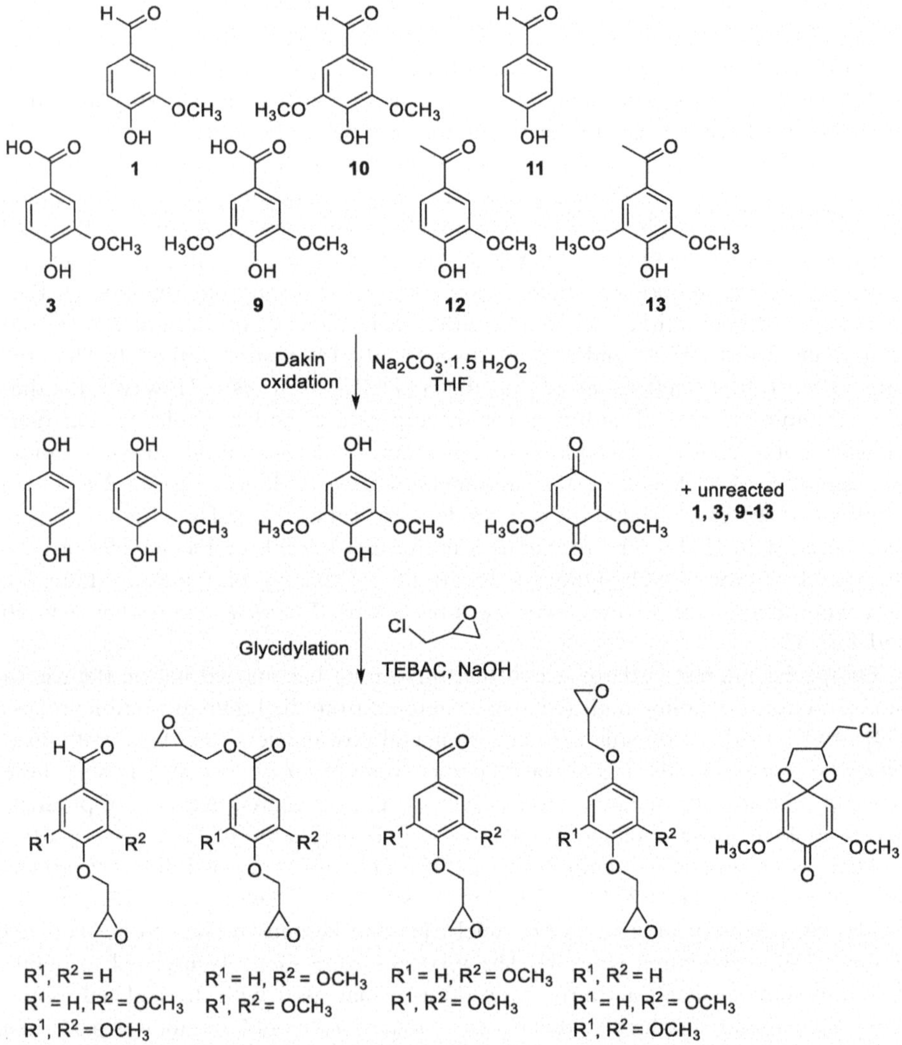

Fig. 3 The different monomers used as a model mixture of a lignin depolymerization product, and their chemical modification (oxidation, glycidylation) for the preparation of epoxy thermosets [34]

homogeneous polymers. The T_g was 99 °C for the polymer formed from the hardwood bio-oil mimic and 113 °C for the mixture mimicking the softwood bio-oil.

Building on their success of converting lignin into methoxyphenol monomers [35], the group of Omar investigated the use of these monomers for epoxy nanocomposites [36]. 4-Propyl guaiacol **14**, chosen as a model compound, was demethylated and the resulting diol **15** glycidylated to yield the bis-epoxide **16** (Fig. 4). The resulting monomer was cured with a diamine; however, its properties were not completely satisfactory (low storage modulus and T_α). The authors attributed this to two effects, namely the formation of the benzodioxane byproduct **17** and the decrease in chain extension caused by the epoxides being in the ortho position. In order to improve the epoxy network, they conducted further research on how to increase the crosslinking density. They applied three strategies: increase of the molecular weight, change of the orientation of the hydroxyl groups, and increase of the number of functional groups [37]. Among the different new monomers synthesized, compound **19** was the most successful. The corresponding epoxy resins had improved performance compared to that based on the previously reported **16** (Fig. 4), where the cross-linking density increased from 0.39 to 9.77 mol dm^{-3}, the T_α rose from 40 to 139 °C, and the statistic heat-resistant index temperature (T_s) went from 125 to 153 °C.

Samec and Plasseraud prepared a diglycidyl ether from iso-eugenol in two steps [38]. They first produced the glycidyl ether by reacting iso-eugenol with epichlorohydrin and then epoxidized the double bond with ozone or H_2O_2. The diepoxide was cured in the presence of several anhydride hardeners, resulting in resins with T_g values in the range 78–120 °C [38].

Samui et al. [39] linked two vanillin fragments by condensing the aldehyde moieties with different ketones. The resulting diols were subsequently reacted with epichlorohydrin. Photoactive liquid crystalline polyester epoxies were successfully

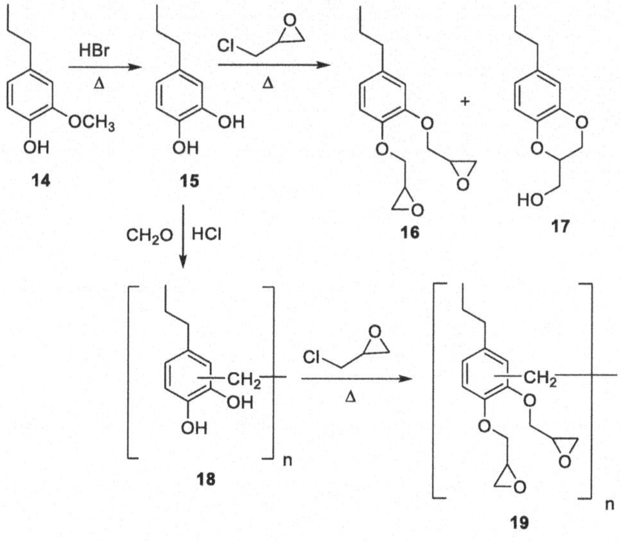

Fig. 4 Modifications of 4-propylguaiacol for the preparation of epoxy monomers [35, 36]

synthesized from these epoxies by reaction with terephthalic or trimesic acid, respectively.

Ma et al. [40] have disclosed an interesting approach to bio-based flame retardant epoxy resins. Two vanillin units **1** were linked with two different diamines and the Schiff base intermediates were reacted in situ with diethyl phosphite (Fig. 5). The resulting molecules **20** were further reacted with epichlorohydrin to produce the epoxy resins **21**. These were cured with 4,4-diaminodiphenylmethane (DDM) to give polymers with excellent flame retardant properties (UL-94 V0 rating and high LOI of 32.8%). The performance exceeded that of the BADGE resin: T_g of 214 °C vs 166 °C; tensile strength of 80.3 MPa vs 76.4 MPa; and tensile modulus of 2709 vs 1893 MPa.

2.2 Polyurethanes

Polyurethanes are synthesized by the condensation between polyols and diisocyanates. Thus, modification of lignin-derived monomers is needed for making polyurethanes, and this might explain the scarcity of research in this field. Allais et al. [41] made a dimer based on ferulic acid by reacting it with different diols from renewable sources. The resulting diol was subjected to polymerization with two commercially available isocyanates. The polymers had low molecular weights, but the authors claimed their importance as a proof of concept for this new class of partially renewable poly(esterurethane) materials.

Wadgaonkar et al. [42] also started from lignin-derived phenolic acids, but in this case they chose vanillic acid and syringic acid. They linked two units and converted the diacid into the diisocyanate via Curtius rearrangement. The diisocyanates were converted into poly(ether urethane) polymers by reacting them with aliphatic diols that could be bio-derived. The T_g of these polymers ranged between 49 °C and 74 °C, and was found to be dependent on both the methoxy substituents on the aromatic rings of the diisocyanate and the number of methylene units in the diols.

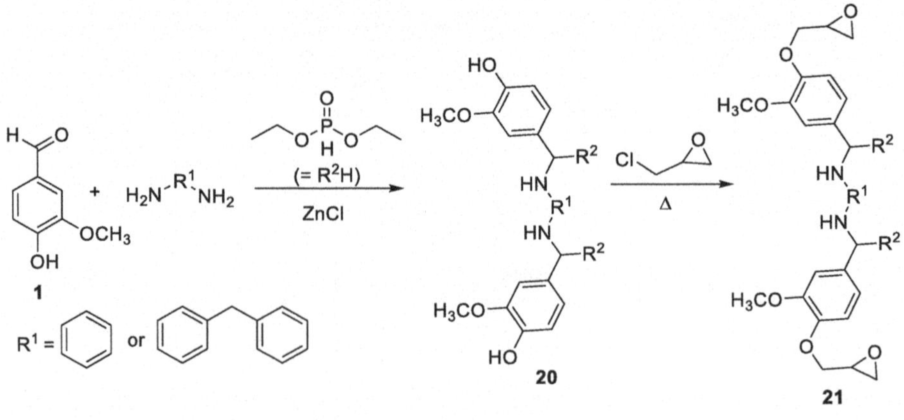

Fig. 5 Preparation of bio-based flame retardant epoxy resins from vanillin [40]

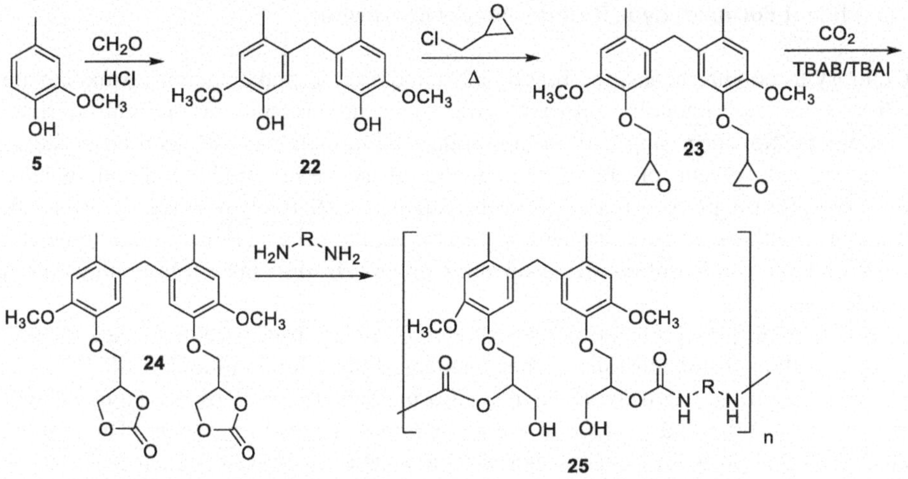

Fig. 6 Synthesis of polyhydroxyurethanes from creosol without using isocyanates [43]

Xie et al. [43] also showed interest in these types of polymers. In order to avoid the toxic isocyanates, they prepared a bis(cyclic carbonate) **24** (Fig. 6). Starting from the bisphenol **22** derived from creosol **5**, the bis-epoxide **23** was synthesized by reaction with epichlorohydrin, and converted to **24** by cycloaddition in the presence of CO_2. By further reacting the resulting bis(cyclic carbonate) **24** with traditional diamines they were able to produce different polyhydroxyurethanes **25**. The T_g of these materials ranged from 44 °C to 90 °C, and was shown to be greatly influenced by the type of amine. By comparison, the BPA analogue of these polymers had a T_g of only 34 °C.

Kim et al. [44] has recently reported on the preparation of polyurethane elastomers based on vanillin. The lignin-derived monomer was dimerized and then reacted with ethanolamine to form a tetraol, which was used to partially replace (i.e., 5–20%) 1,4-butanediol that is often used as a chain extender in polyurethane synthesis. The different mixtures of divanillin-ethanolamine (DV-EA) tetraol and 1,4-butanediol were reacted with a prepolymer derived from methylene diisocyanate. The resulting polyurethanes retained their thermal stability and strength, while strain and Young's modulus were increased compared with the control polyurethane (Table 2). The authors hypothesized that the enhancement comes from either the polyol functionality (that enables the formation of hydrogen bonds between polymeric chains) or the presence of aromatic rings in the vanillin structure.

Table 2 Mechanical properties of vanillin-based polyurethanes [44]

Amount of DV-EA (%)	0 (control)	5	10	20
Stress (MPa)	32.9	29.1	26.4	30.4
Strain (%)	522.6	644.8	655.4	770.9
Young's modulus (MPa)	7.5	9.1	9.7	8.0

2.3 Phenol-Formaldehyde Resins—Polybenzoxazines

Contrary to polyurethanes, the functionality required to make phenol-formaldehyde (PF) resins (i.e., phenolic hydroxyl groups) is already present in lignin-derived monomers. Not surprisingly, these lignin-derived monomers have been proposed as the ideal replacement for phenol in phenolic resins. Lignin model compounds have been used for the preparation of polybenzoxazines, a relatively new type of PF resin. They are synthesized from a phenolic derivative, a primary amine and an aldehyde by a one-pot condensation. Ring-opening polymerization takes place under high temperatures [45].

Ronda et al. [46] prepared benzoxazine derivatives from *p*-coumaric and ferulic acid, as well as from their esters. They condensed these lignin models with 1,3,5-triphenylhexahydro-1,3,5-triazine and paraformaldehyde to form monomers, but these were not thermally stable at the temperatures needed for curing. The authors addressed this problem by using boron trifluoride as a catalyst, which allowed the polymerization to take place at a lower temperature and thus avoided degradation of the resin. The polymers had comparable properties to the petroleum-based analogs.

The research group of Liu [47] prepared two fully bio-based monomers **26a–b** using guiacol **6**, *p*-formaldehyde and either furfurylamine or stearylamine (Fig. 7). Importantly, the procedure was solventless. The two monomers were mixed prior to curing, and it was observed that when the ratio of furfuryl to stearyl exceeded 1:2 the copolymerization was homogeneous. Furthermore, the furan moiety was found to be beneficial for accelerating the curing process, and enhancing the cross-linking density and thermal properties. The T_g increased with increasing amounts of furan benzoxazine monomer **26a** in the mixture. The authors attributed this effect to the higher crosslinking density of the copolymer (due to electrophilic aromatic substitution of the furan), which led to higher rigidity.

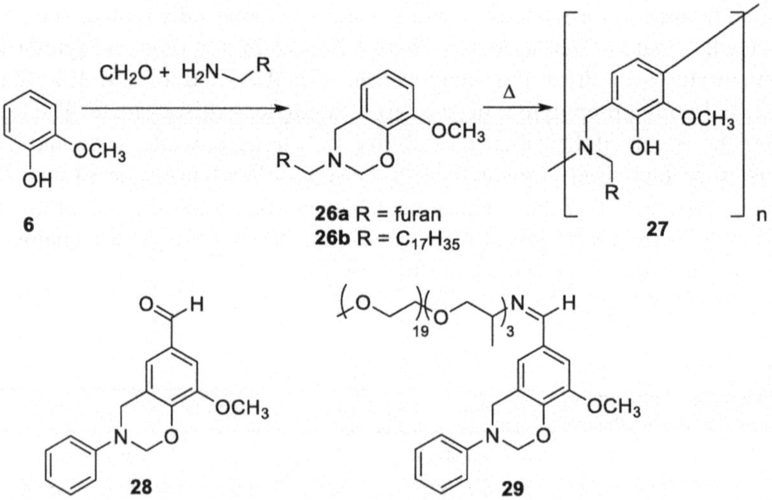

Fig. 7 Benzoxazine monomers for the preparation of polybenzoxazine polymers [47]

Mhaske et al. also started from guaiacol, but in their case the benzoxazine monomer was synthesized using ethanolamine, which allowed the introduction of a hydroxyl group [48]. This was reacted with an isocyanate hardener to prepare a poly(benzoxazine-urethane) coating for corrosion protection of mild steel. It was concluded that increased concentration of urethane moieties in the coatings improved the mechanical, chemical and thermal properties compared to the pure polybenzoxazine coating.

Lomonaco et al. [49] have contributed to this field by designing a better synthetic procedure for preparing benzoxazines. They developed a microwave-assisted synthesis of benzoxazine monomers from guaiacol and different amines. The protocol reduced the reaction time from hours to minutes and eliminated the need for a solvent.

Varma et al. [50] reported on the synthesis of benzoxazine **28** using vanillin instead of guaiacol (Fig. 7). The polymer arising from monomer **28** showed higher T_g (270 °C) and char yield than petroleum-based benzoxazines. The authors hypothesized that the superior properties arose from the presence of the formyl group. Following up on this research, they reported on the synthesis of bis-benzoxazine monomers from vanillin, *p*-formaldehyde and five petroleum-based diamines. Vanillin imparted a better thermal stability to the polymer, and the diamine structures had an influence on both the thermal properties and curing behavior. The cured resins had a T_g ranging from 170 °C to 255 °C, which is significantly higher than the BPA-based analogue. Moreover, the resins showed good thermal stability and char yield, and importantly, had good adhesive strength at 200 °C, making them suitable for high-temperature adhesives. Ishida et al. [51] used vanillin for synthesizing benzoxazine resins. The monomer **28** was formed by condensation of aniline with *p*-formaldehyde. These monomers were further functionalized with an amine terminated poly(ethyleneoxide) to produce a surfactant **29** (Fig. 7), which was shown to stabilize polystyrene for two weeks in mini-emulsions.

2.4 Other Polymeric Materials

Wool et al. [52] prepared vinyl ester resins by reacting vanillin **1** with a methacrylic anhydride **30** to yield vanillin methacrylate **31** (Fig. 8). The methacrylic acid **32,** produced as a byproduct in the first reaction, was consumed by reacting it in situ with glycidyl methacrylate **33**. In this way, the glycerol dimethacrylate **34** was also produced as a cross-linker for the vanillin methacrylate **31**. The sustainability of this non-waste generating process was further supported by the solvent-free procedure.

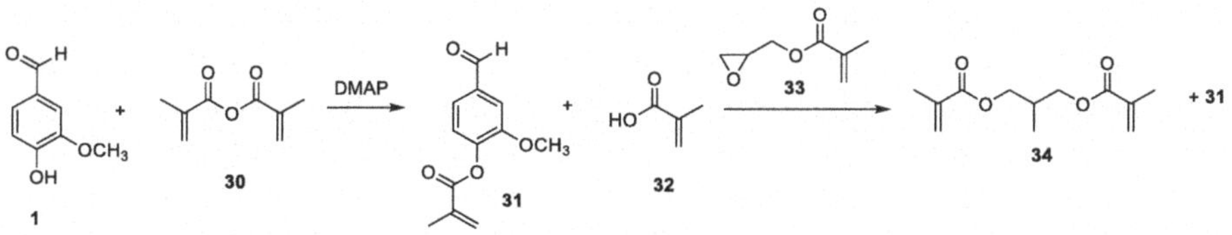

Fig. 8 Preparation of potentially fully bio-based vinyl ester resins [52]

The vinyl ester resin had properties similar to those of commercial resins. The resin was cured via free radical polymerization reaching 66% conversion, which could be increased up to 78% after post-curing. The formed resin was a hard transparent thermoset with a T_g of 155 °C and a storage modulus of 3.6 GPa at 25 °C. This resin had comparable thermal stability and thermomechanical properties to commercial vinyl ester resins while avoiding the use of styrene, which is non-renewable and hazardous.

Wool et al. [53] expanded this research to bio-oil mimics, for which they chose nine phenolic compounds (phenol, guaiacols and catechols). As done before with vanillin, these compounds were first methacrylated individually and then blended together to make the mixture. The bio-oil mimic was then self-cured as a vinyl ester resin. This was possible because the catechols present in the mixture allowed for cross-linking. Additionally, due to its low viscosity, the methacrylated bio-oil mixture was used as a reactive diluent (50%) in a standard commercial vinyl ester resin. However, the viscosity of the resulting resin was slightly higher than what is desired for liquid moulding applications. Curing the resins led to hard transparent thermosets that possessed near complete conversion of free radical polymerizable groups. T_g values were 115 °C for the methacrylated bio-oil and 135 °C for the mixture with the commercial vinyl ester resin. These T_g values, although comparable with the corresponding styrene analogues, were lower than expected, leading the authors to suggest that unreacted monomers were acting as plasticizers.

Harvey et al. [54] reported on the preparation of cyanate ester resins from lignin-derived monomers. They used creosol **5**, which can be obtained from vanillin by hydrogenation, to form different bisphenols. Zinc acetate-catalyzed ortho-coupling of formaldehyde with creosol gave **35**, while acidic conditions led to the meta-coupled (with respect to the hydroxyl group) dimer product **36** with different aldehydes (Fig. 9). The resulting bisphenols **35** and **36** were transformed into the cyanates **37** and **38** by treating them with cyanogen bromide and triethylamine. These were cured to yield cyanate ester thermosets **39** and **40** (Fig. 9). They exhibited T_g values ranging from 219 °C to 248 °C. The steric bulk in the bridge lead to a slower but more complete cure. Interestingly, their TGA/FTIR and mass spectrometry results showed that the resins decomposed to phenols, isocyanic acid, and other decomposition products. This led the authors to suggest that these types of resins could potentially be recycled to the parent creosol by pyrolysis.

In contrast to thermosets, there has been extensive research on thermoplastics derived from lignin model compounds, which has been reviewed recently [2]. As already stressed in this review, bio-oils are complex, and their heterogeneity makes working with them more challenging than with pure monomers. Consequently, there is only one example of the preparation of thermoplastics from bio-oil mimics, reported by the group of Epps [55]. Concerned about the practicality of using bio-oil based heteropolymers, the authors investigated the reversible addition fragmentation chain transfer (RAFT) homopolymerization of several methacrylate monomers and compared them with the heteropolymerization of bio-oil mimics. One of the bio-oil mimics contained model compounds of the pyrolysis of softwood kraft lignin: guaiacol **6**, creosol **5**, 4-ethylguaiacol **41** and vanillin **1** (Fig. 10). These monomers share the same structure except for the

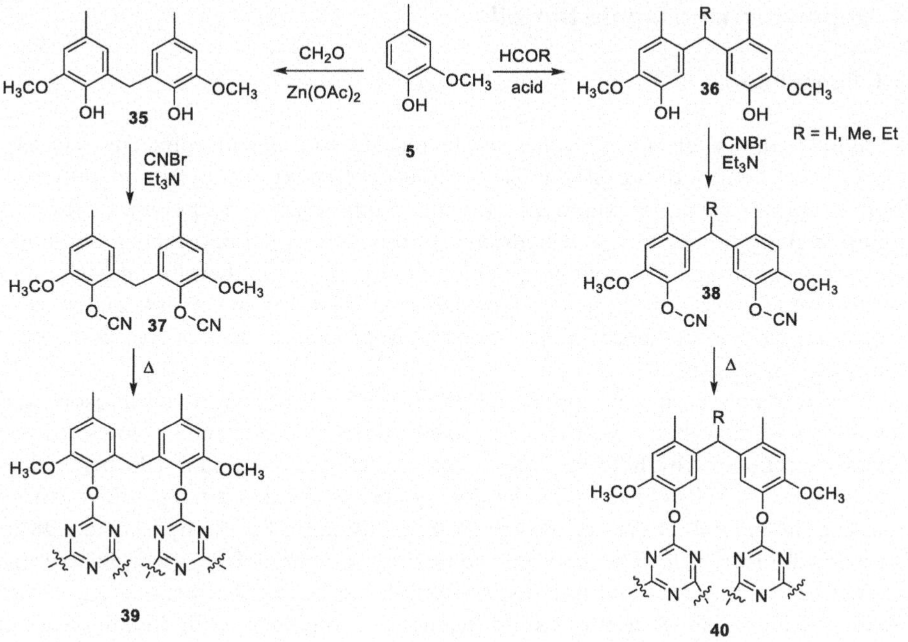

Fig. 9 Synthesis of cyanate ester resins from creosol [54]

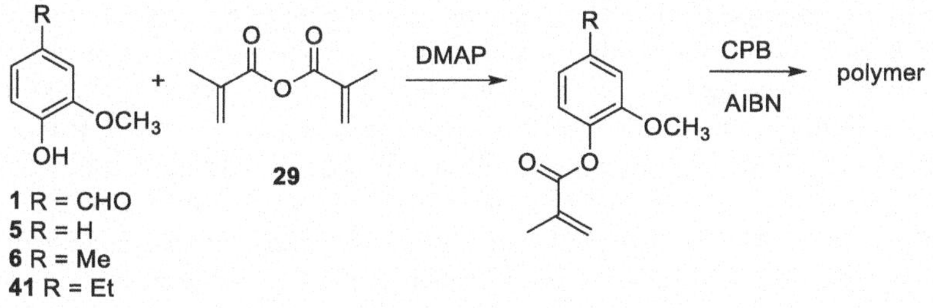

Fig. 10 Reversible addition fragmentation chain transfer (RAFT) polymerization of methacrylated monomers and their mixture mimicking a lignin-derived bio-oil [55]

functionality in the *p*-position. Thus, this was a model of a relatively homogeneous bio-oil. On the other hand, they also prepared a bio-oil mimic with components from different biomass sources: vanillin, phenol, *n*-butanol and lauric acid. In this case, it was intended to model a situation in which the components are less structurally homogeneous. The homogeneous bio-oil, having monomers with similar reactivity, generated a random polymer. The heterogeneous bio-oil yielded a gradient polymer. The authors concluded that extensive kinetic studies will be needed when dealing with heterogenous bio-oils while the polymerization of relatively homogeneous bio-oils will be more straightforward.

3 Depolymerized Lignin Bio-oils

3.1 Epoxy Resins

Phenolic hydroxyl groups on lignin can be reacted with epichlorohydrin to impart epoxy functionality to the lignin. The synthesis of epoxy resins from depolymerized lignin has several advantages over using lignin alone. Depolymerization of lignin leads to higher levels of hydroxyl groups, which is important in obtaining epoxy resins with good levels of epoxide functionality, and therefore crosslinking ability. Depolymerized lignin also has increased solubility in solvents and a lower molecular weight, making it more amenable to processing as a polyol in the synthesis of epoxy resins.

Xu's research group has produced epoxy resins from kraft, organosolv and hydrolysis lignins that were depolymerized by either a proprietary low-pressure alkaline process or by hydrogenolysis using a Ru catalyst under 100 bar hydrogen pressure at 350 °C [56–58]. The depolymerized lignins were glycidylated by reacting the phenolic hydroxyl groups (e.g., 2.3 mmol/g for the hydrolysis lignin) with epichlorohydrin. The epoxide equivalent weight (EEW), weight average molecular weight (M_w) and polydispersity index (PDI) of the resins are given in Table 3. The resins were either solid materials or highly viscous liquids at room temperature that could be processed effectively at elevated temperatures (80 °C) or by using acetone as a solvent. In one study, epoxy resins from depolymerized organosolv and kraft lignin were cured with an aliphatic curing agent [diethylenetriamine (DETA)] and an aromatic curing agent (DDM), and their thermal properties determined [56]. The resins made from depolymerized kraft lignin had greater thermal stability than those made from depolymerized organosolv lignin.

In other studies, epoxy resins from depolymerized organosolv, kraft and hydrolysis lignin were blended with different proportions of BADGE and cured with DDM in producing fiberglass-reinforced epoxy composites [57, 58]. The flexural and tensile strength of the cured resins increased by up to 25% when 25–50% of the BADGE was replaced with depolymerized organosolv or kraft lignin epoxies (Fig. 11). Higher proportions of the lignin-based epoxies lead to reduced strength.

In a recent study, van de Pas and Torr [59] synthesized epoxy resins from depolymerized native softwood lignin. The study made use of the inherent properties of the native lignin by selectively cleaving the ether linkages in the lignin without condensing the lignin as can occur in pulping processes (e.g., kraft and organosolv). The native lignin was depolymerized by mild hydrogenolysis using a Pd catalyst under 34 bar hydrogen pressure at 195 °C. The hydrogenolysis oil product (yield = 78% w/w on starting lignin) consisted mainly of monomers, dimers and oligomers of dihydroconiferyl alcohol (**42**, major component) and 4-propyl guaiacol (**14**, minor component) (Fig. 12) [15]. The hydrogenolysis oil and an oligomeric-rich oil fraction were glycidylated by reacting the phenolic hydroxyl groups (4.1–4.4 mmol/g) with epichlorohydrin. The resulting epoxy resins (LHEP and LHOEP, respectively) had low EEW, M_w and PDI values (Table 3).

Table 3 Selected properties of epoxy resins that incorporate depolymerized lignins. *EEW* Epoxide equivalent weight, *PDI* polydispersity index, *BADGE* diglycidyl ether of bisphenol A, *DDM* 4,4-diaminodiphenylmethane, *DETA* diethylenetriamine, *DSC* differential scanning calorimetry

Lignin	EEW (g/eq)	M_w (g/mol)	PDI	Resin formulation	T_g (°C)[a]	References
Kraft	768	2800	3.5	BADGE blends; cured with DDM; reinforced with fiberglass	101[b] (129[c])	[56, 58]
Organosolv	537	1400	2.8		85[2]	
Hydrolysis	838	5530	4.6			[57]
Native softwood						
Whole oil	359	1084	1.8	BADGE blends; cured with DETA	80[b] (117[c])	[59]
Oligomer-rich	452	1514	1.8		104[4]	
Native hardwood						
Annulated	352			Cured with phenol novolac	134 (95[3])	[60]
Transacetalized	376				94	

[a]Based on DSC measurements

[b]50% lignin-based prepolymer in blended resins

[c]BADGE control resin

[d]25% lignin-based prepolymer in blended resin

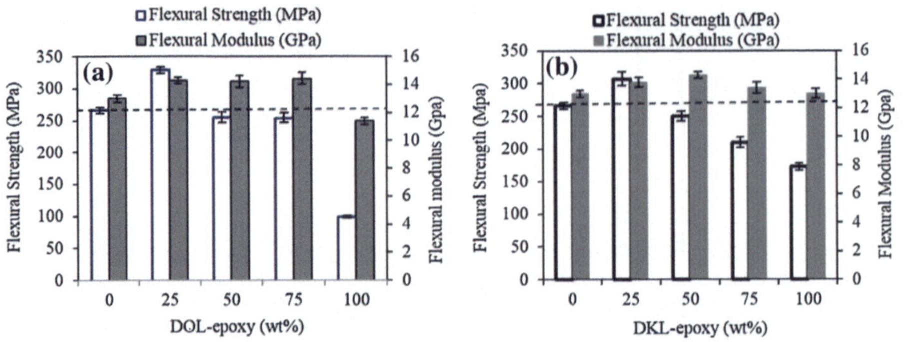

Fig. 11 Flexural strength and flexural modulus of fiberglass-reinforced epoxy composites prepared with different proportions of BADGE and **a** depolymerized organosolv lignin (DOL) epoxy resin, or **b** depolymerized kraft lignin (DKL) epoxy resin. Reprinted from Ferdosian et al. [58], with permission from Elsevier

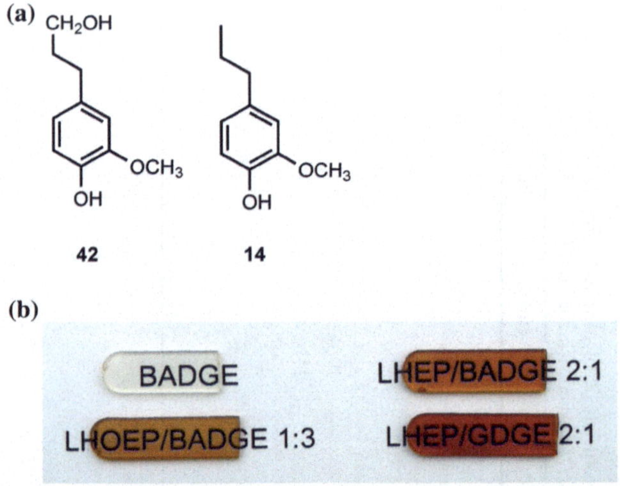

Fig. 12 Chemical structures of **a** dihydroconiferyl alcohol and 4-propyl guaiacol, and **b** transparent epoxy resins cured with diethylenetriamine (*LHEP* whole oil, *LHOEP* oligomer-rich fraction; reprinted with permission from van de Pas and Torr [59], American Chemical Society)

The resins were blended with BADGE and cured with an aliphatic (DETA) or a cycloaliphatic (isophorone diamine) curing agent (Fig. 12).

Thermal and flexural properties of the cured resins were determined. When 25–75% of the BADGE was replaced with the lignin-based resins, the flexural strength increased up to 38% (163 MPa) and the flexural modulus increased up to 52% (4.6 GPa) relative to the BADGE-based control resin (Fig. 13). The lignin-based resins were also blended with glycerol diglycidyl ether (GDGE) to produce 100% bio-based resins that, when cured, also exhibited increased flexural properties compared to the BADGE-based control resin (Figs. 12, 13). Improvements in

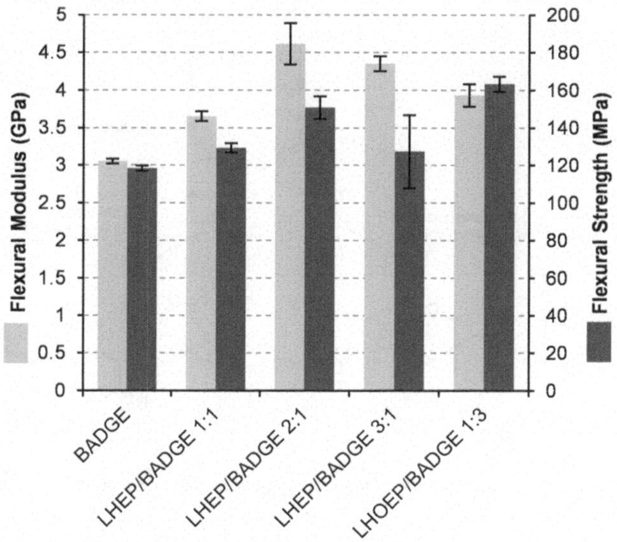

Fig. 13 Flexural properties of lignin hydrogenolysis epoxy resins (*LHEP* unfractionated, *LHOEP* oligomer-rich fraction) blended with BADGE and cured with DETA (*error bars* 95% confidence intervals of at least four replicates) [59]

the mechanical properties were attributed to the oligomers present in the hydrogenolysis products.

Watanabe et al. [60] synthesized epoxy resins from depolymerized hardwood lignin with the aim of chemically modifying the products to tailor the thermal properties of the epoxy resins. An acid-catalyzed process was used to depolymerize the lignin at 140 °C, with methanol used to stabilize the radicals to prevent condensation of the lignin. The depolymerized products were monomeric acetals obtained in a yield of 33% w/w on starting lignin. These products were reacted in one of two ways to give flexible (transacetalization) and rigid (annulation) difunctional intermediates that were subsequently glycidylated with epichlorohydrin to give **43** and **44**, respectively (Fig. 14). Key properties are given in Table 3. The resins were cured with a phenol novolac curing agent. The rigid annulated product had a high T_g of 134 °C while the flexible transacetalized product had a T_g of 94 °C, which was similar to the BADGE-based resin (Table 3).

Lignin can also be utilized as a curing agent for epoxy resins by imparting suitable functionality to the lignin. Zhang et al. [61] synthesized a curing agent for epoxy resins from depolymerized kraft lignin. The lignin was partially depolymerized under base-catalysis in supercritical methanol at 97 bar and 250 °C. The hydroxyl groups on the depolymerized lignin were reacted with succinic anhydride to generate terminal carboxylic acid functionality. The resulting lignin polycarboxylic acid product was used to cure a BADGE-based resin, either alone or with other curing agents [glycerol tris(succinate monoester) and hexahydrophthalic anhydride]. The glass transition and initial degradation temperatures of the

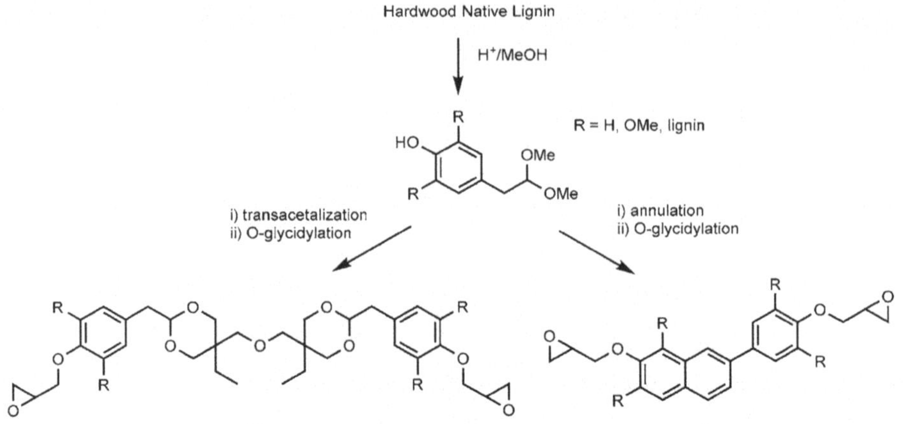

43 **44**

Fig. 14 Preparation of lignin-based epoxy resins from depolymerized hardwood lignin [60]

BADGE resin cured with the lignin-based curing agent were lower than for the same resin cured with the other two curing agents.

3.2 Polyurethanes

Technical lignins can be used as a component of polyurethane formulations to produce coatings, films, and flexible and rigid foams. Using lignins can lead to improved mechanical properties, thermal stability, and moisture and flame resistance [62]. However, levels of lignin incorporation are generally low, at around 20–30%. This is because high molecular weights, lower solubility in polyol systems and low hydroxyl group content can lead to low strength foams at higher incorporation levels [62]. To improve the level of lignin incorporation, chemical modifications such as oxypropylation are commonly used to generate a liquid lignin-based polyol with more reactive aliphatic hydroxyl groups [5, 63]. An alternative strategy is to depolymerize the lignin to lower its molecular weight and increase the total number of hydroxyl groups available to react with isocyanates.

Xu et al. [64] depolymerized kraft lignin using direct hydrolysis with aqueous NaOH at 250°C to give a low molecular weight lignin (M_w = 1700 g/mol) with a high aliphatic hydroxyl number (365 mg KOH/g) in a yield of 77 wt%. Rigid polyurethane foams were produced using this depolymerized lignin by replacing up to 50% of the polypropylene glycol and sucrose polyols. This lead to improvements in compression modulus and compression strength in both foam types (Fig. 15). Additionally, a polyurethane foam produced from oxypropylated depolymerized kraft lignin showed the best mechanical properties (i.e., compression modulus = 10,986 kPa, compression strength 515 kPa at 10% deformation) and lowest thermal conductivity (0.029 W/m K).

Xu et al. [65] also depolymerized an enzymatic hydrolysis lignin in a 1:1 water–ethanol mixture at 250°C and used this depolymerized lignin in up to 50 wt% content with polypropylene glycol or sucrose polyols to prepare rigid polyurethane

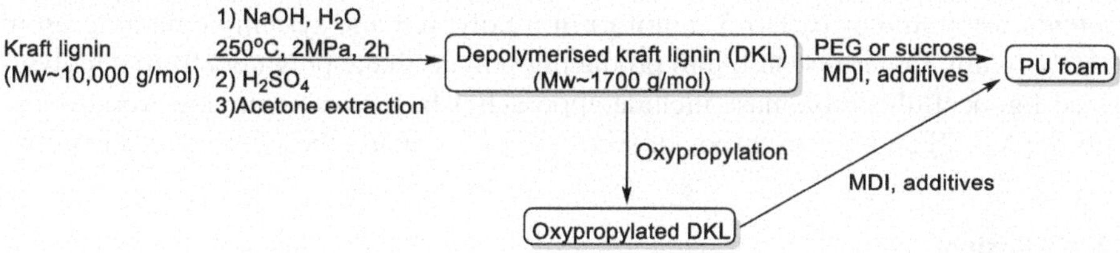

Fig. 15 Production of polyurethane foams using depolymerized kraft lignin (DKL) and oxypropylated DKL [64]

foams. The lignin-based foams had good compression strengths (e.g., 216 kPa at 10% deformation) and low thermal conductivities (e.g., 0.036 W/m K). Mamiński et al. [66] hydrolyzed a lignosulfonate hydrolyzed under alkaline conditions and used the polyol product to prepare polyurethane foams. However, they found that the original lignosulfonate gave polyurethane foams with better mechanical properties. Xue et al. [67] reported on the hydrolytic depolymerization of corncob lignin in an isopropanol-water mixture that was catalyzed by NaOH. They produced a modest yield (22%) of low molecular weight (860 g/mol) lignin-derived polyols (hydroxyl number = 5.10 mmol/g) with the view of producing bio-based rigid polyurethane foams.

Recently, Feghali and Torr [15] prepared polyurethane foams from depolymerized native softwood lignin mixed with sorbitol and sucrose polyols, and methylene diphenyl diisocyanate. The depolymerized lignin was produced as an oil product by hydrogenolysis of native lignin from pine wood. Viable foams were prepared with up to 50 wt% replacement of the sugar-derived polyols with the depolymerized lignin hydrogenolysis oil (LHO) (Fig. 16). Improvements in compressive modulus and thermal stability were observed at 25 wt% LHO replacement.

Polyurethane materials have also been made from low molecular weight lignins isolated by solvent fractionation. Vanderlaan and Thring [68] found no benefit in using low molecular weight fractions of Alcell lignins ($M_w = 720$ g/mol) compared to medium and high molecular weight fractions (2410 and 6950 g/mol, respectively) in preparing polyurethane films containing up to 35 wt% lignin content. By contrast, Yoshida et al. [69] reported that low molecular weight lignin fractions ($M_w = 580$, 1090 g/mol) gave tougher and more flexible polyurethane films than medium and high molecular weight lignins (2140 and > 10,000 g/mol) when the lignin hydroxyl

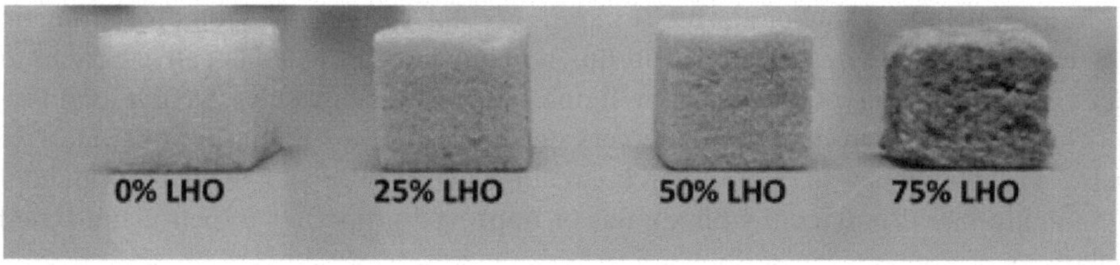

Fig. 16 Polyurethane foams prepared from depolymerized native softwood lignin [lignin hydrogenolysis oil (LHO)] with 0–75 wt% replacement of sorbitol/sucrose polyols

contents were similar (6.1–7.3 mmol/g) in a polyurethane system containing up to 30 wt% lignin. Other research into producing polyurethane polymers from depolymerized lignocellulosic biomass include approaches based on lignin and wood liquefaction [70–72] and fast pyrolysis of straw [73] are outside the scope of this review.

3.3 PF Resins

Depolymerized lignins have been used to make PF resins and foams. Xu et al. [74] compared pure PF resins with resins made using organosolv lignin ($M_w = 1157$ g/mol) and catalytically depolymerized organosolv lignin ($M_w = 512$ g/mol). The lignin depolymerization was achieved in 1:1 water–ethanol at 340°C in the presence of a Ni catalyst and a hydrogen atmosphere. The organosolv and depolymerized lignin PF resins were tested as plywood adhesives at phenol replacement levels of 50% and 75%. The two different lignin-based adhesives were found to have similar average dry and wet tensile bond strengths (1.9–2.2 and 1.6–1.8 MPa, respectively), which surpassed the Japanese Industry standard minimum requirements. The depolymerized lignin PF resins appeared to have an advantage in terms of lower free formaldehyde content. Xu et al. [75] produced PF foams from a depolymerized enzymatic hydrolysis lignin ($M_w = 1910$ g/mol, PDI ~ 3). PF foams with phenol substitution ratios up to 50% were produced and shown to have higher compressive strength and elastic modulus than the reference PF foam.

Gramlich et al. [76] investigated formate assisted fast pyrolysis of kraft lignin as a means of producing low molecular weight alkylated phenols for use in the synthesis of PF resins. They demonstrated that resins with properties similar to pure PF resin and superior to the parent bio-oil resin could be produced using a mixture of model phenol compounds to mimic a purified extract of the pyrolysis bio-oil. Other approaches to producing PF resins from depolymerized biomass feedstocks include using phenolic bio-oils produced by biomass liquefaction in hot-compressed phenol-water or ethanol-water mixtures [77–79].

4 Conclusion and Perspectives

Lignin is produced as a byproduct from the pulp and paper industry and is mostly burnt to produce energy. This is anticipated to change in the future as new biorefinery operations look to extract greater value from the lignin. Lignin has the potential to transition from a low-grade fuel to an important sustainable source of aromatic chemicals. The research discussed in this review shows it is technically feasible to make thermosetting polymers from lignin-based low molecular weight building blocks such as lignin-derived monomers or depolymerized lignin bio-oils. Despite being a relatively new area of research, great advances have been made in producing new lignin-based polymers with similar or improved properties to petroleum-based polymers.

There are still some major barriers to overcome before these new polymers can be produced commercially. Currently, petroleum-based polymers are synthesized from high purity monomers. The replacement of traditional polymers by lignin-derived renewable polymers cannot be foreseen in the short-term as it is not yet technically possible to access individual monomers from lignin in high yield and purity. This review suggests a potential nearer-term approach is to use depolymerized lignin bio-oils comprising mixtures of monomers, dimers and oligomers. Furthermore, new sustainable lignin-derived polymers do not need to be constrained by the need to provide renewable replacements to currently existing polymers. Lignin's inherently rich chemical functionality is currently underutilized. The research described in this review shows that this functionality can impart unique properties to the polymers that could lead to new materials and applications in the future.

Acknowledgments The review was supported by the New Zealand Ministry of Business, Innovation and Employment via Scion funding from the Strategic Science Investment Fund. VITO would like to acknowledge the province of Noord-Brabant (The Netherlands) for the financial support in the framework of the activities at the Shared Research Center Biorizon.

References

1. Delidovich I, Hausoul PJ, Deng L, Pfutzenreuter R, Rose M, Palkovits R (2016) Alternative monomers based on lignocellulose and their use for polymer production. Chem Rev 116(3):1540–1599
2. Llevot A, Grau E, Carlotti S, Grelier S, Cramail H (2016) From lignin-derived aromatic compounds to novel biobased polymers. Macromol Rapid Commun 37(1):9–28
3. Lora JH (2016) Lignin: A platform for renewable aromatic polymeric materials. In: Lau PCK (ed) Quality living through chemurgy and green chemistry. Springer, Berlin, pp 221–260
4. Doherty WOS, Mousavioun P, Fellows CM (2011) Value-adding to cellulosic ethanol: lignin polymers. Ind Crops Prod 33(2):259–276
5. Xu C, Ferdosian F (2017) Conversion of lignin into bio-based chemicals and materials. Green chemistry and sustainable technology. Springer, Berlin
6. Ma S, Li T, Liu X, Zhu J (2016) Research progress on bio-based thermosetting resins. Polym Int 65(2):164–173
7. Duval A, Lawoko M (2014) A review on lignin-based polymeric, micro- and nano-structured materials. React Funct Polym 85:78–96
8. Laurichesse S, Avérous L (2014) Chemical modification of lignins: towards biobased polymers. Prog Polym Sci 39(7):1266–1290
9. Koike T (2012) Progress in development of epoxy resin systems based on wood biomass in Japan. Polym Eng Sci 52(4):701–717
10. Westwood NJ, Panovic I, Lancefield CS (2016) Chemical modification of lignin for renewable polymers and chemicals. In: Fang Z, Smith RLJ (eds) Production of biofuels and chemicals from lignin. Springer, Singapore, pp 183–216
11. Xu C, Arancon RA, Labidi J, Luque R (2014) Lignin depolymerisation strategies: towards valuable chemicals and fuels. Chem Soc Rev 43(22):7485–7500
12. Zakzeski J, Bruijnincx PCA, Jongerius AL, Weckhuysen BM (2010) The catalytic valorization of lignin for the production of renewable chemicals. Chem Rev 110(6):3552–3599
13. Sun Z, Fridrich B, de Santi A, Elangovan S, Barta K (2018) Bright side of lignin depolymerization: toward new platform chemicals. Chem Rev 118(2):614–678. https://doi.org/10.1021/acs.chemrev.7b00588
14. Yan N, Zhao C, Dyson PJ, Wang C, Liu LT, Kou Y (2008) Selective degradation of wood lignin over noble-metal catalysts in a two-step process. Chemsuschem 1(7):626–629
15. Torr KM, van de Pas DJ, Cazeils E, Suckling ID (2011) Mild hydrogenolysis of in situ and isolated pinus radiata lignins. Bioresour Technol 102(16):7608–7611
16. Feghali E, Carrot G, Thuéry P, Genre C, Cantat T (2015) Convergent reductive depolymerization of wood lignin to isolated phenol derivatives by metal-free catalytic hydrosilylation. Energy Environ Sci 8(9):2734–2743

17. Van den Bosch S, Schutyser W, Vanholme R, Driessen T, Koelewijn SF, Renders T, De Meester B, Huijgen WJJ, Dehaen W, Courtin CM, Lagrain B, Boerjan W, Sels BF (2015) Reductive lignocellulose fractionation into soluble lignin-derived phenolic monomers and dimers and processable carbohydrate pulps. Energy Environ Sci 8(6):1748–1763
18. Ragauskas AJ, Williams CK, Davison BH, Britovsek G, Cairney J, Eckert CA, Frederick WJ, Hallett JP, Leak DJ, Liotta CL, Mielenz JR, Murphy R, Templer R, Tschaplinski T (2006) The path forward for biofuels and biomaterials. Science 311(5760):484–489
19. Esposito D, Antonietti M (2015) Redefining biorefinery: the search for unconventional building blocks for materials. Chem Soc Rev 44(16):5821–5835
20. Fache M, Darroman E, Besse V, Auvergne R, Caillol S, Boutevin B (2014) Vanillin, a promising biobased building-block for monomer synthesis. Green Chem 16(4):1987–1998
21. Fache M, Boutevin B, Caillol S (2015) Vanillin, a key-intermediate of biobased polymers. Eur Polym J 68:488–502
22. Jegers HE, Klein MT (1985) Primary and secondary lignin pyrolysis reaction pathways. Ind Eng Chem Process Des Dev 24(1):173–183
23. Santos SG, Marques AP, Lima DLD, Evtuguin DV, Esteves VI (2011) Kinetics of eucalypt lignosulfonate oxidation to aromatic aldehydes by oxygen in alkaline medium. Ind Eng Chem Res 50(1):291–298
24. Villar JC, Caperos A, García-Ochoa F (1997) Oxidation of hardwood kraft-lignin to phenolic derivatives. Nitrobenzene and copper oxide as oxidants. J Wood Chem Technol 17(3):259–285
25. Rodrigues Pinto PC, Borges da Silva EA, Rodrigues AE (2011) Insights into oxidative conversion of lignin to high-added-value phenolic aldehydes. Ind Eng Chem Res 50(2):741–748
26. Upton BM, Kasko AM (2016) Strategies for the conversion of lignin to high-value polymeric materials: review and perspective. Chem Rev 116(4):2275–2306
27. Isikgor FH, Becer CR (2015) Lignocellulosic biomass: a sustainable platform for the production of bio-based chemicals and polymers. Polym Chem 6(25):4497–4559
28. Vandenberg LN, Maffini MV, Sonnenschein C, Rubin BS, Soto AM (2009) Bisphenol-a and the great divide: a review of controversies in the field of endocrine disruption. Endocr Rev 30(1):75–95
29. Hernandez ED, Bassett AW, Sadler JM, La Scala JJ, Stanzione JF (2016) Synthesis and characterization of bio-based epoxy resins derived from vanillyl alcohol. ACS Sustain Chem Eng 4(8):4328–4339
30. Mauck JR, Yadav SK, Sadler JM, La Scala JJ, Palmese GR, Schmalbach KM, Stanzione JF (2017) Preparation and characterization of highly bio-based epoxy amine thermosets derived from lignocellulosics. Macromol Chem Phys 218(14):1700013
31. Auvergne R, Caillol S, David G, Boutevin B, Pascault JP (2014) Biobased thermosetting epoxy: present and future. Chem Rev 114(2):1082–1115
32. Fache M, Auvergne R, Boutevin B, Caillol S (2015) New vanillin-derived diepoxy monomers for the synthesis of biobased thermosets. Eur Polym J 67:527–538
33. Fache M, Viola A, Auvergne R, Boutevin B, Caillol S (2015) Biobased epoxy thermosets from vanillin-derived oligomers. Eur Polym J 68:526–535
34. Fache M, Boutevin B, Caillol S (2016) Epoxy thermosets from model mixtures of the lignin-to-vanillin process. Green Chem 18(3):712–725
35. Parsell T, Yohe S, Degenstein J, Jarrell T, Klein I, Gencer E, Hewetson B, Hurt M, Kim JI, Choudhari H, Saha B, Meilan R, Mosier N, Ribeiro F, Delgass WN, Chapple C, Kenttämaa HI, Agrawal R, Abu-Omar MM (2015) A synergistic biorefinery based on catalytic conversion of lignin prior to cellulose starting from lignocellulosic biomass. Green Chem 17(3):1492–1499
36. Zhao S, Abu-Omar MM (2015) Biobased epoxy nanocomposites derived from lignin-based monomers. Biomacromol 16(7):2025–2031
37. Zhao S, Abu-Omar MM (2016) Renewable epoxy networks derived from lignin-based monomers: effect of cross-linking density. ACS Sustain Chem Eng 4(11):6082–6089
38. François C, Pourchet S, Boni G, Fontaine S, Gaillard Y, Placet V, Galkin MV, Orebom A, Samec J, Plasseraud L (2016) Diglycidylether of iso-eugenol: a suitable lignin-derived synthon for epoxy thermoset applications. RSC Adv 6(73):68732–68738
39. Rao VS, Samui AB (2008) Molecular engineering of photoactive liquid crystalline polyester epoxies containing benzylidene moiety. J Polym Sci Part A Polym Chem 46(23):7637–7655
40. Wang S, Ma S, Xu C, Liu Y, Dai J, Wang Z, Liu X, Chen J, Shen X, Wei J, Zhu J (2017) Vanillin-derived high-performance flame retardant epoxy resins: facile synthesis and properties. Macromolecules 50(5):1892–1901

41. Oulame MZ, Pion F, Allauddin S, Raju KVSN, Ducrot P-H, Allais F (2015) Renewable alternating aliphatic-aromatic poly(ester-urethane)s prepared from ferulic acid and bio-based diols. Eur Polym J 63:186–193
42. Kuhire SS, Nagane SS, Wadgaonkar PP (2017) Poly(ether urethane)s from aromatic diisocyanates based on lignin-derived phenolic acids. Polym Int 66(6):892–899
43. Chen Q, Gao K, Peng C, Xie H, Zhao ZK, Bao M (2015) Preparation of lignin/glycerol-based bis(cyclic carbonate) for the synthesis of polyurethanes. Green Chem 17(9):4546–4551
44. Gang H, Lee D, Choi K-Y, Kim H-N, Ryu H, Lee D-S, Kim B-G (2017) Development of high performance polyurethane elastomers using vanillin-based green polyol chain extender originating from lignocellulosic biomass. ACS Sustain Chem Eng 5(6):4582–4588
45. Endo T, Sudo A (2009) Development and application of novel ring-opening polymerizations to functional networked polymers. J Polym Sci Part A Polym Chem 47(19):4847–4858
46. Comí M, Lligadas G, Ronda JC, Galià M, Cádiz V (2013) Renewable benzoxazine monomers from "lignin-like" naturally occurring phenolic derivatives. J Polym Sci Part A Polym Chem 51(22):4894–4903
47. Wang C, Sun J, Liu X, Sudo A, Endo T (2012) Synthesis and copolymerization of fully bio-based benzoxazines from guaiacol, furfurylamine and stearylamine. Green Chem 14(10):2799–2806
48. Phalak GA, Patil DM, Mhaske ST (2017) Synthesis and characterization of thermally curable guaiacol based poly(benzoxazine-urethane) coating for corrosion protection on mild steel. Eur Polym J 88:93–108
49. Oliveira JR, Kotzebue LRV, Ribeiro FWM, Mota BC, Zampieri D, Mazzetto SE, Ishida H, Lomonaco D (2017) Microwave-assisted solvent-free synthesis of novel benzoxazines: a faster and environmentally friendly route to the development of bio-based thermosetting resins. J Polym Sci Part A Polym Chem 55(21):3534–3544
50. Sini NK, Bijwe J, Varma IK (2014) Renewable benzoxazine monomer from vanillin: synthesis, characterization, and studies on curing behavior. J Polym Sci Part A Polym Chem 52(1):7–11
51. Van A, Chiou K, Ishida H (2014) Use of renewable resource vanillin for the preparation of benzoxazine resin and reactive monomeric surfactant containing oxazine ring. Polymer 55(6):1443–1451
52. Stanzione JF, Sadler JM, La Scala JJ, Reno KH, Wool RP (2012) Vanillin-based resin for use in composite applications. Green Chem 14(8):2346–2352
53. Stanzione JF, Giangiulio PA, Sadler JM, La Scala JJ, Wool RP (2013) Lignin-based bio-oil mimic as biobased resin for composite applications. ACS Sustain Chem Eng 1(4):419–426
54. Meylemans HA, Harvey BG, Reams JT, Guenthner AJ, Cambrea LR, Groshens TJ, Baldwin LC, Garrison MD, Mabry JM (2013) Synthesis, characterization, and cure chemistry of renewable bis(cyanate) esters derived from 2-methoxy-4-methylphenol. Biomacromolecules 14(3):771–780
55. Holmberg AL, Karavolias MG, Epps TH (2015) Raft polymerization and associated reactivity ratios of methacrylate-functionalized mixed bio-oil constituents. Polym Chem 6(31):5728–5739
56. Ferdosian F, Yuan Z, Anderson M, Xu C (2015) Sustainable lignin-based epoxy resins cured with aromatic and aliphatic amine curing agents: curing kinetics and thermal properties. Thermochim Acta 618:48–55
57. Ferdosian F, Yuan Z, Anderson M, Xu C (2016) Synthesis and characterization of hydrolysis lignin-based epoxy resins. Ind Crops Prod 91:295–301
58. Ferdosian F, Zhang Y, Yuan Z, Anderson M, Xu C (2016) Curing kinetics and mechanical properties of bio-based epoxy composites comprising lignin-based epoxy resins. Eur Polym J 82:153–165
59. van de Pas DJ, Torr KM (2017) Biobased epoxy resins from deconstructed native softwood lignin. Biomacromol 18(8):2640–2648
60. Kaiho A, Mazzarella D, Satake M, Kogo M, Sakai R, Watanabe T (2016) Construction of the di(trimethylolpropane) cross linkage and the phenylnaphthalene structure coupled with selective β-O-4 bond cleavage for synthesizing lignin-based epoxy resins with a controlled glass transition temperature. Green Chem 18(24):6526–6535
61. Qin J, Woloctt M, Zhang J (2014) Use of polycarboxylic acid derived from partially depolymerized lignin as a curing agent for epoxy application. ACS Sustain Chem Eng 2(2):188–193
62. Mahmood N, Yuan Z, Schmidt J, Xu C (2016) Depolymerization of lignins and their applications for the preparation of polyols and rigid polyurethane foams: a review. Renew Sustain Energy Rev 60:317–329
63. Li Y, Ragauskas AJ (2012) Kraft lignin-based rigid polyurethane foam. J Wood Chem Technol 32(3):210–224

64. Mahmood N, Yuan Z, Schmidt J, Xu C (2015) Preparation of bio-based rigid polyurethane foam using hydrolytically depolymerized kraft lignin via direct replacement or oxypropylation. Eur Polym J 68:1–9
65. Mahmood N, Yuan Z, Schmidt J, Xu C (2013) Valorization of hydrolysis lignin for polyols and rigid polyurethane foam. J Sci Technol For Prod Processes 3(5):26–31
66. Wyosocka K, Szymona K, McDonald AG, Maminski M (2016) Characterisation of thermal and mechanical properties of ligninsulfonate- and hydrolysed lignosulfonate-based polyurethane foams. BioResources 11(3):7355–7364
67. Xue B-L, Huang P-L, Sun Y-C, Li X-P, Sun R-C (2017) Hydrolytic depolymerization of corncob lignin in the view of a bio-based rigid polyurethane foam synthesis. RSC Adv 7(10):6123–6130
68. Vanderlaan MN, Thring RW (1998) Polyurethanes from alcell® lignin fractions obtained by sequential solvent extraction. Biomass Bioenergy 14(5–6):525–531
69. Yoshida H, Mörck R, Kringstad KP, Hatakeyama H (1990) Kraft lignin in polyurethanes. Ii. Effects of the molecular weight of kraft lignin on the properties of polyurethanes from a kraft lignin–polyether triol–polymeric mdi system. J Appl Polym Sci 40(11–12):1819–1832
70. Cinelli P, Anguillesi I, Lazzeri A (2013) Green synthesis of flexible polyurethane foams from liquefied lignin. Eur Polym J 49(6):1174–1184
71. Bernardini J, Anguillesi I, Coltelli M-B, Cinelli P, Lazzeri A (2015) Optimizing the lignin based synthesis of flexible polyurethane foams employing reactive liquefying agents. Polym Int 64(9):1235–1244
72. Wei Y, Cheng F, Li H, Yu J (2004) Synthesis and properties of polyurethane resins based on liquefied wood. J Appl Polym Sci 92(1):351–356
73. Li H, Mahmood N, Ma Z, Zhu M, Wang J, Zheng J, Yuan Z, Wei Q, Xu C (2017) Preparation and characterization of bio-polyol and bio-based flexible polyurethane foams from fast pyrolysis of wheat straw. Ind Crops Prod 103:64–72
74. Cheng S, Yuan Z, Leitch M, Anderson M, Xu C (2013) Highly efficient de-polymerization of organosolv lignin using a catalytic hydrothermal process and production of phenolic resins/adhesives with the depolymerized lignin as a substitute for phenol at a high substitution ratio. Ind Crops Prod 44:315–322
75. Li B, Wang Y, Mahmood N, Yuan Z, Schmidt J, Xu C (2017) Preparation of bio-based phenol formaldehyde foams using depolymerized hydrolysis lignin. Ind Crops Prod 97:409–416
76. Vithanage AE, Chowdhury E, Alejo LD, Pomeroy PC, DeSisto WJ, Frederick BG, Gramlich WM (2017) Renewably sourced phenolic resins from lignin bio-oil. J Appl Polym Sci 134(19):44827
77. Wang M, Leitch M, Xu CC (2009) Synthesis of phenolic resol resins using cornstalk-derived bio-oil produced by direct liquefaction in hot-compressed phenol–water. J Ind Eng Chem 15(6):870–875
78. Cheng S, D'Cruz I, Yuan Z, Wang M, Anderson M, Leitch M, Xu C (2011) Use of biocrude derived from woody biomass to substitute phenol at a high-substitution level for the production of biobased phenolic resol resins. J Appl Polym Sci 121(5):2743–2751
79. Yan L, Cui Y, Gou G, Wang Q, Jiang M, Zhang S, Hui D, Gou J, Zhou Z (2017) Liquefaction of lignin in hot-compressed water to phenolic feedstock for the synthesis of phenol-formaldehyde resins. Compos B Eng 112:8–14

Affiliations

Elias Feghali[1] · Kirk M. Torr[2] · Daniel J. van de Pas[2] · Pablo Ortiz[3] · Karolien Vanbroekhoven[3] · Walter Eevers[3,4] · Richard Vendamme[3]

✉ Elias Feghali
elias.feghali@ndu.edu.lb

✉ Kirk M. Torr
kirk.torr@scionresearch.com

✉ Richard Vendamme
richard.vendamme@vito.be

[1] Chemical Engineering Program, Notre Dame University-Louaize, PO Box: 72, Zouk Mikael, Zouk Mosbeh, Lebanon

[2] Scion, Private Bag 3020, Rotorua 3046, New Zealand

[3] Flemish Institute for Technological Research (VITO), Boeretang 200, 2400 Mol, Belgium

[4] Department of Chemistry, University of Antwerp, Groenenborgerlaan 171, 2020 Antwerp, Belgium

Topics in Current Chemistry (2018) 376:33
https://doi.org/10.1007/s41061-018-0210-7

REVIEW

Carbon Materials from Technical Lignins: Recent Advances

Alexander M. Puziy[1] · **Olga I. Poddubnaya**[1] · **Olena Sevastyanova**[2]

Received: 23 February 2018 / Accepted: 2 July 2018 / Published online: 11 July 2018

Abstract

Lignin, a major component of lignocellulosic biomass, is generated in enormous amounts during the pulp production. It is also a major coproduct of second generation biofuels. The effective utilization of lignin is critical for the accelerated development of the advanced cellulosic biorefinery. Low cost and availability of lignin make it attractive precursor for preparation of a range of carbon materials, including activated carbons, activated carbon fibers (CF), structural CF, graphitic carbons or carbon black that could be used for environmental protection, as catalysts, in energy storage applications or as reinforcing components in advanced composite materials. Technical lignins are very diverse in terms of their molecular weight, structure, chemical reactivity, and chemical composition, which is a consequence of the different origin of the lignin and the various methods of lignin isolation. The inherent heterogeneity of lignin is the main obstacle to the preparation of high-performance CF. Although lignin-based CF still do not compete with polyacrylonitrile-derived CF in mechanical properties, they nevertheless provide new markets through high availability and low production costs. Alternatively, technical lignin could be used for production of carbon adsorbents, which have very high surface areas and pore volumes comparable to the best commercial activated carbons. These porous carbons are useful for purifying gas and aqueous media from organic pollutants or adsorption of heavy metal ions from aqueous solutions. They also could be used as catalysts or electrodes in electrochemical applications.

Keywords Lignin · Activated carbon · Carbon fibers · Carbon catalyst · Carbon electrodes

---

Chapter 5 was originally published as Puziy, A. M., Poddubnaya, O. I. & Sevastyanova, O. Topics in Current Chemistry (2018) 376: 33. https://doi.org/10.1007/s41061-018-0210-7.

✉ Alexander M. Puziy
alexander.puziy@gmail.com

[1] Institute for Sorption and Problems of Endoecology, NAS of Ukraine, Naumov Street 13, Kiev 03164, Ukraine

[2] Department of Fiber and Polymer Technology, The Royal Institute of Technology (KTH), Stockholm, Sweden

1 Introduction

Lignin is the second most abundant natural polymer after cellulose and is a main source of aromatic compounds in nature [1, 2]. A huge amount of technical lignins is generated as a by-product in pulp and paper manufacturing and in production of second generation of biofuels using kraft, sulfite, soda, organosolv and hydrolysis methods. The estimated global annual production of lignin in the pulp making industry is ca. 70 million tons [3–5]. The amount of lignin waste will grow due to regulations directed to replacing fossil-derived fuels and chemical commodities by biomass.

Lignin is commonly used as a low-grade fuel for pulping operations to recover energy and chemicals. However, the utilization of lignin as a fuel is economically ineffective—the value of lignin as fuel (US$ 0.18/kg) is much less than a conservative estimate for lignin used in chemical conversion (US$ 1.08–1.51/kg) [6]. The potential of lignin as a raw material for value-added applications is underexploited—only 1–2% of technical lignins are used in non-fuel high-value applications such as oil well drilling additives, concrete additives, dyestuff dispersants, agricultural chemicals, animal feed and other industrial binders [1, 7, 8].

Lignin is treated as a waste product in biorefinery processes that focus on the valorization of cellulose and hemicelluloses within the sugar platform [6]. Currently, biorefineries use harsh processing conditions for delignification, leading to highly heterogeneous, complex and polydispersed lignin with uncertain reactivity, which constitutes a major obstacle to its wide-scale commercial valorization [6]. However, an emerging tendency for switching to the lignin-first biorefinery, which treats lignin as a primary valorization target without compromising the carbohydrate fraction, could improve the quality of lignin and facilitate its conversion to marketable value-added products [9, 10].

Lignin as an abundant low-cost biopolymer with high carbon content and high aromaticity provides the industrially important precursor for the production of both functional (activated carbons, activated carbon fibers (ACF), carbon catalysts, electrodes for electrochemical applications) and structural (carbon fibers (CF), carbon black) carbon materials [11–14]. CF and activated carbons represent the highest value-added products from technical lignins and are located in the upper part of the potential lignin applications pyramid (Fig. 1) [15].

The production of lignin-derived carbon materials represents a rapidly growing research area. Many reviews have been published on the conversion of lignin to valuable carbon materials, including CF [3, 13, 16–23], carbon adsorbents [11, 20, 24–26], catalysts [4] or electrodes for electrochemical devices [4, 20]. The objective of this review is to integrate the current knowledge available on the transformation of lignin to value-added carbon materials, with a focus on the most recent advances in this area.

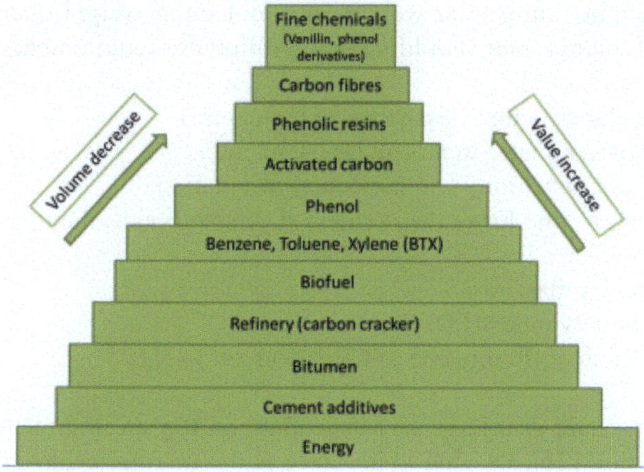

Fig. 1 Potential lignin applications. Reproduced from [15]

2 Carbon fibers

CF are one of the most value-added products produced from lignin (Fig. 1) and have attracted the particular attention of researchers. CF are lightweight materials with excellent mechanical properties and broad applications in sporting goods, in the automotive, aerospace and other industries, and in wind turbine blades [5, 12, 20]. Commercially, CF are produced mostly from polyacrylonitrile (PAN) (90%) [16, 18]. The major limitation for CF applications is the high cost of PAN (~US$ 33/kg), which accounts for more than half of CF production costs [12, 17]. As a bio-renewable, abundant, and low-cost waste, lignin is an attractive precursor to replace PAN, and hence reduce the price and promote the broader utilization of CF in the auto industry. The major advantages of lignin as a CF precursor are related to its high carbon (>60%) content and high carbon yield during carbonization, the absence of toxic elimination products, such as HCN or nitrous gases released during PAN carbonization, as well as its availability and low cost [16].

CF are classified by mechanical properties as high-performance CF or general-purpose CF. High-performance CF are used in aerospace and sporting goods applications, and as reinforcing components in advanced composite materials, while general-purpose CF are used as activated CF for adsorption applications (see section: Porous carbon fibers) and substrates for catalytic (Carbon catalysts) and electrochemical applications (Electrodes for electrochemical applications).

Lignin is an aromatic polymer composed of molecules with different molecular weights, various functional groups and diverse chemical linkages. These intrinsic heterogeneities of lignin could account for the poor mechanical performance of lignin-based CF. A major challenge in the development of lignin-based CFs is the spinnability of lignin. As with other polymers, the processability and properties of lignins are influenced by factors such as atomic and molecular bonds, molecular chain conformation and configuration, degree of order and disorder in the molecular chains, molecular weight, and molecular weight distribution [21]. To form a fiber, the polymer should meet the following requirements:

- Linear molecular structure (less-branched structure).
- High degree of symmetry in molecular backbone.
- High degree of crystallinity and molecular orientation.
- Flexible structure (the lower the number of double bonds, the higher is the flexibility).
- Higher molecular mass.
- Low polydispersity index (PDI).
- High carbon content (in the case of CF precursors) [21].

Other properties are also important depending on the specific method of precursor fiber production [21]. Of vital importance is the thermal behavior of the lignin for the melt spinning method. Hardwood lignin is a more suitable precursor than softwood lignin, which did not show a softening point and thus could not be melt spun. The dissolution of lignin in solvents, molecular weight, rheological properties of lignin solutions, solvent type, and solvent volatility are critical in wet spinning and electrospinning methods.

The manufacturing CF from lignin has attracted particular interest, which is reflected in the exponentially growing number of publications on this topic [17]. The preparation of CF from lignin has been described in many excellent reviews [3, 13, 16–18, 20–22]. Different strategies were used to enhance the spinnability of lignin, such as (1) polymer blending with PEO [27–30], PET [31] or PAN [32–35]; (2) reinforcement with clay [36, 37] or carbon nanotubes [38, 39]; (3) chemical modifications such as cross-linking using hexamethylenetetramine [40], hydrogenation [41], acetylation [42, 43], phenolation [44], butyration [45], derivatization of lignin via attachment of ester, ether or urethane groups [46]; and (4) fractionation by membrane filtration [47, 48] or via solvent extraction [41, 49–51]. Recently several approaches to improve the quality of CF has emerged.

2.1 Lignin–lignin blends

Improving the processing and performance of pure lignin CF through hardwood and herbaceous lignin blends has been proposed [52]. The spinnability of switchgrass lignin markedly improved by blending it with yellow poplar lignin. In contrast, switchgrass lignin significantly improved the thermostabilization performance of yellow poplar fibers, preventing the fusion of fibers during fast stabilization and improving the mechanical properties of fibers. The CF obtained had diameters in the range 16–32 μm, tensile strength 230–750 MPa and Young's modulus 30.4–41.8 GPa. The mechanical properties of CF were highest in the samples with the lowest switchgrass lignin content and were stabilized at the lowest rate. However, samples with the highest switchgrass lignin content had consistent mechanical properties across all stabilization rates evaluated. These results suggested a route towards a 100% renewable CF with a significant decrease in production time and improved mechanical performance.

2.2 Lignin–cellulose blends

Addition of small amount of nanocrystalline cellulose (up to 5%) allowed preparation of lignin-based CF with homogeneous cross section by direct carbonization without an oxidative thermal stabilization step [53]. Nanocrystalline cellulose interact with lignin preventing the fusion of precursor fibers during heat treatment. Thus, unique interconnected carbon mats with increased electrical conductivity (5–35 S/cm) were formed. Inclusion of nanocrystalline cellulose during electrospinning provides a route to reduce processing time and energy cost associated with CF production.

2.3 Reinforcement

The reinforcement of lignin-derived CF with lignin grafted to carbon nanotubes (CNTs-g-L) allowed the preparation of well-orientated CF with increased mechanical properties [39]. The tensile strength increased from 171.2 to 289.3 MPa when 0.5% CNTs-g-L was incorporated, which exceeded the tensile strength of lignin/CNT-based CF with 1% unmodified CNTs. However, many voids appeared in the carbonized lignin/CNTs-g-L fibers, which were generated by the breakage of chemical links between CNTs and lignin, thus decreasing the tensile strength of the obtained CF. It has been noted that the generation of voids can be controlled by optimizing the stabilization and carbonization processes.

2.4 Fractionation

Solid and smooth CF were fabricated from soft wood kraft lignin [48]. CF were produced by direct spinning of the permeate obtained by ultrafiltration of the black liquor through a 15-kD ceramic membrane, followed by oxidative stabilization and carbonization. In another instance, the permeate of hard wood kraft lignin was added to unfractionated soft wood kraft lignin as a softening agent. Thus, good quality kraft lignin-based CF with a carbon content ranging from 93 at% to 97 at% were produced.

Quality lignin-based CF were fabricated by developed enzyme-mediator system to fractionate and modify lignin [5]. It has been shown that laccase-1-hydroxybenzotriazole hydrate treatment not only enhanced crystallite microstructure of the CF but also led to a high reduced elastic modulus of 21.8 GPa compared with 20.7 GPa for PAN-based CF. The water-insoluble lignin fraction derived from the enzymatic processing had higher molecular weights, fewer functional groups and more β-*O*-4 interunitary linkages, which served as a better CF precursor.

The impact of molecular weights (MW) and lignin uniformity on the performance of CF has been studied [54]. Fractionating lignin into fractions with different MW and PDI using an enzyme-mediator-based method and a dialysis method were applied to obtain lignin fractions with an increased MW and decreased PDI. Lignin fractions were electro-spun into fibers after blending with polyacrylonitrile (PAN) at a 1:1 (w/w) ratio. Fractionation in general improved the spinnability of lignin and allowed the acquisition of finer lignin-based CF. The elastic modulus of lignin CF, as measured by nanoindentation, increased as the lignin MW increased and as PDI decreased. The higher lignin MW and lower PDI led to an increase in crystallite size and content of the pre-graphitic carbon, improving the CF mechanical performance. The study revealed the impact of lignin MW and uniformity on the mechanical properties of CF, and pawed a way toward the development of lignin processing technologies for quality lignin-based CF.

2.5 Chemical modification

2.5.1 Acetylation

The impact of acetylation on production of CF from an industrial corn stover lignin has been investigated [55]. First, fractionation with methanol was applied, which removed most of the impurities in the raw lignin and selectively removed the molecules with high melting points. However, neither methanol fractionation nor thermal treatment resulted in melt-spinnable precursors. A two-step acetylation of methanol-fractionated lignin was used, which greatly improved the mobility of lignin while enhancing the thermal stability of the precursor during melt spinning. It has been shown that the contents of phenolic and aliphatic hydroxyls, as well as the hydroxycinnamates, decreased in the acetylated precursors. The optimum precursor was a partially acetylated lignin with a glass transition temperature of 85 °C. Upon oxidative stabilization and carbonization, CF with an average tensile strength of 454 MPa and modulus of 62 GPa were obtained.

2.5.2 Iodine pretreatment

Using iodine pretreatment, the thermodynamic stability of the lignin-based precursor fibers increased significantly, and thus energy consumption during the preparation of CF was reduced [56]. Moreover, CF produced from iodine pretreated lignin precursor fiber showed a higher yield of 59–61% compared with 22–38% for iodine-free CFs, a higher tensile strength 85–89 MPa compared with 20–56 MPa for iodine-free CFs and a higher modulus of 5.2–5.3 GPa compared with 2–3.2 GPa for iodine-free CFs.

2.6 New types of lignin

A recently discovered new type of lignin, poly-(caffeyl alcohol) (PCFA, also known as C-lignin), in the seeds of vanilla orchid (*Vanilla planifolia*) has been used for the preparation of CF without additional modification or blending with polymers [57]. PCFA lignin, in contrast to all known lignins, which are comprised of polyaromatic networks, is a linear polymer derived almost totally from caffeyl alcohol monomers linked head to tail into benzodioxane chains. CF obtained by electrospinning followed by carbonization at 900 °C had a smaller diameter (~ 10 vs ~ 50 μm) and were more ordered (I_D/I_G 1.15 vs 1.92) compared to CF from kraft lignin. CF from PCFA and kraft lignin had comparable axial moduli measured by nanoindentation compared to commercial carbon fibers, while the transverse moduli of the PCFA were better than kraft and commercial CF.

3 Carbon adsorbents

Carbon adsorbents are highly porous materials with carbon as the main component [58]. Carbon adsorbents are most commonly used in a wide range of applications, such as cleaning gas and fluid streams, gas separation, catalyst supports, filtration systems and electrochemical applications [59, 60]. World consumption of carbon adsorbents was estimated at approximately 1.1 million tons per annum in 2010, and driven by environmental regulations, it is growing at a rate of 9% yearly [61].

Activated carbon can be produced from a number of organic, carbon-rich sources, including coal, wood or coconut shells, by physical or chemical activation processes [58, 62, 63]. During physical activation, the carbonaceous precursor is first carbonized in an inert atmosphere and then activated with oxidizing gases such as oxygen (air), H_2O or CO_2. Chemical activation is performed by impregnating a carbonaceous precursor with an activating agent such as KOH, H_3PO_4, and $ZnCl_2$, followed by heat treatment.

Approximately 80–85% (ca. 0.9 million tons) of the total production of activated carbon is derived from non-renewable coal-based resources [61]. As a polyaromatic macromolecule with a carbon content greater than 60%, lignin may provide the high carbon yield required for the commercial manufacturing of activated carbon and can partially substitute for non-renewable coal-based resources. An additional advantage of using renewable precursors, such as lignin, is a reduction of the carbon footprint for the production of activated carbon. Virtually all kinds of preparation methods could be used for production of carbon adsorbents from lignin.

3.1 Carbonization

In some cases, activation in not necessary step for preparation of porous carbon materials. Nitrogen-containing hollow carbon nanospheres were prepared by direct carbonization of the polyaniline–lignosulfonate composite spheres at different temperatures under a nitrogen atmosphere [64, 65]. Hollow carbon nanospheres were uniform with an average diameter of 135 nm and developed a micro-mesoporous structure-BET (Brunauer-Emmett-Teller) surface area in the range 382–700 m^2/g, micropore volume in the range 0.12–0.28 cm^3/g and mesopore volume in the range 0.12–0.16 cm^3/g. The hollow carbon nanospheres showed a high adsorption capacity towards papain up to 1.16 g/g, which holds much promise for the isolation of proteins or enzyme immobilization.

3.2 Physical activation

The preparation of activated carbons from eucalyptus kraft lignin by physical activation with CO_2 at 800–850 °C has been reported [66, 67]. It has been noted that ash content plays a key role in the carbonization behavior of kraft lignin. Low ash kraft lignin showed a plastic and swelling behavior during pyrolysis under a nitrogen atmosphere even at heating rates as low as 1°/min, while no fusion or swelling was observed when kraft lignin with an ash content above 8% was used. To avoid serious operating problems in practice, the as received high ash (12.4%) kraft lignin was pre-carbonized at 350 °C and demineralized with 1% H_2SO_4 prior to activation with CO_2. The obtained carbons had developed a porous structure with the contribution of mesoporosity increased with burnoff (Table 1). Development of the porous structure with activation allowed the preparation of a range of products with distinct potential uses in gas and liquid phase operations. The efficiency of the removal of organic compounds associated with environmental concerns, such as dyes [68–71], different aromatic compounds [72–75], surfactant (sodium dodecyl benzene sulfonate) [76] from aqueous solutions and volatile organic compounds from the gas phase [77] was demonstrated using activated carbons from kraft lignin. The large amount of Na observed in the carbons from kraft lignin increased the adsorption of water vapor and thus offered the prospect of using such activated carbons as desiccants [78].

A review was recently published on preparation of carbon adsorbents from hydrolysis lignin, a large-scale by-product of the hydrolysis industry, operated in the Soviet Union and several eastern European countries from the 1930s to the end of the 1990s [26]. Although percolation hydrolysis and its variants that were developed in the former Soviet Union are not currently in use, the large amounts of hydrolysis lignin accumulated during the twentieth century represent a good source for processing into value-added products. Hydrolysis lignin has been shown to be a suitable precursor for the production of activated carbon. Numerous studies have demonstrated that

Table 1 Preparation conditions and porosity characteristics of activated carbons from eucalyptus kraft lignin. Data compiled from [66]

Activation temperature (°C)	Activation time (h)	Yield (%)	BET surface area (m^2/g)	Micropore volume (cm^3/g)	Mesopore volume (cm^3/g)
800	4	35.4	747	0.29	0.02
800	12	26.3	947	0.34	0.31
800	19	19.5	1223	0.36	0.63
800	30	13.7	1434	0.43	0.73
800	40	11.8	1613	0.47	0.77
850	4	31.5	822	0.31	0.60
850	12	18.2	1273	0.39	0.48
850	15	14.6	1348	0.42	0.78
850	20	9.3	1853	0.57	0.86

highly porous carbon adsorbents can be produced from hydrolysis lignin using physical activation with steam (specific surface area of 1200–1260 m^2/g), carbon dioxide (specific surface area of 1640 m^2/g) or chemical activation with H_3PO_4 (specific surface area of 900–1500 m^2/g), K_2CO_3 (specific surface area of 1480–1730 m^2/g), NaOH (specific surface area of 2100–2500 m^2/g) or KOH (specific surface area of 2400–2750 m^2/g). Activated carbons from hydrolysis lignin showed comparable or even superior characteristics in numerous environmental and other applications compared to commercially available carbon adsorbents.

3.3 Chemical activation

Chemical activation with H_3PO_4 was used to prepare highly porous carbons with a well-developed micro- and meso-porous structure from eucalyptus kraft lignin [79]. The maximum surface area of 1460 m^2/g was obtained at 425 °C and impregnation (acid/precursor) ratio of 2. The obtained carbon showed a high adsorption capacity towards the three important target compounds representing water toxic pollutants (phenol, 2,4,5-trichlorophenol and Cr(VI)) that was comparable, or higher than, other activated carbons, thus demonstrating their potential use as adsorbents for water pollutant removal.

Activation of kraft lignin with ortho-phosphoric acid in air at temperatures ranging from 400 to 650 °C has been reported using an impregnation ratio of 0.7–1.75 [80]. An increasing carbonization temperature has led to a substantial decline in yield (Fig. 2). With an increasing activation temperature, the BET surface area first increased, reached a maximum of 1305 m^2/g at 600 °C, and considerably decreased at higher temperatures (Fig. 2). The steady increase in surface area with a temperature rise up to 600 °C was ascribed to the evolution of compounds produced from cross-linking reactions, while the decrease in surface area at higher temperatures was due to the degradation of phosphate and polyphosphate bridges accompanied by the oxidation of carbon material [81, 82].

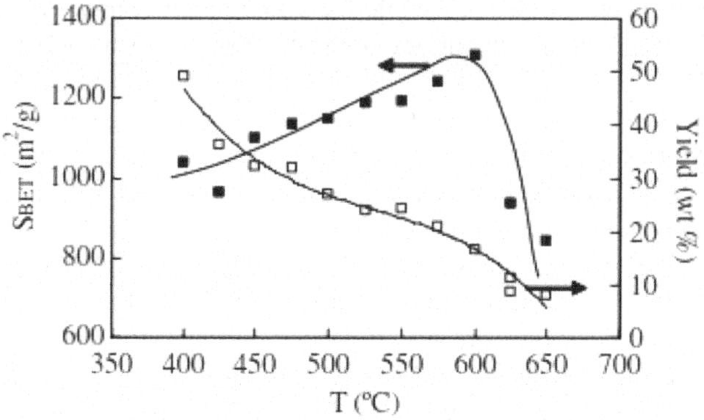

Fig. 2 Brunauer-Emmett-Teller (BET) surface area and carbon yield of the activated carbons as a function of the activation temperature (P/L = 1.4 and 1 h impregnation time). Reproduced with permission from [80]

The impact of the demineralization of kraft lignin on the properties of activated carbons obtained by chemical activation with phosphoric acid was examined [83]. Porosity characteristics (surface area and pore volume) and methylene blue adsorption have been shown to be higher for carbons prepared from raw lignin. This effect was explained by a decrease in the number of hydroxyl groups during demineralization, which in turn decreased the reaction of lignin with phosphoric acid and subsequent cross-linking. It was concluded that since superior active carbons were produced from raw lignin, the demineralization, if required, should be carried out using activated carbon rather than precursors.

Chemical activation with $ZnCl_2$ of eucalyptus kraft lignin to produce highly porous activated carbons has been reported [84]. The maximum surface area ($\approx$1800 m^2/g) was obtained at 500 °C and an impregnation ratio of 2.3. As received, eucalyptus kraft lignin was demineralized with 1% H_2SO_4, which reduced the ash content from 12.4% to 0.1%. It was noted that the plastic phase of kraft lignin with an ash content lower than 4% occurs at relatively low temperatures ($\approx$180–280 °C) and has no effect on the development of porosity. Lignin-derived activated carbons feature reasonably good mechanical properties, showing very acceptable bulk density values despite their high porosity. Chemical activation with $ZnCl_2$ developed a substantially higher microporosity and much lower mesoporosity compared with CO_2-activated carbons [66, 67], while maintaining an equivalent BET-specific surface area. Moreover, the yield of chemically activated carbons is higher: 1800 m^2/g of BET surface area was reached through $ZnCl_2$ activation with approximately 40% yield [84], in contrast to only 10% with CO_2 activation [66, 67]. The efficiency of $ZnCl_2$-activated carbon in SO_2 removal was demonstrated [85]. The presence of oxygen and water vapor in the inlet stream increased the SO_2 adsorption capacity of the carbon due to the surface oxidation of the SO_2 to SO_3 and subsequent hydration to H_2SO_4. This holds much promise for cleaning gas streams from SO_2 to produce H_2SO_4.

The preparation of carbon adsorbents by chemical activation of kraft lignin with $ZnCl_2$, H_3PO_4, K_2CO_3, Na_2CO_3, KOH and NaOH has been described [86]. The maximum surface areas of the activated carbons prepared by $ZnCl_2$ and H_3PO_4 activation was observed at 600 °C, while for alkali metal compounds it occurred at 800 °C (Fig. 3). The maximum surface area of 2000 m^2/g was achieved for activation with K_2CO_3.

A favorable effect of microwave treatment during impregnation of kraft lignin with $ZnCl_2$ for the preparation of activated carbons at 500–800 °C has been reported [87]. The use of microwave pretreatment increased the BET specific surface area, total pore volume and total amount of acid surface groups (Table 2). The maximum development of porosity and total acid surface groups was obtained at an activation temperature of 600 °C and impregnation ratio of 1:1.5. Higher total amount of acid surface groups for microwave-pretreated carbon were responsible for an increased adsorption of copper ions via ion exchange mechanism.

Chemical activation of kraft lignin NaOH and KOH using different preparation conditions, such as various temperatures, impregnation ratios, activation times, flow rates of inert gas, and heating rates, has been thoroughly studied [88]. KOH produced a more developed porous structure than NaOH under all conditions. The

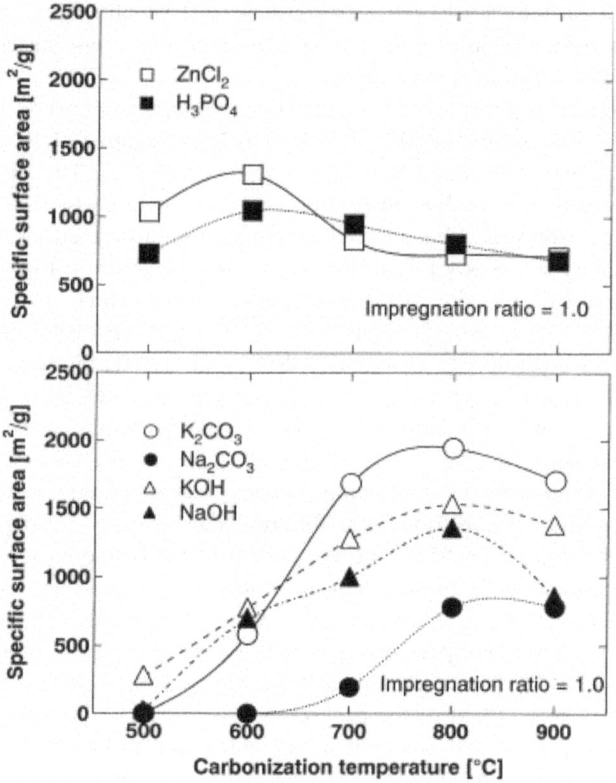

Fig. 3 Influence of carbonization temperature and activating reagent on the surface area of prepared activated carbon Reproduced with permission from [86]

Table 2 Surface area, total pore volume and amount of surface groups in activated carbon from kraft lignin obtained by $ZnCl_2$ activation with and without microwave treatment. Data compiled from [87]

Carbon	BET surface area (m^2/g)	Total pore volume (cm^3/g)	Total amount of acid surface groups (mmol/g)	Maximum adsorption capacity for copper (mmol/g)
With microwave pretreatment	1172	0.64	3.14	1.54
Without microwave pretreatment	918	0.51	2.66	1.15

maximum BET surface area (3100 and 2400 m^2/g for KOH and NaOH, respectively) and micropore volumes (1.5 and 1 cm^3/g for KOH and NaOH, respectively) were attained at 700 °C and an impregnation ratio of 3. Such highly porous carbons were obtained with a good yield of 30%, which is lower than that obtained by activation with H_3PO_4 [80, 89] and $ZnCl_2$ [84] and slightly lower than that obtained for pure charring of lignin at the same temperatures [90, 91]. The drawback of alkali-activated carbons is the low packing density (typically 0.2 g/cm^3) of carbon materials, which significantly reduces the advantage of a high surface area when volume-based characteristics are required.

The preparation of phosphorus-functionalized nanoporous carbon by KOH activation of de-alkaline lignin at 800 °C followed by heat treatment with triphenylphosphine at 800 °C under nitrogen flow has been reported [92]. The obtained carbon had a BET surface area of 837 m^2/g and total pore volume of 0.41 cm^3/g along with 0.9% of phosphorus. Phosphorus-containing lignin-derived carbon showed a very high capacity for neodymium (356 mg/g) and dysprosium (345 mg/g), while the adsorption of iron was one order of magnitude lower, which suggests a possible separation of these rare earth elements from Fe(III). The XPS study revealed a crucial role of surface phosphate groups in Nd(III) and Dy(III) binding, with a minor contribution of carboxylic groups. The study showed the potential of lignin-derived phosphorus-functionalized nanoporous carbon for the enrichment and separation of Nd(III) and Dy(III).

Highly porous phosphorus-containing carbons were prepared by phosphoric acid activation of sodium lignosulfonate at different temperatures (Table 3) [93]. The yield was in the range of 49–55% for carbons obtained at temperatures below 800 °C and was approximately 9% higher than that for acid-free carbons. Lignosulfonate-based carbons contained a significant amount of phosphorus (Table 3). With an increasing carbonization temperature up to 800 °C, the phosphorus content increased to a maximum of 10.7% and decreased at higher temperatures. The increasing phosphorus content was ascribed to the progressive formation of polyphosphate esters, while the decreasing phosphorus content was attributed to both the evaporation of phosphorus compounds due to the thermal instability of the C–O–P linkage above 800 °C and phosphate reduction by carbon [94, 95], which is supported by thermodynamic calculations [93, 95]. Previous investigations have shown that the main phosphorus species in phosphoric acid activated carbons are phosphates/polyphosphates [96], which impart acidic properties to the carbon [97–102]. As a result,

Table 3 Chemical composition and porous structure of lignosulfonate-based activated carbons obtained by phosphoric acid activation. Adapted from [93]

Carbonization temperature (°C)	Yield (%)	C (%)	O (%)	P (%)	P/O	A_{BET} (m^2/g)	V_{tot} (cm^3/g)	V_{mi} (cm^3/g)
400	55	80.6	17	2.1	15.3	766	0.45	0.25
500	49.2	81.3	16.7	3	10.7	587	0.46	0.17
600	51.4	78.6	17.4	7	4	319	0.3	0.08
700	53.4	70.4	18.99	10.6	3.7	179	0.26	0.04
800	48.6	73.8	15.3	10.7	2.8	230	0.29	0.06
900	36	73.2	14.4	8.7	3.2	904	0.65	0.31
1000	26.2	82.6	11.2	4.4	5.8	1373	0.97	0.41
800[a]	39.6					142	0.14	0.05

[a]Obtained without phosphoric acid

phosphorus-containing carbon adsorbents show a high capacity for metal ions in aqueous solutions [94, 95, 99, 100, 102–105]. Lignosulfonate-based activated carbons exhibited pronounced adsorption of copper ions (Fig. 4) [93]. With an increasing carbonization temperature up to 900 °C, copper binding increased and decreased for carbon obtained at 1000 °C, which was ascribed to the combined effect of the increasing phosphorus content in temperature range 400–800 °C and increasing specific surface area at 900 °C. It is interesting that all lignosulfonate-based carbons could bind a significant amount of copper from very acidic solutions ($pH < 2$), which was attributed to an ion-exchange reaction between the delocalized protonated π-electrons of the graphene layer ($-C\pi-H_3O^+$) and the metal ion [106, 107]. Thus, lignosulfonate-based carbon adsorbents offer a promise for use in water treatment methods.

Preparation of highly porous (specific surface area 2640–2750 m^2/g) carbon adsorbents from hydrolysis lignin by charring followed by chemical activation has been reported [108]. It was found that preliminary charring is a necessary step for the preparation of high-surface-area carbon adsorbents from hydrolysis lignin. The most effective activating agent was KOH and NaOH, while K_2CO_3 and Na_2CO_3 produced carbons with almost no activity towards methylene blue. Increasing amounts of KOH, activation temperature and time led to an increase in adsorption activity and decrease in yield.

Microporous carbons with high specific surface areas of 2000–3100 m^2/g, large micropore volumes, and narrow pore size distributions were prepared from hydrolysis lignin by activation with H_2O/CO_2 mixture or KOH [109]. It has been shown that KOH-activated carbon exhibits considerable cryogenic hydrogen uptake capacities

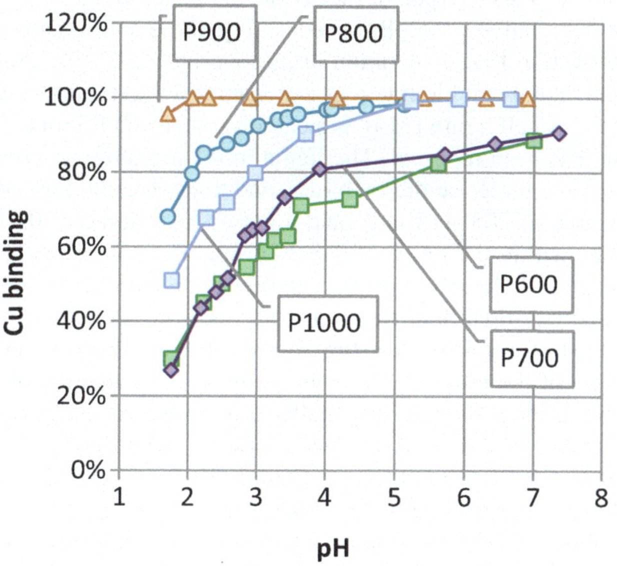

Fig. 4 Copper binding by lignosulfonate-based carbons obtained by phosphoric acid activation at different temperatures. Adapted from [93]

exceeding 5% at 10 bar, and attaining 2.55% at 1 bar with an adsorption heat of 4.1–7.5 kJ mol^{-1}.

3.4 Template Synthesis

Microporous–mesoporous carbons were synthesized by the template method using microwave-pretreated kraft lignin in sodium bicarbonate solution as a carbon precursor and colloidal silica as a hard template [110]. A characteristic feature of the carbons is the appearance of uniform spherical mesopores after dissolving colloidal silica. Depending on the types of template, the mesopore size of carbon was 14–15 nm for 20 nm colloidal silica and 21–25 nm for 13 nm colloidal silica. A microporous structure was developed by post-synthesis activation with CO_2. The activated carbons obtained were highly porous (specific surface area up to 2000 m^2/g, total pore volume up to 2.2 cm^3/g) and an easily tunable porous structure was achieved by adjusting the activation conditions and optimizing the amount and particle size of the colloidal silica used as template.

The potential of Alcell lignin as a low cost carbon precursor for the preparation of porous carbons with a hierarchical pore structure by the template method using different zeolites has been demonstrated [111]. Microporosity develops from both replication of the hard template and carbonization of lignin, whereas mesoporosity can be adjusted by selection of the hard template and infiltration conditions. The BET surface area of porous carbons was in the range of 620–950 m^2/g, which could be increased up to 1380 m^2/g by additional CVD of propylene. Lignin-derived porous carbons contained a relatively high amount of nitrogen (pyrrolic, pyridinic) and oxygen (hydroxyl, carbonyl) surface groups, presumably originating from the surface ammonia ions and oxygen of the zeolite template. The hydroxyl groups were responsible for the relatively high Brönsted acidity of the porous carbons.

Chemical-activation-free preparation of mesoporous carbon by the soft template method using methanol-soluble fractions extracted from kraft hardwood lignin [112, 113] or hardwood Alcell lignin [114] as a carbon source and Pluronic F127 as a sacrificial template has been reported. The lignin in the mixture was cross-linked with formaldehyde (37%) under acidic or basic conditions. Lignin-derived carbons had a BET surface area of 205–418 m^2/g and a mesopore volume of 0.11–0.34 cm^3/g, with pore widths within the range of 2.5–12.0 nm. The controlled release of three model drugs, captopril, furosemide and ranitidine hydrochloride in simulated gastric fluid or simulated intestinal fluid from mesoporous lignin-derived carbons has been studied [112, 113]. The release kinetics showed that the diffusivities of the drugs from carbon were of the order of 10^{-22}–10^{-24} m^2/s. The reduction of pore width in the carbon reduced the diffusivity of smaller drug molecule by an order of magnitude. Another study [114] showed a longer release of antipyrine (1–7 h) from lignin-derived carbon as compared to phloroglucinol-derived carbon (0.5–3 h), which was attributed to lower porosity due to the polydispersity and hyperbranched structure of lignin that inhibited the formation of a perfect self-assembled morphology of the carbon precursor. Thus, the potential use of lignin-based mesoporous carbon for controlled drug delivery was demonstrated.

A combination of hydrothermal treatment and a soft template method, using sodium lignosulfonate as a carbon precursor and cetyltrimethylammonium bromide as sacrificial template for the production of mesoporous carbons has been reported [115]. To remove the templating agent, biochar after hydrothermal treatment was carbonized at 800 °C in a nitrogen atmosphere. The carbon obtained exhibited a spherical shape and developed pore structure with specific surface area of 633 m^2/g and pore size of 4.6 nm. Lignosulfonate-based carbon showed pronounced binding of U(VI)—the maximum adsorption capacity in the simulated aqueous solution and actual radioactive wastewater was 109.5 and 195.6 mg/g, respectively, at pH 5.5.

An ice templating method [116] was used to prepare two-dimensional (2D) carbon nanosheets (CNSs) by freeze-drying a liquid-nitrogen-quenched alkali lignin dispersion followed by direct carbonization without any activation treatment [117]. Increasing the lignin concentration in the aqueous dispersion increased the size and stacking compactness of the obtained lignin CNSs, while the specific surface area decreased from 855 to 97 m^2/g. This study provided a simple, inexpensive, and environmentally friendly method for the preparation of CNSs from various water-soluble biomass carbon precursors.

3.5 Hydrothermal Carbonization

Owing to the necessity for the optimal and sustainable use of natural resources, hydrothermal treatment of biomass and biomass waste for energy and resource recovery has received increasing attention [25]. Several excellent reviews on the hydrothermal treatment of biomass including lignin are available [20, 25, 118, 119].

Lignin provides an attractive alternative as a raw material for hydrothermally treated carbons (HTC) because it is widely available and has a high carbon content. Lignin shows a higher thermal stability compared with cellulose, resulting in less conversion in the HTC and higher solid yield [119]. Hydrothermal treatment slightly increases carbon content and removes ashes. However, biochar obtained by hydrothermal treatment shows rather low porosity and requires further activation to obtain carbon adsorbents with high adsorption ability [25]. The special convenience of hydrothermal carbonization is that it allows treating liquid lignin wastes without evaporation.

Hydrothermal treatment of lignin waste was used to enrich carbon content prior to KOH activation for the production of highly porous activated carbons [120]. The advantage of hydrothermal treatment was a decrease in ash content from approximately 20% for lignin waste to almost zero for hydrochar. Activation of ash-free hydrochar at an impregnation ratio of 2 generated highly microporous carbons ($\geq$97% micropore surface area and 93% micropore volume), which exhibited excellent CO_2 uptake capacity up to 4.6 mmol/g at 1 bar and 25 °C and 17.3 mmol/g at 20 bar. An increase in the impregnation ratio to 4 produced activated carbons with a surface area and pore volume up to 3235 m^2/g and 1.77 cm^3/g, respectively, revealing hydrogen uptake of up to 6.2% at −196 °C and 20 bar.

HP-L™ lignin, Lignol's proprietary product, was treated hydrothermally followed by chemical activation with KOH (IR 4) [121]. Hydrothermal chars exhibited agglomerated structures with some vesicles at elevated temperatures. Major structural changes in the HP-L™ lignin occurred at temperatures higher than 300 °C, with significant aromatization observed at 350 °C. The activated carbon produced from lignin hydrochar was predominantly microporous with a BET surface area 1341 m^2/g.

4 Porous CF

Carbon adsorbents in the form of fibers offer advantage of rapid adsorption–desorption rate and high adsorption capacity. Although activated CF are much in demand, their widespread use is limited due to relatively high price. To lower production costs, lignin was studied as relatively cheap and renewable material for preparation of porous CF.

4.1 Carbonization

Porous CF could be produced by carbonization without activation. By exploring idea of using two polymers with different thermal properties—one crosslinking at high temperature giving carbon material and the other melting and acting as a pore-forming agent—porous CF were produced from immiscible polymer blends of hardwood kraft lignin and polypropylene (PP) [122, 123]. Porosity was achieved by a two-step process—oxidative degradation of the PP component followed by pyrolysis gasification. It was shown that gasification was the main factor for development of porosity. The advantage of the method is that a relatively simple processes can produce CF with a high specific surface area up to 444 m^2/g without activation. Furthermore, such porous CF could be activated with steam or CO_2 to make them suitable for commercial applications.

The preparation of filled and hollow CF with well-developed porous structure from Alcell lignin has been reported [124, 125]. Lignin fibers were prepared by electrospinning of ethanol solution of Alcell lignin using a coaxial spinneret that created a thin sheath of ethanol on the fiber surface to prevent solidification of the precursor fiber due to evaporation of the solvent. For preparation of hollow fibers, a tri-coaxial spinneret was used, with a third needle supplying glycerol as templating fluid. CF were obtained by oxidative stabilization of lignin fibers at 200 °C in air, followed by carbonization at 600–1000 °C. CF had diameters as small as 200 nm and smooth surfaces with no defects. It is interesting that CF showed a well-developed porous structure with a BET surface area up to 1200 m^2/g and total pore volume up to 0.48 cm^3/g without the activation step. Addition of platinum acetyl acetonate to Alcell lignin solution allowed the production of CF containing 0.2–0.6% platinum [126]. The obtained CF showed a high oxidation resistance despite the well-developed porous structure or even the presence of platinum.

Nitrogen-doped free-standing fused carbon fibrous mats were prepared from an Alcell lignin-polyethylene oxide (90:10) blend by electrospinning followed by carbonization and thermal annealing at 900 °C in the presence of urea [127]. The obtained CF mats showed a relatively large BET-specific surface area of 473 and 381 m^2/g with a total pore volume of 0.31 and 0.28 cm^3/g for N-free and N-containing samples, respectively. The research demonstrated an economic and straightforward fabrication process that holds much promise for large-scale production.

Mechanically flexible mats consisting of carbon nanofibers were prepared by electrospinning aqueous mixtures of alkali kraft lignin with polyvinyl alcohol into composite nanofiber mats followed by stabilization in air and carbonization in an inert atmosphere [128]. Carbon nanofiber mats derived from the composite nanofibers with 70% lignin had an average fiber diameter of ~ 100 nm and BET specific surface area of 583 m^2/g. It has been shown that, with an increasing amount of lignin in the precursor nanofiber, the average pore size decreases and the volume and specific surface area increases.

4.2 Physical Activation

Activated CF (ACFs) were prepared from acetic acid lignin-based CF by steam activation [129]. The ACFs showed a BET specific surface area of 1250 m^2/g, more rapid adsorption kinetics and higher iodine and methylene blue adsorption capacities, compared to commercially available activated carbon. Rapid uptake by ACFs has been ascribed to adsorption occurring on the surface, but not inside the fibers due to the absence of macropores.

4.3 Chemical Activation

Addition of phosphoric acid to Alcell lignin/ethanol solution for electrospinning allowed shortening of the conventional stabilization process from more than 90 h to only 2 h, or even enabled the stabilization of the lignin fibers in an inert atmosphere [130]. Phosphoric acid as a chemical activating agent generated phosphate and/or polyphosphate esters with lignin molecules, which were cross-linked during the stabilization step, increasing the glass transition temperature of the lignin fibers and reducing the time needed for stabilization. Moreover, phosphoric acid played the role of a chemical activating agent generating porosity in CF from lignin. Due to the presence of phosphorus species, CF showed greatly increased oxidation resistance, avoiding complete combustion during carbonization under a low concentration of oxygen at temperatures as high as 900 °C. The resulting CF had sub-micron diameters (≤ 1 µm), a large surface area (≈ 2000 m^2/g), relatively high mechanical performance with Young's modulus 2–4 GPa and a tensile strength ranging from 0.25 to 0.30 GPa. Phosphorus-containing CF showed high adsorption of phenol from aqueous solutions, reaching a capacity of approximately 180 mg/g at 25 °C and concentration of 25 mg/L.

5 Carbon Catalysts

Lignin-based carbon materials were used as effective catalysts in esterification, hydrolysis, etherification or dehydration reactions.

5.1 Esterification

A simple and environmentally friendly method for the preparation of a low-cost solid base catalyst for biodiesel production has been described [131]. A solid base catalyst was prepared by mixing kraft lignin with K_2CO_3, followed by carbonization at 800 °C. The highest biodiesel yield of 99.6% was achieved with a catalyst content of 3.0%, methanol to oil molar ratio of 15:1, reaction time of 2 h and reaction temperature of 65 °C. The catalyst could be recycled and reused four times without marked deactivation.

The preparation of a solid acid catalyst by direct sulfonation of lignin residue obtained after hydrolysis of *Xanthoceras sorbifolia* Bunge hulls has been reported [132]. The solid acid catalyst showed high catalytic activity and cycling ability in the esterification of acidified soybean soap with methanol. The maximum free fatty acid conversion of 97.2% was achieved with a catalyst content of 7%, reaction temperature of 70 °C and molar ratio of methanol to acidified oil of 9:1. The study demonstrated not only a decreased cost and simplified process but also full use of the biomass waste. The activity of lignin-derived catalyst was similar to that of 1.5% of H_2SO_4.

Lignin residue of pine wood hydrolysis was directly sulfonated with sulfuryl chloride at 50 °C for 5 h [133]. The obtained lignin-derived solid acid catalyst showed high catalytic activity in the esterification of acetic acid with ethanol and hydration of 2,3-dimethyl-2-butene. Recycling catalytic tests showed the superior stability of the lignin-based solid acid.

The potential of bio-refinery by-product lignin to produce a mesoporous solid acid catalyst has been demonstrated [134]. Lignin was obtained as a by-product of 5-(chloromethyl)furfural production from Swedish spruce sawdust. The advantage of this type of lignin is a moderate level of mesoporosity with a surface area of 63 m^2/g that allowed direct sulfonating without pre-carbonization steps. It has been shown that the obtained solid acid catalysts were efficient in the esterification of levulinic acid with excess ethanol with the highest ester yield of up to 92–95 and 100% selectivity.

5.2 Cellulose Hydrolysis

The preparation of lignin-derived solid acid catalysts for microcrystalline cellulose hydrolysis has been reported [135]. Three carbon catalysts were prepared by carbonization at 425 °C, followed by sulfonation at 180 °C of differently pretreated (impregnation with H_3PO_4 at 80 °C, no pretreatment, hydrothermal treatment with 35% H_2SO_4 at 140 °C) alkali kraft lignin. A solid acid catalyst prepared by pretreatment with phosphoric acid showed superior catalytic activity due to its higher densities of –COOH groups (1.68 mmol/g) as binding sites for cellulose and $–SO_3H$ groups (0.88 mmol/g) as catalytic sites for cellulose hydrolysis as well as higher specific surface area (488.4 m^2/g) than for other carbon catalysts. A total reducing sugar yield of 50.8% was achieved during cellulose hydrolysis in water at 180 °C for 3 h. Reaction kinetic analysis demonstrated that lignin-derived solid acid catalyst could promote the hydrolysis of cellulose by reducing the activation energy of the reaction to 83.3 kJ mol^{-1}, compared to 170 kJ mol^{-1} for sulfuric acid.

In an attempt to increase the amount of carboxylic groups, acrylic acid was added to alkali lignin during hydrothermal treatment (240 °C), followed by sulfonation (180 °C) [136]. The obtained lignin-derived solid acid catalyst showed high catalytic activity in the hydrolysis of cellulose to reducing sugar in an ionic liquid ([BMIM] Cl) containing a small amount of water. The high catalytic activity was ascribed to a high density of –COOH groups (2.85 mmol/g) contributing to the synergistic effect of surface oxygen-containing functional groups including $–SO_3H$ groups (1.58 mmol/g) as main catalytic sites. The lignin-derived solid acid catalyst could be recycled five times without notable loss of catalytic activity.

The preparation of solid acid catalyst by direct sulfonation of alkali lignin with sulfuric acid at 170 °C or fuming sulfuric acid (15% SO_3) at 100 °C has been reported [131]. The obtained catalysts showed high catalytic activity in the hydrolysis of cellulose, delignified sugarcane bagasse and cassava starch. Lignin catalyst prepared by direct sulfonation with fuming sulfuric acid without pyrolysis had equal or higher catalytic performance than catalysts prepared by pyrolysis of lignin and subsequent sulfonation. It was concluded that the cost and energy of catalyst synthesis could be reduced by using the direct sulfonation method.

A magnetic carbonaceous solid acid catalyst containing chlorine and sulfonic groups for the hydrolysis of microcrystalline cellulose has been prepared [137]. Carbonaceous solid acid catalyst containing chlorine (Cl) groups as cellulose-binding sites and sulfonic (SO_3H) groups as cellulose-hydrolyzing sites was synthesized by the impregnation of enzymatic hydrolysis lignin with $FeCl_3$, co-carbonization of Fe-loaded lignin with chloromethylated polystyrene at 500 °C, followed by sulfonation with H_2SO_4 at 150 °C. The obtained magnetic solid acid catalyst showed a satisfactory yield of total reducing sugars (TRS) of 78.5% in the hydrolysis of microcrystalline cellulose in the presence of ionic liquid [BMIM]Cl at a moderate reaction temperature of 130 °C for 4 h. The magnetic properties of solid acid catalyst allowed easy separation from the reaction mixture by using an external magnet. Lignin-derived solid acid catalyst demonstrated good stability—the TRS yield did not decrease during five successive reaction runs. Using the same catalytic system of lignin-derived solid acid catalyst and [BMIM]Cl, the rice straw-derived cellulose could also be directly hydrolyzed into TRS with a good yield of 73.2%, which is very beneficial for the comprehensive utilization of rice straw.

5.3 Ethyl Tert-Butyl Ether Synthesis (Etherification)

Preparation of acid carbon catalysts by phosphoric acid activation followed by sulfonation of different lignocellulosic precursors including lignosulfonate has been reported [138]. Lignosulfonate-based carbon catalyst was among of the most efficient catalysts for the synthesis of ethyl tert-butyl ether (ETBE) from ethanol and iso-butene. The activity of lignosulfonate-based carbon catalyst was comparable to sulfonic resin Amberlyst-15. Modeling catalytic activity revealed a crucial role of relatively weak polyphosphate groups in ETBE synthesis (Fig. 5). The turnover frequency of phosphate groups (0.0102 s^{-1}) was almost an order of magnitude higher than that of sulfonic groups (0.00116 s^{-1}). The study revealed a key role of relatively weak phosphate groups in ETBE synthesis, which were induced during activation with H_3PO_4.

5.4 Decomposition of 2-Propanol

The preparation of solid acid carbon catalyst by phosphoric acid activation at 500 °C of Alcell lignin has been reported [139]. Despite the washing step, a significant amount of phosphorus (3.7%) remained in the carbon catalyst, imparting thermally stable surface acid groups. The XPS P2p study revealed the presence of C–O–PO_3 and C–PO_3 groups on the surface of the catalyst. The total concentration of acid surface groups determined by NH_3-TPD was 0.86 mmol/g, which was distributed between weak (0.41 mmol/g), moderate (0.24 mmol/g) and strong acid (0.22 mmol/g) groups. The solid acid carbon catalyst showed high catalytic activity in the decomposition of 2-propanol (Fig. 6). Total conversion of 2-propanol was achieved at 350 °C. The catalyst yielded only dehydration products, mainly propylene, with a very low amount of di-iso-propyl ether at low temperatures. It was concluded that the acid carbon catalysts dehydrated 2-propanol to propylene over strong Brønsted acid sites formed during the activation step. Based on the assumption that all the acid sites are available for the 2-propanol dehydration reaction, the catalyst showed a turnover frequency value of 0.0026 s^{-1} at 225 °C. The increase in the amount and strength of acid sites increased the catalytic activity, but it did not significantly affect the value of the apparent activation energy of the dehydration reaction.

High catalytic activity of phosphorus-containing CF from Alcell lignin in 2-propanol decomposition has been shown [130]. The temperature of the total conversion decreased almost linearly from 660 to 300 °C with increasing phosphorus content from 0 to 3.4% (Fig. 7a). The selectivity of the P-containing CF

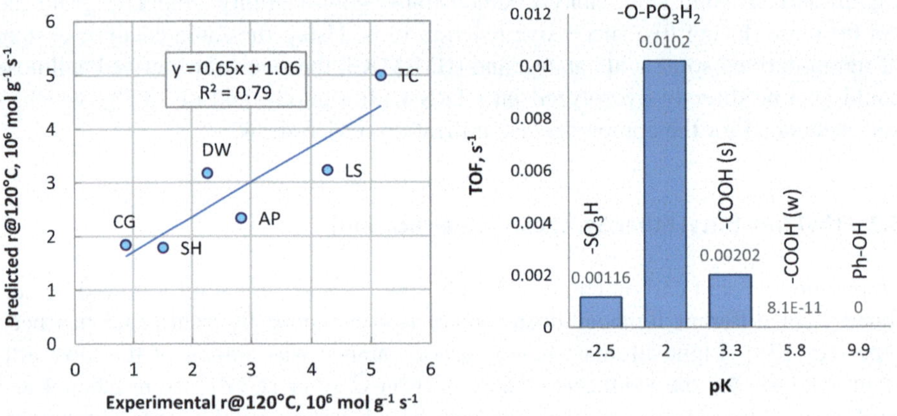

Fig. 5 Predicted vs experimental reaction rate of ethyl tert-butyl ether (ETBE) synthesis at 120 °C (left) and turnover frequency of surface groups with different pK (right) Adapted from [138]

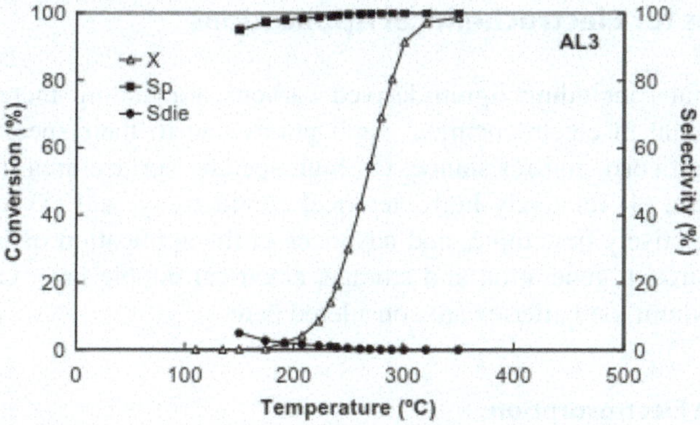

Fig. 6 Evolution of the conversion of 2-propanol and selectivity to propylene (Sp) and di-isopropyl ether (Sdie) with the temperature for AL3 catalyst. Reproduced with permission from [139]

towards the production of polypropylene increased rapidly with temperature, reaching more than 90% above 200 °C (Fig. 7a), indicating the presence of acidic characteristics that were attributed to the functionality of phosphorus. In contrast, P-free CF showed high selectivity to acetone, suggesting the presence of basic sites. In addition, phosphorus-containing CF showed a high oxidation resistance, which is necessary for use as catalysts and catalyst supports in several reactions involving oxygen at moderate temperatures. P-free CF started to lose weight (oxidize) slowly from 350–400 to 500 °C, a temperature at which they quickly combusted, while P-containing CF did not show any weight-loss until 525–550 °C, indicating increased oxidation resistance induced by phosphorus species.

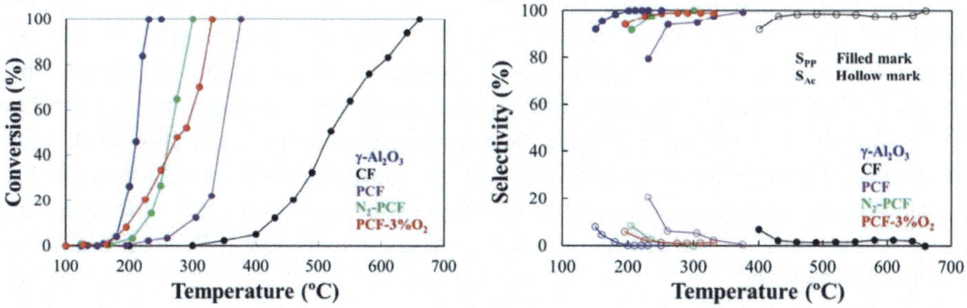

Fig. 7 **a** Evolution of 2-propanol conversion and **b** selectivity to propylene (S_{PP}) and acetone (S_{Ac}) as a function of temperature for various lignin-based carbon fibers (CF) and a commercial reference material (γ-Al_2O_3) used as catalysts Reproduced with permission from [130]

6 Electrodes for Electrochemical Applications

Carbon materials, including lignin-derived carbons are finding increasing use as electrode material in electrochemical applications due to their specific properties such as (1) high corrosion resistance, (2) high specific surface area, (3) controlled porous structure, (4) relatively high electrical conductivity, and (5) low cost. This represents a relatively new topic, and advances in the application of lignin-derived carbons in hydrogen generation and storage, electrical double layer capacitors and lithium and sodium ion batteries are considered here.

6.1 Hydrogen Electrosorption

A high hydrogen electrosorption of 510 mA h/g (1.89%) for KOH activated carbon from kraft lignin has been reported [140]. Concurrently, the carbon revealed very good electrooxidation of stored hydrogen during discharge and, thus, showed promise as the negative electrode material in rechargeable hydride batteries. The favorable behavior of the carbon was ascribed to specific pore size distribution, with high content of pores below 0.8 nm resulting from a specific lignin precursor and technological parameters of treatment, particularly KOH activation.

6.2 Electrical Double Layer Capacitors (EDLC)

Electrochemical performance as electrodes in EDLC of mesoporous lignin-derived carbon obtained by the soft-template method (Pluronic F127 as structure-directing agent) followed by physical (CO_2) or chemical (KOH) activation has been reported [141]. The gravimetric capacitance in 6 M KOH of the non-activated carbon was 77.1 F/g, which was enhanced after activation to 102.3 and 91.7 F/g for CO_2 and KOH-activated carbons, respectively. The study demonstrated the potential of biomass-derived mesoporous carbon materials for use in specific electrochemical applications.

Zeolite-templated carbons with a hierarchical porous structure [111] showed good electrochemical performance as carbon electrode in supercapacitors [142]. The gravimetric capacitance was as high as 250 F/g and the volumetric capacitance 140 F/cm^3 at a current load of 50 mA/g in 1 M H_2SO_4 electrolyte. With an increasing current load to 2 A/g, these values decreased to 170 F/g and 107 F/cm^3, but the materials retained more than 70% of that capacitance at currents higher than 20 A/g due to a hierarchical porous structure. A symmetric capacitor working at 1.2 V showed energy densities of 4.2 W h/kg at a power density of 1.3 kW/kg [143]. The asymmetric in mass design of the electrodes allowed extension of the operating window safely to 1.4 V, which increased the energy density to 6.3 W h/kg at a power density of 1.3 kW/kg.

A high-performance monolithic carbon electrode for an electric double layer supercapacitor with three-dimensional (3D) interconnected hierarchical porous structure has been synthesized from naturally abundant lignin via a dual-template

(Pluronic P123 and mesoporous silica KIT-6) approach [144]. The carbon monolith delivered simultaneously excellent areal and volumetric capacitances of 3.0 F/cm^2 and 97.1 F/cm^3, respectively, at a high mass loading of 14.4 mg/cm^2. The carbon monolith-based symmetric supercapacitor achieved a high areal energy density of 0.16 mW h/cm^2 at 1.75 mW/cm^2, with excellent cycling performance (~95% retention after 10,000 cycles). The study demonstrated the successful utilization of biomass waste as high-performance binderless electrode material for energy storage systems.

The special convenience of CF in electrochemical applications is that they allow the fabrication of binderless and flexible electrodes that exhibit ultrafast charge–discharge and excellent energy density (power density up to 61 kW/kg and energy density 10 W h/kg) and cyclability performance (95–100% retention after 100,000 cycles) [145]. Self-standing phosphorus-containing lignin-derived CF showed stable performance in a wide potential window of 1.60, 1.65 or 1.90 V in aqueous H_2SO_4, NaOH and Na_2SO_4, respectively [130]. The observed enhanced electro-oxidation resistance has been attributed to the presence of surface phosphorus functionalities as in other phosphorus-containing carbons [146–150]. The electrical capacity of phosphorus-containing CF was in the range of 150–200 F/g, while energy density due to enhanced potential window reached 15–20 W h/kg, which made them promising low-cost electrode candidates for energy storage.

A high electrochemical performance in supercapacitors of free-standing and mechanically flexible mats consisting of electrospun carbon nanofibers made from a kraft lignin/PVA blend has been reported [128]. More lignin in the precursor nanofibers produced CF with a smaller average pore size, larger pore volume, and higher specific surface area. The carbon nanofibers derived from the precursor nanofibers with 70% lignin had an average diameter of ~ 100 nm and BET specific surface area of 583 m^2/g. The lignin-based carbon nanofiber mats demonstrated high performance as free-standing and/or binder-free electrode materials for supercapacitors, showing a gravimetric capacitance of 64 F/g at the current density of 400 mA/g and 50 F/g at the current density of 2000 mA/g in 6 M KOH aqueous solution. The gravimetric capacitance was reduced by ~ 10% after 6000 cycles of charge/discharge, indicating electrochemical stability of the carbon nanofiber mat electrode.

The great potential of lignin-derived carbon nanofibers for electrical energy storage has been demonstrated [137]. Highly porous submicron activated CF were prepared from low sulfonated alkali kraft lignin/PEO blend by electrospinning, followed by carbonization at 850 °C. Activation via KOH or NaOH has led to hydrophilic and readily water wettable ACFs with larger basal plane sizes and more defective sp^2 amorphous carbon. Activated CF showed a remarkable gravimetric capacitance of 344 F/g at a low electrode mass loading of 1.8 mg and 10 mV/s scan rate and maintained capacitance of 196 F/g at an even higher loading of 10 mg and scan rate of 50 mV/s, yielding an energy density of 8.1 W h/kg. The constructed supercapacitor cells showed rather low equivalent series resistance and outstanding cycling stability, indicating that lignin-based carbons are excellent carbonaceous electrodes for supercapacitors in aqueous electrolytes.

The 2D CNSs obtained by the ice templating method from alkali kraft lignin exhibited good conductivity (2.5 S/cm) and excellent electrochemical performance as supercapacitor electrodes [117]. The capacitance was as high as 281 F/g at the current density of 0.5 A/g in 1 M H_2SO_4 aqueous electrolyte with good capacitance retention (91%) after cycling tests for 5000 cycles. Moreover, CNSs showed an energy density of 14.3 W h/kg at a power density of 28.6 kW/kg. Thus, the potential for preparation of CNSs from various water-soluble biomass for energy storage was demonstrated.

6.3 Li-Ion Batteries

The nitrogen-containing hollow carbon nanospheres obtained by direct pyrolysis of lignosulfonate-polyethylaniline polymer nanospheres exhibited excellent charge/discharge characteristics in Li-ion batteries [65]. The first charge and discharge capacities of carbon nanospheres at current densities of 60 and 100 mA/g were 980 and 432 mA h/g, and 764 and 342 mA h/g, respectively. Carbon nanospheres showed a high coulombic efficiency of 92–94% and good recyclability. The study showed that lignin-derived nitrogen-containing carbon nanospheres could be used as a promising candidate for electrode materials of lithium-ion batteries.

A lignin-derived free-standing fused carbon fibrous mat showed high specific capacity as an electrode in a Li-ion battery up to 445 mA h/g at a current density of 30 mA/g, which is comparable to polyacrylonitrile-derived carbon nanofibers, and good cyclic stability at different current rates [127]. Nitrogen doping via thermal annealing in the presence of urea increased the charge capacity to 576 mA h/g. A good capacity of approximately 200 mA h/g was maintained even at a high current rate of 2000 mA/g. The research demonstrated the great promise of lignin-derived nanocarbon materials for applications in energy storage systems.

Lignin-derived macro/microporous carbon for lithium-sulfur batteries was prepared by direct activation of alkali lignin with KOH at 700 °C [151]. The obtained carbon with a high BET specific surface area of 1211.6 m^2/g and pore volume of 0.59 cm^3/g was used after sulfur encapsulation as a cathode for LiS batteries. Sulfur-loaded carbon delivered a high discharge capacity of 1241.0 mA h/g in the 2nd cycle and maintained a specific value of 791.6 mA h/g in the 100th cycle at 0.5 °C. The study demonstrated high electrochemical performance of lignin-derived carbon and its potential application as Li–S battery cathode.

A carbonaceous cathode for Li–Se battery was prepared by chemical activation of alkaline lignin with KOH followed by heat treatment with Se [152]. The high specific surface area, large pore volume and good electron conductivity of Se/carbon composite facilitated the reversible electrochemical reaction of selenium towards metallic Li, and at 0.5 C, a Se/carbon composite electrode exhibited a reversible capacity of 596 mA h/g in the 2nd cycle and a capacity retention of 453 mA h/g over 300 cycles with an average decay of 0.08% per cycle. The obtained Se/carbon composite also showed a high-rate performance 363 mA h/g at 4 °C, indicating its effectiveness in renewable green energy sources.

6.4 Na-Ion Batteries

Recently sodium ion batteries have attracted the attention as a promising electrical energy storage device due to abundance of sodium resources and their cost advantage. Low-cost and bio-based carbon nanofibrous webs fabricated from hardwood lignosulfonate/polyacrylonitrile (PAN) blends showed excellent performance in a sodium-ion battery [153]. Carbon fibrous webs obtained from 50% lignin composition at 1300 °C exhibited a reversible capacity of approximately 292.6 mA h/g and high initial efficiency of 70.5% at a current density of 20 mA/g with good rate capability (80 mA h/g at 1 A/g), and excellent cycling stability (247 mA h/g over 200 cycles at 0.1 A/g). Due to the excellent performance, combined with the green and inexpensive precursor sources and economic synthesis method the carbon nanofibrous webs hold much promise as a sustainable electrode material for sodium-ion batteries.

7 Other Carbon Materials

7.1 Graphitic Carbons

In search for efficient strategies to minimize waste production, the thermal conversion of kraft lignin to carbon materials with a significant degree of structural ordering has been studied [154]. XRD and Raman studies have shown progressive formation of the 3D order as the heat treatment temperature increased up to 2800 °C. The inorganic impurities of the precursor (mainly Na) enhanced the onset of structural ordering, allowing utilization of the high-ash lignin precursor for the production of graphitic materials. The obtained high-temperature carbons showed high oxidation resistance comparable to graphite SP-1, which holds much promise for producing components of carbon composites from lignin.

7.2 Glassy Carbon

Lignin is a suitable precursor for the production of bulky samples of glassy carbon. The preparation of glassy carbon from lignosulfonate and alkali lignin has been reported [155]. The lignins were condensed with H_2O_2, NaOH + HCl, formaldehyde and phenol–formaldehyde. Glassy carbons were obtained by pyrolyzing up to 1000 °C ground and compressed (cylinders with diameter of 13 mm and height typically of 3 mm) lignins. Both lignins provided carbon material with a relatively high yield ranging from 42% to 47%, apparent porosity of 33–40% and comparatively high compression strength 70–110 N/mm^2. The compressive strengths of lignosulfonate carbon samples were lower than those of alkali-lignin carbon samples, which was attributed to the deteriorating effect of high ash of lignosulfonate. Although the compressive strength is lower than that for phenol–formaldehyde-based glassy carbon, the optimization of compression and pyrolysis conditions would give carbons with still better properties. Decreasing temperature ramp from 18.5°/h to 6°/h increased compressive strength to 155 N/mm^2. The study showed that unmodified alkali-lignin is a suitable raw material for production of glassy carbon.

7.3 Carbon Black

The potential of bio-based carbon black from hydrolysis lignin as an alternative to fossil fuel-derived commercial carbon black has been demonstrated [156]. It has been shown that optimized carbonized ball milled lignin had a carbon content greater than 90% and exhibited a surface area 8.8 times larger and a thermal conductivity 36% greater, while the electrical conductivity (9.5 S/m) was lower in comparison to conductive carbon black (C-NERGY Super P Li carbon black). The study demonstrated the feasibility of bio-based carbon black in nonelectrical applications as non-conductive black ink, toner, paint, thermal paste, and thermally conductive filler.

8 Conclusions

Lignin, a major component of the lignocellulosic biomass, is generated in enormous amounts during pulp production. It is also a major coproduct of second generation biofuels. Effective utilization of lignin is critical for the accelerated development of advanced cellulosic biorefineries. Lignin as renewable, abundant, low cost, and high content of aromatics is an attractive precursor for the preparation of a range of carbon materials, including activated carbons, ACF, structural CF, graphitic carbons or carbon black that could be used for environmental protection, as catalysts, in energy storage applications or as reinforcing components in advanced composite materials.

The most economically attractive application of lignin-derived carbon materials is structural CF. However, taking into account the inherent heterogeneity of lignin, which is the main obstacle for the preparation of high-performance materials, the production of CF from lignin should be treated as long term prospect. More research is needed in preparation of high quality, sufficiently robust lignin to yield good CF. Although lignin-based CF still do not compete with PAN-derived CF in mechanical properties, they nevertheless provide a new potential markets through high availability and low production costs.

More elaborated is preparation of carbon adsorbents with very high surface areas and pore volumes comparable to the best commercial activated carbons. These porous materials as universal adsorbents are useful for purifying gas and aqueous media from organic pollutants or adsorption of heavy metal ions from aqueous solutions. They could also be used as catalysts or electrodes in electrochemical applications. While many methods for preparation of carbon adsorbents from lignin were studied, the challenges remain. The key issue in manufacture of lignin-derived carbon adsorbents is thermoplastic foaming behavior, which leads to low density and low mechanical strength of the final product. More research is needed in this field. Another issue is a high ash content of lignin-derived carbons that limits their applications. Some progress in this field was achieved and laboratory methods for diminishing the ash content in carbon adsorbents (acid treatment or hydrothermal carbonization) have been proposed.

Alternatively, technical lignin could be used for the production of bulky carbon materials like graphitic carbons for carbon composites, glassy carbon for electrodes or sensors and carbon black for nonelectrical applications.

Acknowledgements The Knut and Alice Wallenberg Foundation in connection with the Wallenberg Wood Science Centre Program (WWSC) is gratefully acknowledged for the financial support of the work of Dr. Olena Sevastyanova. The Cost Action FP 1306 LIGNOVAL is acknowledged for the initiation of the current work.

References

1. Gosselink RJA, de Jong E, Guran B, Abächerli A (2004) Co-ordination network for lignin—standardisation, production and applications adapted to market requirements (EUROLIGNIN). Ind Crops Prod 20:121–129. https://doi.org/10.1016/j.indcrop.2004.04.015
2. Gargulak JD, Lebo SE, McNally TJ (2015) Lignin. Kirk–Othmer encyclopedia of chemical technology. Wiley, Hoboken, pp 1–26
3. Norgren M, Edlund H (2014) Lignin: recent advances and emerging applications. Curr Opin Colloid Interface Sci 19:409–416. https://doi.org/10.1016/j.cocis.2014.08.004
4. Liu W-J, Jiang H, Yu H-Q (2015) Thermochemical conversion of lignin to functional materials: a review and future directions. Green Chem 17:4888–4907. https://doi.org/10.1039/C5GC01054C
5. Li Q, Xie S, Serem WK et al (2017) Quality carbon fibers from fractionated lignin. Green Chem 19:1628–1634. https://doi.org/10.1039/C6GC03555H
6. Vishtal A, Kraslawski A (2011) Challenges in industrial applications of technical lignins. BioResources 6:3547–3568
7. Gargulak JD, Lebo SE (1999) Commercial use of lignin-based materials. In: Glasser WG, Northey RA, Schultz TP (eds) Lignin: historical, biological and materials perspectives. ACS Symposium Series, American Chemical Society, Washington DC, pp 304–320
8. Satheesh Kumar MN, Mohanty AK, Erickson L, Misra M (2009) Lignin and its applications with polymers. J Biobased Mater Bioenergy 3:1–24. https://doi.org/10.1166/jbmb.2009.1001
9. Rinaldi R, Jastrzebski R, Clough MT et al (2016) Paving the way for lignin valorisation: recent advances in bioengineering, biorefining and catalysis. Angew Chem Int Ed 55:8164–8215. https://doi.org/10.1002/anie.201510351
10. Renders T, Van den Bosch S, Koelewijn S-F et al (2017) Lignin-first biomass fractionation: the advent of active stabilisation strategies. Energy Environ Sci 10:1551–1557. https://doi.org/10.1039/C7EE01298E
11. Suhas Carrott PJM, Ribeiro Carrott MML (2007) Lignin—from natural adsorbent to activated carbon: a review. Bioresour Technol 98:2301–2312. https://doi.org/10.1016/j.biortech.2006.08.008
12. Mainka H, Täger O, Körner E et al (2015) Lignin—an alternative precursor for sustainable and cost-effective automotive carbon fiber. J Mater Res Technol 4:283–296. https://doi.org/10.1016/j.jmrt.2015.03.004
13. Chatterjee S, Saito T (2015) Lignin-derived advanced carbon materials. Chemsuschem 8:3941–3958. https://doi.org/10.1002/cssc.201500692
14. Graichen FHM, Grigsby WJ, Hill SJ et al (2017) Yes, we can make money out of lignin and other bio-based resources. Ind Crops Prod 106:74–85. https://doi.org/10.1016/j.indcrop.2016.10.036
15. Gosselink RJA (2011) Lignin as a renewable aromatic resource for the chemical industry. PhD Thesis, Wageningen University, Wageningen, the Netherlands
16. Huang X (2009) Fabrication and properties of carbon fibers. Materials 2:2369–2403. https://doi.org/10.3390/ma2042369
17. Baker DA, Rials TG (2013) Recent advances in low-cost carbon fiber manufacture from lignin. J Appl Polym Sci 130:713–728. https://doi.org/10.1002/app.39273
18. Frank E, Steudle LM, Ingildeev D et al (2014) Carbon fibers: precursor systems, processing, structure, and properties. Angew Chem Int Ed 53:5262–5298. https://doi.org/10.1002/anie.201306129

19. Ragauskas AJ, Beckham GT, Biddy MJ et al (2014) Lignin valorization: improving lignin processing in the biorefinery. Science 344:1246843. https://doi.org/10.1126/science.1246843
20. Titirici M-M, White RJ, Brun N et al (2015) Sustainable carbon materials. Chem Soc Rev 44:250–290. https://doi.org/10.1039/c4cs00232f
21. Ko FK, Goudarzi A, Lin L-T, et al (2016) Lignin-based composite carbon nanofibers. In: Lignin in polymer composites. Elsevier, Amsterdam, pp 167–194
22. Dias OAT, Negrão DR, Gonçalves DFC et al (2017) Recent approaches and future trends for lignin-based materials. Mol Cryst Liq Cryst 655:204–223. https://doi.org/10.1080/15421406.2017.1360713
23. Fang W, Yang S, Wang X-L et al (2017) Manufacture and application of lignin-based carbon fibers (LCFs) and lignin-based carbon nanofibers (LCNFs). Green Chem 19:1794–1827. https://doi.org/10.1039/C6GC03206K
24. Calvo-Flores FG, Dobado JA (2010) Lignin as renewable raw material. Chemsuschem 3:1227–1235. https://doi.org/10.1002/cssc.201000157
25. Jain A, Balasubramanian R, Srinivasan MP (2016) Hydrothermal conversion of biomass waste to activated carbon with high porosity: a review. Chem Eng J 283:789–805. https://doi.org/10.1016/j.cej.2015.08.014
26. Rabinovich ML, Fedoryak O, Dobele G et al (2016) Carbon adsorbents from industrial hydrolysis lignin: the USSR/Eastern European experience and its importance for modern biorefineries. Renew Sustain Energy Rev 57:1008–1024. https://doi.org/10.1016/j.rser.2015.12.206
27. Kadla JF, Kubo S, Venditti RA et al (2002) Lignin-based carbon fibers for composite fiber applications. Carbon 40:2913–2920. https://doi.org/10.1016/S0008-6223(02)00248-8
28. Kadla JF, Kubo S, Gilbert RD, Venditti RA (2002) Lignin-based carbon fibers. In: Hu TQ (ed) Chemical modification, properties, and usage of lignin. Springer, New York, pp 121–137
29. Kubo S, Kadla JF (2004) Poly(ethylene oxide)/organosolv lignin blends: relationship between thermal properties, chemical structure, and blend behavior. Macromolecules 37:6904–6911. https://doi.org/10.1021/ma0490552
30. Kubo S, Kadla JF (2005) Kraft lignin/poly(ethylene oxide) blends: effect of lignin structure on miscibility and hydrogen bonding. J Appl Polym Sci 98:1437–1444. https://doi.org/10.1002/app.22245
31. Kubo S, Kadla JF (2005) Lignin-based carbon fibers: effect of synthetic polymer blending on fiber properties. J Polym Environ 13:97–105. https://doi.org/10.1007/s10924-005-2941-0
32. Maradur SP, Kim CH, Kim SY et al (2012) Preparation of carbon fibers from a lignin copolymer with polyacrylonitrile. Synth Met 162:453–459. https://doi.org/10.1016/j.synthmet.2012.01.017
33. Liu HC, Chien AT, Newcomb BA et al (2016) Stabilization kinetics of gel spun polyacrylonitrile/lignin blend fiber. Carbon 101:382–389. https://doi.org/10.1016/j.carbon.2016.01.096
34. Oroumei A, Fox B, Naebe M (2015) Thermal and rheological characteristics of biobased carbon fiber precursor derived from low molecular weight organosolv lignin. ACS Sustain Chem Eng 3:758–769. https://doi.org/10.1021/acssuschemeng.5b00097
35. Youe WJ, Lee SM, Lee SS et al (2016) Characterization of carbon nanofiber mats produced from electrospun lignin-g-polyacrylonitrile copolymer. Int J Biol Macromol 82:497–504. https://doi.org/10.1016/j.ijbiomac.2015.10.022
36. Sevastyanova O, Qin W, Kadla JF (2010) Effect of nanofillers as reinforcement agents for lignin composite fibers. J Appl Polym Sci 117:2877–2881. https://doi.org/10.1002/app.32198
37. Qin W, Kadla JF (2011) Effect of organoclay reinforcement on lignin-based carbon fibers. Ind Eng Chem Res 50:12548–12555. https://doi.org/10.1021/ie201319p
38. Baker FS, Baker DA, Menchhofer PA (2011) Carbon nanotube (CNT)-enhanced precursor for carbon fiber production and method of making a CNT-enhanced continuous lignin fiber. US Patent 2011285049, 24 Nov 2011
39. Wang S, Zhou Z, Xiang H et al (2016) Reinforcement of lignin-based carbon fibers with functionalized carbon nanotubes. Compos Sci Technol 128:116–122. https://doi.org/10.1016/j.compscitech.2016.03.018
40. Lin J, Koda K, Kubo S et al (2014) Improvement of mechanical properties of softwood lignin-based carbon fibers. J Wood Chem Technol 34:111–121. https://doi.org/10.1080/02773813.2013.839707
41. Sudo K, Shimizu K (1992) A new carbon fiber from lignin. J Appl Polym Sci 44:127–134. https://doi.org/10.1002/app.1992.070440113
42. Uraki Y, Kubo S, Nigo N et al (1995) Preparation of carbon fibers from organosolv lignin obtained by aqueous acetic acid pulping. Holzforschung 49:343–350. https://doi.org/10.1515/hfsg.1995.49.4.343

43. Eckert RC, Abdullah Z (2008) Carbon fibers from kraft softwood lignin. US Patent 2008317661, 25 Dec 2008
44. Sudo K, Shimizu K, Nakashima N, Yokoyama A (1993) A new modification method of exploded lignin for the preparation of a carbon fiber precursor. J Appl Polym Sci 48:1485–1491. https://doi.org/10.1002/app.1993.070480817
45. Ding R, Wu H, Thunga M et al (2016) Processing and characterization of low-cost electrospun carbon fibers from organosolv lignin/polyacrylonitrile blends. Carbon 100:126–136. https://doi.org/10.1016/j.carbon.2015.12.078
46. Wohlmann B, Woelki M, Ebert A et al (2010) Lignin derivative, shaped body comprising the derivative, and carbon-fibres produced from the shaped body. World Patent 2010081775, 22 Jul 2010
47. Brodin I, Sjöholm E, Gellerstedt G (2009) Kraft lignin as feedstock for chemical products: the effects of membrane filtration. Holzforschung 63:290–297. https://doi.org/10.1515/HF.2009.049
48. Nordström Y, Norberg I, Sjöholm E, Drougge R (2013) A new softening agent for melt spinning of softwood kraft lignin. J Appl Polym Sci 129:1274–1279. https://doi.org/10.1002/app.38795
49. Yoshida H, Mörck R, Kringstad KP, Hatakeyama H (1987) Fractionation of kraft lignin by successive extraction with organic solvents. II. Thermal Properties of kraft lignin fractions. Holzforschung 41:171–176. https://doi.org/10.1515/hfsg.1987.41.3.171
50. Mörck R, Reimann A, Kringstad KP (1988) Fractionation of kraft lignin by successive extraction with organic solvents. III. Fractionation of kraft lignin from birch. Holzforschung 42:111–116. https://doi.org/10.1515/hfsg.1988.42.2.111
51. Kubo S, Uraki Y, Sano Y (1998) Preparation of carbon fibers from softwood lignin by atmospheric acetic acid pulping. Carbon 36:1119–1124. https://doi.org/10.1016/S0008-6223(98)00086-4
52. Hosseinaei O, Harper DP, Bozell JJ, Rials TG (2017) Improving processing and performance of pure lignin carbon fibers through hardwood and herbaceous lignin blends. Int J Mol Sci 18:1410. https://doi.org/10.3390/ijms18071410
53. Cho M, Karaaslan M, Chowdhury S et al (2018) Skipping oxidative thermal stabilization for lignin-based carbon nanofibers. ACS Sustain Chem Eng 6:6434–6444. https://doi.org/10.1021/acssuschemeng.8b00209
54. Li Q, Serem WK, Dai W et al (2017) Molecular weight and uniformity define the mechanical performance of lignin-based carbon fiber. J Mater Chem A 5:12740–12746. https://doi.org/10.1039/C7TA01187C
55. Qu W, Liu J, Xue Y et al (2018) Potential of producing carbon fiber from biorefinery corn stover lignin with high ash content. J Appl Polym Sci 135:1–11. https://doi.org/10.1002/app.45736
56. Dai Z, Shi X, Liu H et al (2018) High-strength lignin-based carbon fibers via a low-energy method. RSC Adv 8:1218–1224. https://doi.org/10.1039/C7RA10821D
57. Nar M, Rizvi HR, Dixon RA et al (2016) Superior plant based carbon fibers from electrospun poly-(caffeyl alcohol) lignin. Carbon 103:372–383. https://doi.org/10.1016/j.carbon.2016.02.053
58. Marsh H, Rodríguez-Reinoso F (2006) Activated carbon. Elsevier, Oxford
59. Yang RT (2003) Adsorbents: fundamentals and applications. Wiley, Hoboken
60. Bansal RC, Goyal M (2005) Activated carbon adsorption. CRC, Boca Raton
61. Ragan S, Megonnell N (2011) Activated carbon from renewable resources—lignin. Cellul Chem Technol 45:527–531
62. Rodríguez-Reinoso F (2002) Production and applications of activated carbons. In: Schüth F, Sing KSW, Weitkamp J (eds) Handbook of porous solids. Wiley, Weinheim, pp 1766–1827
63. González-García P (2018) Activated carbon from lignocellulosics precursors: a review of the synthesis methods, characterization techniques and applications. Renew Sustain Energy Rev 82:1393–1414. https://doi.org/10.1016/j.rser.2017.04.117
64. Lü Q-F, He Z-W, Zhang J-Y, Lin Q (2011) Preparation and properties of nitrogen-containing hollow carbon nanospheres by pyrolysis of polyaniline–lignosulfonate composites. J Anal Appl Pyrolysis 92:152–157. https://doi.org/10.1016/j.jaap.2011.05.009
65. He Z-W, Lü Q-F, Lin Q (2013) Fabrication, characterization and application of nitrogen-containing carbon nanospheres obtained by pyrolysis of lignosulfonate/poly(2-ethylaniline). Bioresour Technol 127:66–71. https://doi.org/10.1016/j.biortech.2012.09.132
66. Rodríguez-Mirasol J, Cordero T, Rodriguez JJ (1993) Activated carbons from carbon dioxide partial gasification of eucalyptus kraft lignin. Energy Fuels 7:133–138. https://doi.org/10.1021/ef00037a021

67. Rodríguez-Mirasol J, Cordero T, Rodríguez JJ (1993) Preparation and characterization of activated carbons from eucalyptus kraft lignin. Carbon 31:87–95. https://doi.org/10.1016/0008-6223(93)90160-C
68. Cotoruelo LM, Marqués MD, Díaz FJ et al (2007) Activated carbons from lignin: their application in liquid phase adsorption. Sep Sci Technol 42:3363–3389
69. Cotoruelo LM, Marqués MD, Díaz FJ et al (2010) Equilibrium and kinetic study of congo red adsorption onto lignin-based activated carbons. Transp Porous Media 83:573–590. https://doi.org/10.1007/s11242-009-9460-8
70. Cotoruelo LM, Marqués MD, Rodríguez-Mirasol J et al (2011) Cationic dyes removal by multilayer adsorption on activated carbons from lignin. J Porous Mater 18:693–702. https://doi.org/10.1007/s10934-010-9428-7
71. Cotoruelo LM, Marqués MD, Díaz FJ et al (2012) Lignin-based activated carbons as adsorbents for crystal violet removal from aqueous solutions. Environ Prog Sustain Energy 31:386–396. https://doi.org/10.1002/ep.10560
72. Cotoruelo LM, Marqués MD, Rodríguez-Mirasol J et al (2007) Adsorption of aromatic compounds on activated carbons from lignin: equilibrium and thermodynamic study. Ind Eng Chem Res 46:4982–4990. https://doi.org/10.1021/ie061415h
73. Cotoruelo LM, Marqués MD, Rodríguez-Mirasol J et al (2007) Adsorption of aromatic compounds on activated carbons from lignin: kinetic study. Ind Eng Chem Res 46:2853–2860. https://doi.org/10.1021/ie061445k
74. Cotoruelo LM, Marqués MD, Leiva A et al (2011) Adsorption of oxygen-containing aromatics used in petrochemical, pharmaceutical and food industries by means of lignin based active carbons. Adsorption 17:539–550. https://doi.org/10.1007/s10450-010-9319-x
75. Cotoruelo LM, Marqués MD, Díaz FJ et al (2012) Adsorbent ability of lignin-based activated carbons for the removal of p-nitrophenol from aqueous solutions. Chem Eng J 184:176–183. https://doi.org/10.1016/j.cej.2012.01.026
76. Cotoruelo LM, Marqués MD, Rodríguez-Mirasol J et al (2009) Lignin-based activated carbons for adsorption of sodium dodecylbenzene sulfonate: equilibrium and kinetic studies. J Colloid Interface Sci 332:39–45. https://doi.org/10.1016/j.jcis.2008.12.031
77. Rodríguez-Mirasol J, Bedia J, Cordero T, Rodríguez JJ (2005) Influence of water vapor on the adsorption of VOCs on lignin-based activated carbons. Sep Sci Technol 40:3113–3135. https://doi.org/10.1080/01496390500385277
78. Bedia J, Rodríguez-Mirasol J, Cordero T (2007) Water vapour adsorption on lignin-based activated carbons. J Chem Technol Biotechnol 82:548–557. https://doi.org/10.1002/jctb.1698
79. Gonzalez-Serrano E, Cordero T, Rodríguez-Mirasol J et al (2004) Removal of water pollutants with activated carbons prepared from H_3PO_4 activation of lignin from kraft black liquors. Water Res 38:3043–3050. https://doi.org/10.1016/j.watres.2004.04.048
80. Fierro V, Torné-Fernández V, Celzard A (2006) Kraft lignin as a precursor for microporous activated carbons prepared by impregnation with ortho-phosphoric acid: synthesis and textural characterisation. Micropor Mesopor Mater 92:243–250. https://doi.org/10.1016/j.micromeso.2006.01.013
81. Montané D, Torné-Fernández V, Fierro V (2005) Activated carbons from lignin: kinetic modeling of the pyrolysis of kraft lignin activated with phosphoric acid. Chem Eng J 106:1–12. https://doi.org/10.1016/j.cej.2004.11.001
82. Fierro V, Torné-Fernández V, Montané D, Celzard A (2005) Study of the decomposition of kraft lignin impregnated with orthophosphoric acid. Thermochim Acta 433:142–148. https://doi.org/10.1016/j.tca.2005.02.026
83. Fierro V, Torné-Fernández V, Celzard A, Montané D (2007) Influence of the demineralisation on the chemical activation of kraft lignin with orthophosphoric acid. J Hazard Mater 149:126–133. https://doi.org/10.1016/j.jhazmat.2007.03.056
84. Gonzalez-Serrano E, Cordero T, Rodríguez-Mirasol J, Rodríguez JJ (1997) Development of porosity upon chemical activation of kraft lignin with $ZnCl_2$. Ind Eng Chem Res 36:4832–4838. https://doi.org/10.1021/ie970261q
85. Rosas JM, Ruiz-Rosas R, Rodríguez-Mirasol J, Cordero T (2017) Kinetic study of SO_2 removal over lignin-based activated carbon. Chem Eng J 307:707–721. https://doi.org/10.1016/j.cej.2016.08.111
86. Hayashi J, Kazehaya A, Muroyama K, Watkinson AP (2000) Preparation of activated carbon from lignin by chemical activation. Carbon 38:1873–1878. https://doi.org/10.1016/S0008-6223(00)00027-0

87. Maldhure AV, Ekhe JD (2011) Preparation and characterizations of microwave assisted activated carbons from industrial waste lignin for Cu(II) sorption. Chem Eng J 168:1103–1111. https://doi.org/10.1016/j.cej.2011.01.091
88. Fierro V, Torné-Fernández V, Celzard A (2007) Methodical study of the chemical activation of kraft lignin with KOH and NaOH. Micropor Mesopor Mater 101:419–431. https://doi.org/10.1016/j.micromeso.2006.12.004
89. Guo Y, Rockstraw DA (2006) Physical and chemical properties of carbons synthesized from xylan, cellulose, and kraft lignin by H_3PO_4 activation. Carbon 44:1464–1475. https://doi.org/10.1016/j.carbon.2005.12.002
90. Sharma RK, Wooten JB, Baliga VL et al (2004) Characterization of chars from pyrolysis of lignin. Fuel 83:1469–1482. https://doi.org/10.1016/j.fuel.2003.11.015
91. Li J, Li B, Zhang X (2002) Comparative studies of thermal degradation between larch lignin and manchurian ash lignin. Polym Degrad Stab 78:279–285. https://doi.org/10.1016/S0141-3910(02)00172-6
92. Saha D, Akkoyunlu SD, Thorpe R et al (2017) Adsorptive recovery of neodymium and dysprosium in phosphorous functionalized nanoporous carbon. J Environ Chem Eng 5:4684–4692. https://doi.org/10.1016/j.jece.2017.09.009
93. Myglovets M, Poddubnaya OI, Sevastyanova O et al (2014) Preparation of carbon adsorbents from lignosulfonate by phosphoric acid activation for the adsorption of metal ions. Carbon 80:771–783. https://doi.org/10.1016/j.carbon.2014.09.032
94. Puziy AM, Tascón JMD (2012) Adsorption by phosphorus-containing carbons. In: Tascón JMD (ed) Novel carbon adsorbents. Elsevier, Amsterdam, pp 245–267
95. Puzii AM (2011) Methods of production, structure, and physicochemical characteristics of phosphorylated carbon adsorbents. Theor Exp Chem 47:277–291. https://doi.org/10.1007/s11237-011-9216-8
96. Puziy AM, Poddubnaya OI, Socha RP et al (2008) XPS and NMR studies of phosphoric acid activated carbons. Carbon 46:2113–2123. https://doi.org/10.1016/j.carbon.2008.09.010
97. Puziy AM, Poddubnaya OI (1998) The properties of synthetic carbon derived from nitrogen- and phosphorus-containing polymer. Carbon 36:45–50. https://doi.org/10.1016/S0008-6223(97)00149-8
98. Puziy AM, Poddubnaya OI (1999) Characterization of surface heterogeneity of carbon-composite adsorbents. Mater Sci Forum 308–311:908–916. https://doi.org/10.4028/www.scientific.net/MSF.308-311.908
99. Puziy AM, Poddubnaya OI, Martínez-Alonso A et al (2002) Synthetic carbons activated with phosphoric acid I. Surface chemistry and ion binding properties. Carbon 40:1493–1505. https://doi.org/10.1016/S0008-6223(01)00317-7
100. Puziy AM, Poddubnaya OI, Martínez-Alonso A et al (2003) Synthetic carbons activated with phosphoric acid. III. Carbons prepared in air. Carbon 41:1181–1191. https://doi.org/10.1016/S0008-6223(03)00031-9
101. Puziy AM, Poddubnaya OI, Martínez-Alonso A et al (2005) Surface chemistry of phosphorus-containing carbons of lignocellulosic origin. Carbon 43:2857–2868. https://doi.org/10.1016/j.carbon.2005.06.014
102. Puziy AM, Poddubnaya OI, Martínez-Alonso A et al (2007) Oxygen and phosphorus enriched carbons from lignocellulosic material. Carbon 45:1941–1950. https://doi.org/10.1016/j.carbon.2007.06.014
103. Puziy AM, Poddubnaya OI, Zaitsev VN, Konoplitska OP (2004) Modeling of heavy metal ion binding by phosphoric acid activated carbon. Appl Surf Sci 221:421–429. https://doi.org/10.1016/S0169-4332(03)00956-5
104. Puziy AM, Poddubnaya OI, Gawdzik B et al (2007) Functionalization of carbon and silica gel by phosphoric acid. Adsorpt Sci Technol 25:531–542
105. Sych NV, Trofymenko SI, Poddubnaya OI et al (2012) Porous structure and surface chemistry of phosphoric acid activated carbon from corncob. Appl Surf Sci 261:75–82. https://doi.org/10.1016/j.apsusc.2012.07.084
106. Sánchez-Polo M, Rivera-Utrilla J (2002) Adsorbent–adsorbate interactions in the adsorption of Cd(II) and Hg(II) on ozonized activated carbons. Environ Sci Technol 36:3850–3854. https://doi.org/10.1021/es0255610

107. Rivera-Utrilla J, Sánchez-Polo M (2003) Adsorption of Cr(III) on ozonised activated carbon. Importance of Cπ-cation interactions. Water Res 37:3335–3340. https://doi.org/10.1016/S0043-1354(03)00177-5
108. Zou Y, Han B (2001) Preparation of activated carbons from chinese coal and hydrolysis lignin. Adsorpt Sci Technol 19:59–72. https://doi.org/10.1260/0263617011493971
109. Cheng F, Liang J, Zhao J et al (2008) Biomass waste-derived microporous carbons with controlled texture and enhanced hydrogen uptake. Chem Mater 20:1889–1895. https://doi.org/10.1021/cm702816x
110. Fierro CM, Górka J, Zazo JA et al (2013) Colloidal templating synthesis and adsorption characteristics of microporous–mesoporous carbons from kraft lignin. Carbon 62:233–239. https://doi.org/10.1016/j.carbon.2013.06.012
111. Valero-Romero MJ, Márquez-Franco EM, Bedia J et al (2014) Hierarchical porous carbons by liquid phase impregnation of zeolite templates with lignin solution. Micropor Mesopor Mater 196:68–78. https://doi.org/10.1016/j.micromeso.2014.04.055
112. Saha D, Payzant EA, Kumbhar AS, Naskar AK (2013) Sustainable mesoporous carbons as storage and controlled-delivery media for functional molecules. ACS Appl Mater Interfaces 5:5868–5874. https://doi.org/10.1021/am401661f
113. Saha D, Warren KE, Naskar AK (2014) Soft-templated mesoporous carbons as potential materials for oral drug delivery. Carbon 71:47–57. https://doi.org/10.1016/j.carbon.2014.01.005
114. Saha D, Warren KE, Naskar AK (2014) Controlled release of antipyrine from mesoporous carbons. Micropor Mesopor Mater 196:327–334. https://doi.org/10.1016/j.micromeso.2014.05.024
115. Zhao W, Lin X, Cai H et al (2017) Preparation of mesoporous carbon from sodium lignosulfonate by hydrothermal and template method and its adsorption of uranium(VI). Ind Eng Chem Res 56:12745–12754. https://doi.org/10.1021/acs.iecr.7b02854
116. Deville S, Saiz E, Nalla RK, Tomsia AP (2006) Freezing as a path to build complex composites. Science 311:515–518. https://doi.org/10.1126/science.1120937
117. Liu W, Yao Y, Fu O et al (2017) Lignin-derived carbon nanosheets for high-capacitance supercapacitors. RSC Adv 7:48537–48543. https://doi.org/10.1039/C7RA08531A
118. Funke A, Ziegler F (2010) Hydrothermal carbonization of biomass: a summary and discussion of chemical mechanisms for process engineering. Biofuels Bioprod Biorefining 4:160–177. https://doi.org/10.1002/bbb.198
119. Wikberg H, Grönberg V, Jermakka J et al (2015) Hydrothermal refining of biomass—an overview and future perspectives. Tappi J 14:195–207
120. Sangchoom W, Mokaya R (2015) Valorization of lignin waste: carbons from hydrothermal carbonization of renewable lignin as superior sorbents for CO2 and hydrogen storage. ACS Sustain Chem Eng 3:1658–1667. https://doi.org/10.1021/acssuschemeng.5b00351
121. Atta-Obeng E, Dawson-Andoh B, Seehra MS et al (2017) Physico-chemical characterization of carbons produced from technical lignin by sub-critical hydrothermal carbonization. Biomass Bioenerg 107:172–181. https://doi.org/10.1016/j.biombioe.2017.09.023
122. Kadla JF, Kubo S, Venditti RA, Gilbert RD (2002) Novel hollow core fibers prepared from lignin polypropylene blends. J Appl Polym Sci 85:1353–1355. https://doi.org/10.1002/app.10640
123. Kubo S, Yoshida T, Kadla JF (2007) Surface porosity of lignin/PP blend carbon fibers. J Wood Chem Technol 27:257–271. https://doi.org/10.1080/02773810701702238
124. Lallave M, Bedia J, Ruiz-Rosas R et al (2007) Filled and hollow carbon nanofibers by coaxial electrospinning of alcell lignin without binder polymers. Adv Mater 19:4292–4296. https://doi.org/10.1002/adma.200700963
125. Ruiz-Rosas R, Bedia J, Lallave M et al (2010) The production of submicron diameter carbon fibers by the electrospinning of lignin. Carbon 48:696–705. https://doi.org/10.1016/j.carbon.2009.10.014
126. García-Mateos FJ, Cordero-Lanzac T, Berenguer R et al (2017) Lignin-derived Pt supported carbon (submicron)fiber electrocatalysts for alcohol electro-oxidation. Appl Catal B Environ 211:18–30. https://doi.org/10.1016/j.apcatb.2017.04.008
127. Wang S-X, Yang L, Stubbs LP et al (2013) Lignin-derived fused electrospun carbon fibrous mats as high performance anode materials for lithium ion batteries. ACS Appl Mater Interfaces 5:12275–12282. https://doi.org/10.1021/am4043867
128. Lai C, Zhou Z, Zhang L et al (2014) Free-standing and mechanically flexible mats consisting of electrospun carbon nanofibers made from a natural product of alkali lignin as binder-free electrodes for high-performance supercapacitors. J Power Sources 247:134–141. https://doi.org/10.1016/j.jpowsour.2013.08.082

129. Uraki Y, Kubo S, Kurakami H, Sano Y (1997) activated carbon fibers from acetic acid lignin. Holzforschung 51:188–192. https://doi.org/10.1515/hfsg.1997.51.2.188
130. García-Mateos FJ, Berenguer R, Valero-Romero MJ et al (2018) Phosphorus functionalization for the rapid preparation of highly nanoporous submicron-diameter carbon fibers by electrospinning of lignin solutions. J Mater Chem A 6:1219–1233. https://doi.org/10.1039/C7TA08788H
131. Li X, Zuo Y, Zhang Y et al (2013) In situ preparation of K_2CO_3 supported kraft lignin activated carbon as solid base catalyst for biodiesel production. Fuel 113:435–442. https://doi.org/10.1016/j.fuel.2013.06.008
132. Guo F, Xiu Z-L, Liang Z-X (2012) Synthesis of biodiesel from acidified soybean soapstock using a lignin-derived carbonaceous catalyst. Appl Energy 98:47–52. https://doi.org/10.1016/j.apenergy.2012.02.071
133. Liang F, Song Y, Huang C et al (2013) Preparation and performance evaluation of a lignin-based solid acid from acid hydrolysis lignin. Catal Commun 40:93–97. https://doi.org/10.1016/j.catcom.2013.06.005
134. Budarin VL, Clark JH, Henschen J et al (2015) Processed lignin as a byproduct of the generation of 5-(Chloromethyl)furfural from biomass: a promising new mesoporous material. Chemsuschem 8:4172–4179. https://doi.org/10.1002/cssc.201501319
135. Zhu J, Gan L, Li B, Yang X (2017) Synthesis and characteristics of lignin-derived solid acid catalysts for microcrystalline cellulose hydrolysis. Korean J Chem Eng 34:110–117. https://doi.org/10.1007/s11814-016-0220-5
136. Gan L, Zhu J, Lv L (2017) Cellulose hydrolysis catalyzed by highly acidic lignin-derived carbonaceous catalyst synthesized via hydrothermal carbonization. Cellulose 24:5327–5339. https://doi.org/10.1007/s10570-017-1515-3
137. Hu S, Zhang S, Pan N, Hsieh Y-L (2014) High energy density supercapacitors from lignin derived submicron activated carbon fibers in aqueous electrolytes. J Power Sources 270:106–112. https://doi.org/10.1016/j.jpowsour.2014.07.063
138. Puziy AM, Kochkin YN, Poddubnaya OI, Tsyba MM (2017) Ethyl tert-butyl ether synthesis using carbon catalysts from lignocellulose. Adsorpt Sci Technol 35:473–481. https://doi.org/10.1177/0263617417696091
139. Bedia J, Rosas JM, Márquez J et al (2009) Preparation and characterization of carbon based acid catalysts for the dehydration of 2-propanol. Carbon 47:286–294. https://doi.org/10.1016/j.carbon.2008.10.008
140. Babeł K, Jurewicz K (2008) KOH activated lignin based nanostructured carbon exhibiting high hydrogen electrosorption. Carbon 46:1948–1956. https://doi.org/10.1016/j.carbon.2008.08.005
141. Saha D, Li Y, Bi Z et al (2014) Studies on supercapacitor electrode material from activated lignin-derived mesoporous carbon. Langmuir 30:900–910. https://doi.org/10.1021/la404112m
142. Ruiz-Rosas R, Valero-Romero MJ, Salinas-Torres D et al (2014) Electrochemical performance of hierarchical porous carbon materials obtained from the infiltration of lignin into zeolite templates. Chemsuschem 7:1458–1467. https://doi.org/10.1002/cssc.201301408
143. Salinas-Torres D, Ruiz-Rosas R, Valero-Romero MJ et al (2016) Asymmetric capacitors using lignin-based hierarchical porous carbons. J Power Sources 326:641–651. https://doi.org/10.1016/j.jpowsour.2016.03.096
144. Li H, Yuan D, Tang C et al (2016) Lignin-derived interconnected hierarchical porous carbon monolith with large areal/volumetric capacitances for supercapacitor. Carbon 100:151–157. https://doi.org/10.1016/j.carbon.2015.12.075
145. Berenguer R, García-Mateos FJ, Ruiz-Rosas R et al (2016) Biomass-derived binderless fibrous carbon electrodes for ultrafast energy storage. Green Chem 18:1506–1515. https://doi.org/10.1039/C5GC02409A
146. Hulicova-Jurcakova D, Puziy AM, Poddubnaya OI et al (2009) Highly stable performance of supercapacitors from phosphorus-enriched carbons. J Am Chem Soc 131:5026–5027. https://doi.org/10.1021/ja809265m
147. Huang C, Sun T, Hulicova-Jurcakova D (2013) Wide electrochemical window of supercapacitors from coffee bean-derived phosphorus-rich carbons. Chemsuschem 6:2330–2339. https://doi.org/10.1002/cssc.201300457
148. Huang C, Puziy AM, Sun T et al (2014) Capacitive behaviours of phosphorus-rich carbons derived from lignocelluloses. Electrochim Acta 137:219–227. https://doi.org/10.1016/j.electacta.2014.05.101

149. Berenguer R, Ruiz-Rosas R, Gallardo A et al (2015) Enhanced electro-oxidation resistance of carbon electrodes induced by phosphorus surface groups. Carbon 95:681–689. https://doi.org/10.1016/j.carbon.2015.08.101
150. Huang C, Puziy AM, Poddubnaya OI et al (2018) Phosphorus, nitrogen and oxygen co-doped polymer-based core-shell carbon sphere for high-performance hybrid supercapacitors. Electrochim Acta 270:339–351. https://doi.org/10.1016/j.electacta.2018.02.115
151. Yu F, Li Y, Jia M et al (2017) Elaborate construction and electrochemical properties of lignin-derived macro-/micro-porous carbon-sulfur composites for rechargeable lithium-sulfur batteries: the effect of sulfur-loading time. J Alloys Compd 709:677–685. https://doi.org/10.1016/j.jallcom.2017.03.204
152. Zhang H, Jia D, Yang Z et al (2017) Alkaline lignin derived porous carbon as an efficient scaffold for lithium–selenium battery cathode. Carbon 122:547–555. https://doi.org/10.1016/j.carbon.2017.07.004
153. Jin J, Yu BJ, Shi ZQ et al (2014) Lignin-based electrospun carbon nanofibrous webs as free-standing and binder-free electrodes for sodium ion batteries. J Power Sources 272:800–807. https://doi.org/10.1016/j.jpowsour.2014.08.119
154. Rodríguez-Mirasol J, Cordero T, Rodríguez JJ (1996) High-temperature carbons from kraft lignin. Carbon 34:43–52. https://doi.org/10.1016/0008-6223(95)00133-6
155. Törmälä P, Romppanen M (1981) Preparation of glassy carbon from lignins and lignin condensates. J Mater Sci 16:272–274. https://doi.org/10.1007/BF00552084
156. Snowdon MR, Mohanty AK, Misra M (2014) A study of carbonized lignin as an alternative to carbon black. ACS Sustain Chem Eng 2:1257–1263. https://doi.org/10.1021/sc500086v

Topics in Current Chemistry (2018) 376:36
https://doi.org/10.1007/s41061-018-0214-3

REVIEW

Catalytic Strategies Towards Lignin-Derived Chemicals

S. Van den Bosch[1] · S.-F. Koelewijn[1] · T. Renders[1] · G. Van den Bossche[1] · T. Vangeel[1] · W. Schutyser[1] · B. F. Sels[1]

Received: 29 June 2018 / Accepted: 10 August 2018 / Published online: 27 August 2018

Abstract

Lignin valorization represents a crucial, yet underexploited component in current lignocellulosic biorefineries. An alluring opportunity is the selective depolymerization of lignin towards chemicals. Although challenged by lignin's recalcitrant nature, several successful (catalytic) strategies have emerged. This review provides an overview of different approaches to cope with detrimental lignin structural alterations at an early stage of the biorefinery process, thus enabling effective routes towards lignin-derived chemicals. A first general strategy is to isolate lignin with a better preserved native-like structure and therefore an increased amenability towards depolymerization in a subsequent step. Both mild process conditions as well as active stabilization methods will be discussed. An alternative is the simultaneous depolymerization-stabilization of native lignin towards stable lignin monomers. This approach requires a fast and efficient stabilization of reactive lignin intermediates in order to minimize lignin repolymerization and maximize the envisioned production of chemicals. Finally, the obtained lignin-derived compounds can serve as a platform towards a broad range of bio-based products. Their implementation will improve the sustainability of the chemical industry, but equally important will generate opportunities towards product innovations based on unique biobased chemical structures.

Keywords Biorefinery · Lignin · Lignocellulose · Catalysis · Biobased chemicals

---

Chapter 6 was originally published as Van den Bosch, S., Koelewijn, S.-F., Renders, T., Van den Bossche, G., Vangeel, T., Schutyser, W. & Sels, B. F. Topics in Current Chemistry (2018) 376: 36. https://doi.org/10.1007/s41061-018-0214-3.

✉ S. Van den Bosch
sander.vandenbosch@kuleuven.be

✉ B. F. Sels
bert.sels@kuleuven.be

[1] Center for Surface Chemistry and Catalysis, KU Leuven, Celestijnenlaan 200F, 3001 Heverlee, Belgium

1 Introduction

Today, alternative technologies are being developed to reduce our society's dependency on fossil resources [1–4]. Several renewable resources, viz., hydro, solar, wind, hydrothermal, and biomass, can be applied for the production of electricity or heat, but only biomass possesses a carbon precursor structure that is required to synthesize chemicals, materials, and liquid transportable fuels. In this light, the concept of the biorefinery, in which biomass is converted to marketable products, is receiving a lot of attention [4–6]. Lignocellulose, the most abundant type of biomass on earth, is regarded as a highly promising feedstock, though its sustainable exploitation should be closely managed; this is likely more realistic in the context of medium volume applications such as chemicals and materials [7, 8].

Due to lignocellulose's complex matrix of entangled biopolymers, viz., cellulose, hemicellulose, and lignin, biorefineries often apply a fractionation strategy to lower the complexity and to allow a tailored downstream conversion of each of these fractions to targeted products [4–6]. Whereas traditional biorefineries primarily focus on the carbohydrate fraction, e.g., for paper and pulp or bio-ethanol production, lignin valorization is becoming increasingly recognized as a crucial link in the valorization chain [9–11]. A major obstacle, however, is lignin's strong tendency towards irreversible degradation, resulting in a recalcitrant lignin-derived side-stream that is usually incinerated for energy recuperation or electricity production [12–14]. Since lignin constitutes the largest source of aromatic compounds in nature, its full valorization potential is nonetheless far from reached. An alternative to incineration is the use of lignin for macromolecular applications such as binding and dispersing agents, polymer composites, and carbon nanofibers, which forms the subject of a highly active research field [11, 15–19]. More recently, the utilization of lignin as a feedstock for chemicals came into the spotlight, and this forms the topic of this chapter (see Fig. 1).

After a brief introduction on the compositional and structural features of lignocellulosic biomass (Sect. 2), traditional lignocellulose fractionation methods will be discussed with emphasis on the ongoing lignin chemistry (Sect. 3). Subsequently,

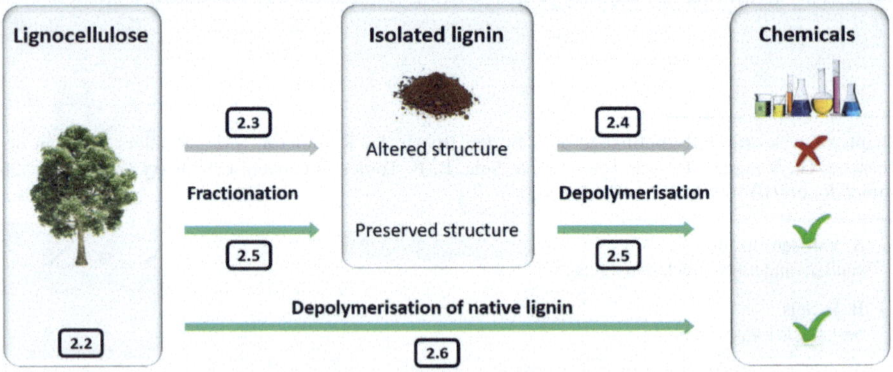

Fig. 1 Schematic overview of major lignin valorization strategies to chemicals discussed in this chapter

major lignin depolymerization strategies will be reviewed, illustrating the recalcitrant nature of lignin upon traditional isolation (Sect. 4). The reactivity of native lignin can be (partially) preserved during milder fractionation processes, hereby enabling a more selective and high-yield production of chemicals from isolated lignin (Sect. 5). Alternatively, the selective depolymerization of lignin into stable products can be performed already during the fractionation step. This strategy will be discussed in Sect. 6. Finally, a perspective on lignin-derived chemicals is provided in Sect. 7.

2 Lignocellulosic Biomass

Lignocellulose, the main constituent in plant cell walls, represents the most abundant source of biomass on earth [20, 21]. Its widespread availability and limited interference with the food industry have turned this biomass feedstock into a favorable renewable candidate for the future chemical industry [22]. The most important sources of lignocellulosic biomass are forestry (hardwoods and softwoods) and agricultural wastes (e.g., corncob, corn stover, rice straw, and wheat straw) next to fast-growing perennial grasses like *Miscanthus* and switchgrass [23]. An impeding factor, however, is lignocellulose's recalcitrance towards chemical and biological conversion, originating from its natural role as a structural and antimicrobial component in plants [24, 25]. This recalcitrant behavior is mainly caused by strong covalent and non-covalent interactions between the three major biopolymers, i.e., cellulose, hemicellulose, and lignin, which together form the lignocellulose matrix, as illustrated in Fig. 2. Briefly, cellulose is organized into highly ordered fibrillary macrostructures, which are supported by a complex intertwined network of hemicellulose and lignin polymers [21, 26]. The fraction represented by each of these three constituents, as well as their chemical composition, varies significantly within different biomass substrates [27–29]. In addition, other compounds such as pectin, proteins, ashes, terpenes, waxes, and fatty acids are also present, albeit in smaller amounts. Altogether, this contributes to the heterogeneous nature of the lignocellulosic feedstock.

Cellulose is the main biopolymer in plant cell walls, constituting roughly 35–55% of the lignocellulosic biomass. It is defined as a linear homo-polysaccharide, composed out of D-glucopyranose units (up to 104), which are linked via β-1,4 glycosidic bonds [30, 31]. The obtained polymer has a strong tendency towards intra- and inter-unit hydrogen bonding, which in combination with van der Waals forces and hydrophobic forces results in the formation of organized crystalline zones [30, 31]. In cellulose, crystalline domains co-exist with less-ordered amorphous zones, leading to a semi-crystalline macrostructure, which is insoluble in water and has a very high tensile strength. These properties are critical to provide both chemical stability and mechanical strength to the plant, but at the same time hamper downstream conversion towards fuels (e.g., bio-ethanol) or chemicals (e.g., levulinic acid, sorbitol) [30].

In contrast to cellulose, hemicellulose is a branched hetero-polysaccharide with a variety of monomer building blocks. Typically, hemicellulose comprises a linear

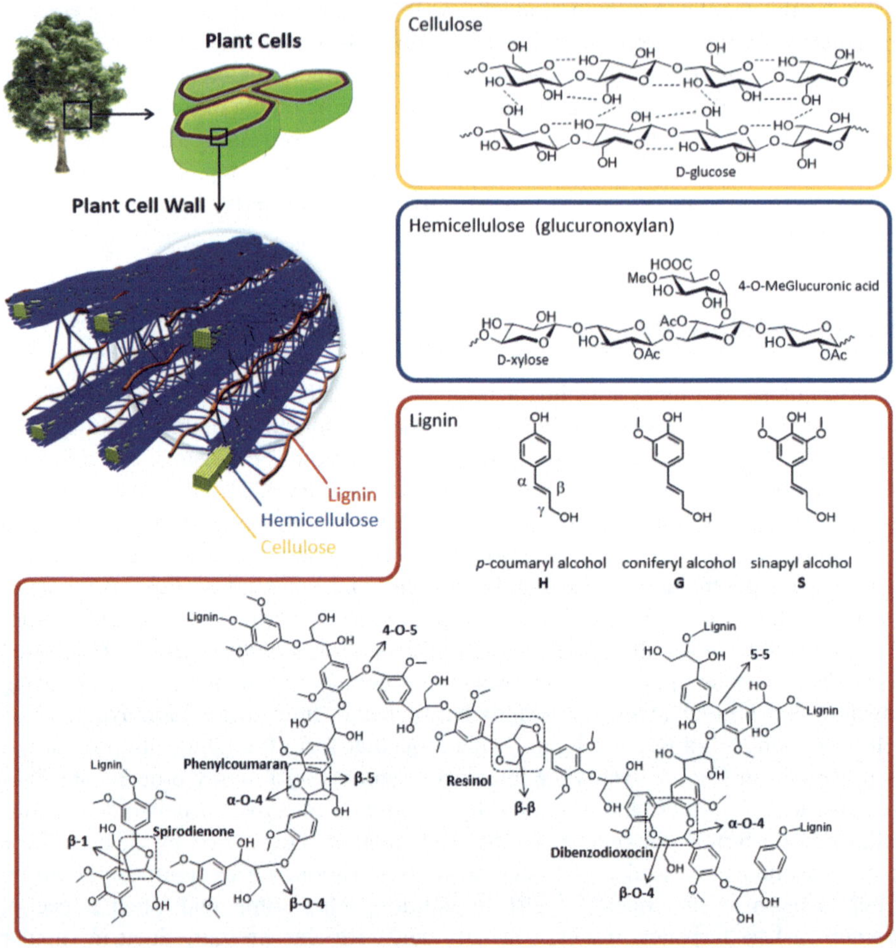

Fig. 2 Schematic illustration of the lignocellulose matrix [35] in plant cell walls, mainly composed of three biopolymers: cellulose, hemicellulose and lignin Adapted from Sels et al. [36]

polymer chain of D-xylose (e.g., glucuronoxylan in hardwoods, illustrated in Fig. 2, or arabinoglucuronoxylan in grasses or softwoods) or D-glucose and D-mannose (e.g., galactoglucomannan in softwoods) linked through β-1,4 glycosidic bonds [32, 33]. The addition of L-arabinose, D-galactose, uronic acids, and acetyl groups on the linear chain generates a branched amorphous structure that is relatively prone to biological or chemical attack. Hemicellulose makes up 10–35% of lignocellulosic biomass and has a lower degree of polymerization compared to cellulose, typically ranging from 50 to 300 units [33, 34]. Hemicellulose's main biological function is to reinforce the cell wall by tethering cellulose microfibrils and participating in the construction of covalently bonded lignin-carbohydrate complexes (LCCs) [21].

Lignin is the third major component, representing about 15–30% of the lignocellulose matrix [14, 37]. This polymer has an amorphous aromatic structure and plays a vital role in plant's antimicrobial defense mechanism [26]. By strongly interacting with hemicellulose, through LCCs, lignin also supports cellulose fibres, thus contributing to the rigidity of the plant cell wall [38, 39].

The lignin polymer is primarily synthesized from three monolignols, i.e., *p*-coumaryl (H), coniferyl (G), and sinapyl alcohol (S), which possess 0, 1, or 2 ortho-methoxy-groups, respectively. Once incorporated in the lignin polymer chain, the corresponding phenolic units are denominated as *p*-hydroxyphenyl (H), guaiacyl (G), and syringyl (S) units [13, 33]. Through enzyme-catalyzed H-abstraction of the phenolic OH-group, resonance-stabilized radicals are formed. The delocalized electrons enable radical coupling reactions at several positions, leading to various types of inter-unit linkages (see Fig. 2), being either C–C bonds or ether bonds [40, 41]. With a relative contribution of 50–80%, the β-O-4 ether bond is by far the most abundant inter-unit linkage in lignin [14]. Other common structures are phenylcoumaran (β-5 + α-O-4), resinol (β–β + α-O-γ), spirodienone (β-1 + α-O-α), biphenyl (5-5), dibenzodioxocin (5-5 + α-O-4 + β-O-4), and diphenyl ether (4-O-5) [37, 41]. From these linkages, the α-O-4 and β-O-4 bonds are most readily cleaved during processing. However, since α-O-4 bonds are mostly present in combination with more resilient C–C linkages, cleavage of the α-O-4 bond has no direct effect on the extent of lignin depolymerization. Due to its high abundancy and readily cleavable nature, the β-O-4 bond is the most investigated inter-unit linkage in lignin research.

The distribution of inter-unit linkages in lignin varies significantly with different sources of lignocellulose. An important variable is the ratio of the H, G, and S building blocks in the lignin polymer. Since methoxy groups occupy either one or both of the *ortho*-positions (relative to the phenolic group) on the aromatic ring, they block the formation of resilient β-5 and 5-5 bonds. For this reason, lignin with a higher content of methoxy groups (S > G > H) contains a lower fraction of resistant C–C bonds and therefore has a higher susceptibility towards extensive depolymerization [42–44]. Hardwood lignin has a high S-content (50–80%) and a smaller fraction of G units, whereas softwood lignin is almost completely composed out of G units [14, 37]. Consequently, hardwood lignin typically contains a higher amount of ether bonds compared to softwood lignin (60 vs. 45–50%) [14, 42]. Finally, lignin in herbaceous crops is predominantly built from G units (55–70%) next to smaller amounts of S and H compounds [45–47].

3 Lignin Chemistry During Biomass Fractionation

The objective of any biorefinery is the conversion of biomass into a spectrum of marketable products. Due to the heterogeneous and complex nature of the lignocellulose matrix, a fractionation strategy is often applied, prior to tailored processing of each individual product stream. As the pulp and paper industry played a dominant role in the landscape of lignocellulosic biorefining during the 19th and 20th centuries, many traditional fractionation processes are geared towards efficient isolation and purification of a fibrous cellulose product stream [48, 49]. With the more recent emergence of the second-generation bio-ethanol industry, again technological efforts are focused on the carbohydrate fraction [50, 51]. In both carbohydrate-oriented contexts, lignin is perceived as an uninteresting constituent due to its contribution to biomass recalcitrance, which impedes carbohydrate valorization [52, 53]. Therefore, lignocellulose is processed under harsh conditions to separate the carbohydrates from lignin, causing irreversible chemical alterations in lignin's chemical structure. The resulting lignin side-stream has long been considered a waste product and is mostly burned for process heat and electricity [13, 37]. In this chapter, the most common (industrial) fractionation processes are summarized and the impact of each of those methods on the chemical structure of lignin is highlighted, especially with regard to the fate of β-O-4 bonds.

3.1 Base-Catalyzed Fractionation

3.1.1 Application and Mechanism

Alkaline media are widely applied for the removal of lignin from lignocellulosic feedstock. Major sources of commercial lignin (e.g., sulphite, kraft, soda) are obtained through alkaline lignocellulose fractionation in the pulp and paper industry [54]. An important reason for this is the base-catalyzed deprotonation of phenolic and benzylic OH-groups in lignin (see Fig. 3), resulting in the formation of ionic intermediates with an improved water-solubility, compared to native lignin [12, 55, 56]. Lignin ionization therefore enhances the extraction of lignin from the lignocellulose matrix. Several other base-catalyzed reactions further impact lignin depolymerization and thus its extraction (Fig. 3). For instance, deprotonation of the benzylic OH-group in non-phenolic units initiates

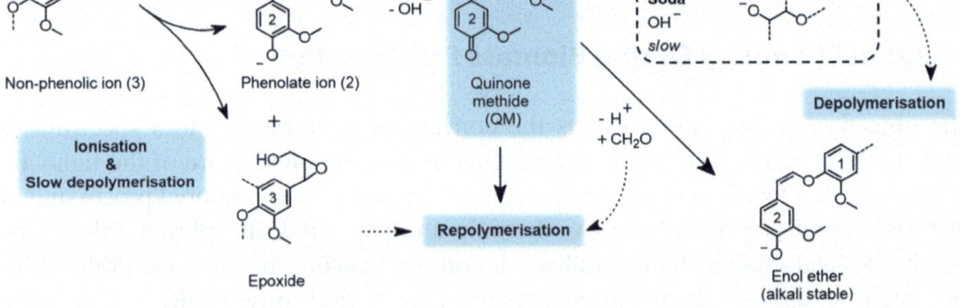

Fig. 3 Fate of the β-O-4 ether bond during base-catalyzed fractionation of lignocellulosic biomass: a balance between ionization (deprotonation), depolymerization (ether bond cleavage), and repolymerization (C–C bond formation) reactions

slow β-O-4 bond cleavage, resulting in additional phenolate ions, but also in the formation of epoxides, which readily participate in repolymerization reactions [56, 57]. Phenolate ions are readily converted to quinone methide (QM) intermediates, provided that a good leaving group (–OH, –OR) is present at the benzylic position [55, 56, 58]. The QM has a strong tendency to restore the aromaticity, which can occur through several reaction pathways. The selectivity to which these pathways occur has a large influence on the final lignin product properties, making the QM a key intermediate during alkaline biomass fractionation. Nucleophilic attack on the QM can induce further lignin depolymerization via ether bond cleavage, but can just as well introduce new C–C bonds via repolymerization with lignin-derived nucleophiles [55, 56, 58]. A third route to restore QM's aromaticity is by breaking the C_β–C_γ bond in the aliphatic side-chain, thus creating an alkali-stable enol-ether. The hereby released formaldehyde might, however, introduce additional C–C linkages, through phenol–formaldehyde condensation reactions [14, 56, 59]. This intricate balance between solubilization, depolymerization and repolymerization determines the lignin product outcome and differs for several important base-catalyzed fractionation processes (e.g., kraft, sulphite, soda), as discussed next.

3.1.2 Kraft Pulping

The kraft process is today's dominant chemical pulping process, reaching a global market share of over 90% and an annual pulp production volume of around 130 MT [14, 60, 61]. During the kraft process, lignocellulose is fractionated in an aqueous solution of NaOH and Na_2S at temperatures around 443 K [59, 61]. This results in a solid cellulose-enriched pulp and a liquid product stream called 'black liquor', which contains a mixture of soluble lignin and hemicellulose components. Although the kraft process produces the largest lignin product stream in industry, at this moment, only a small portion (<2%) is actually available as isolated kraft lignin [13, 54]. The reason for this is the predominant incineration of black liquor for the generation of electricity and heat, as well as the regeneration of Na_2S [14]. Nevertheless, lignin can be readily isolated from the black liquor, for instance via precipitation induced by neutralization of the alkaline liquor [62]. It is important to emphasize that isolated kraft lignin has a strongly altered chemical structure compared to native lignin [47, 55]. Strong nucleophilic HS^- ions, present during the process, react with the α-carbon of QMs, hereby initiating β-O-4 cleavage and the incorporation of sulphur, ultimately ending up in the form of thiol groups. Furthermore, the alkaline conditions promote a complex network of repolymerization reactions [14], briefly discussed in the context of Fig. 3. As a result, kraft processing leads to a condensed, recalcitrant, and water-insoluble lignin product, which has lost most of the native β-O-4 ether bonds and possesses a (relatively low) sulphur content of 1-3%, which is known to act as a catalyst poison [14, 54, 63]. Kraft lignin therefore represents a challenging feedstock for catalytic conversions to chemicals and fuels.

3.1.3 Sulphite Pulping

Sulphite pulping was the dominant process until the 1950s, but has since been superseded by the more flexible kraft process and now represents less than 10% of the chemical pulping industry [55, 64]. By adjusting the cation (Na^+, NH_4^+, Mg^{2+}, Ca^{2+}) of the sulphite or bisulphite salt, the process can proceed under a wide range of pHs, from 1 to 13 [55, 65]. Although the underlying mechanism is pH-dependent, typically the benzylic α-carbon in lignin is sulphonated upon nucleophilic attack of the (bi)sulphite anion, which is followed by the cleavage of β-O-4 bonds [56, 66]. The resulting lignosulphonates have a higher sulphur content (4–8%) compared to kraft lignin and possess a greatly improved solubility in water (even at low pH), which has enabled their implementation as a macromolecular component in multiple applications (e.g., dispersants, adhesives) [65, 66]. Lignosulphonate's good solubility in water slightly complicates their isolation, though several techniques such as ultrafiltration, extraction, or precipitation are being successfully applied [66, 67]. Due to their high applicability and feasible isolation, lignosulphonates represent the largest commercial source of lignin at this moment, despite the smaller production capacity of the sulphite process [54].

3.1.4 Soda Pulping

The soda process is a third important fractionation method used in the pulping industry. It is similar to kraft processing, but uses an aqueous NaOH solution without Na_2S [65]. The absence of a strong nucleophile leads to a less-efficient cleavage of β-O-4 bonds and steers the QM intermediate towards repolymerization pathways, discussed previously (see Fig. 3) [55, 65]. Therefore, soda pulping is primarily used for the fractionation of herbaceous feedstock (e.g., perennial grasses, agricultural residues), which typically possess a more accessible structure, a lower lignin content, and a higher amount of alkali-unstable ester linkages [62, 63]. To increase the fractionation efficiency during soda-pulping, anthraquinone is often used, because it can act as a redox shuttle to promote ether bond cleavages (reductive) and stabilize the carbohydrate fraction (oxidative) [61, 68]. A major advantage of the soda pulping process is the production of a more valuable sulphur-free lignin, thus avoiding catalyst poisoning or bad odors (often linked to sulphur) during downstream lignin processing [63, 69].

3.2 Acid-Catalyzed Fractionation

3.2.1 Application and Mechanism

Acidic media are traditionally applied in lignocellulose fractionation to catalyze the hydrolytic cleavage of glycosidic bonds in hemicellulose and cellulose [70, 71]. This results in the formation of soluble mono- and oligomeric carbohydrates, ready for downstream bio- or chemocatalytic processing. The process severity can also be reduced to perform a selective hydrolysis of hemicellulose while retaining the more stable cellulose as a solid product fraction [12, 70, 72]. In contrast to base-catalyzed fractionation, acidic conditions do not exert a direct promoting effect on the solubilization of lignin [72, 73]. However, a complex network of acid-catalyzed reactions evokes lignin depolymerization and repolymerization, hereby altering the chemical structure of the native lignin polymer [73, 74]. Depending on the process conditions and setup, the final lignin product is either obtained as a highly degraded solid residue or as a solubilized product that in certain cases can be separated from the solubilized carbohydrates through lignin precipitation. Before discussing traditional acid-catalyzed fractionation processes, a general introduction on the underlying acid-catalyzed lignin chemistry is provided.

Due to its high abundance and readily cleavable nature, the fate of the β-O-4 ether bond has a major impact on the final lignin properties. In acidic media, β-O-4 bond cleavage is initiated via protonation and subsequent dehydration of the benzylic OH-group, resulting in a benzylic carbocation (see Fig. 4) [74]. This ion can either undergo (1) proton abstraction or (2) C_β–C_γ bond cleavage, which in both cases leads to the formation of an enol-ether. Acid-catalyzed hydration of the C_α=C_β double bond generates a hemiketal or hemiacetal, which after β-O-4 bond cleavage results in a carbonyl group on the β-carbon [74, 75]. Ketone-substituted phenolics (route I) have a C_3-side-chain and are denominated as Hibbert ketones, whereas the aldehyde-substituted phenolics (route II) have a C_2-side-chain resulting from the release of CH_2O upon C_β–C_γ bond cleavage [75–77]. Besides its conversion to enol-ethers, the carbenium ion is also prone to undergo nucleophilic attack, leading to repolymerization reactions with for

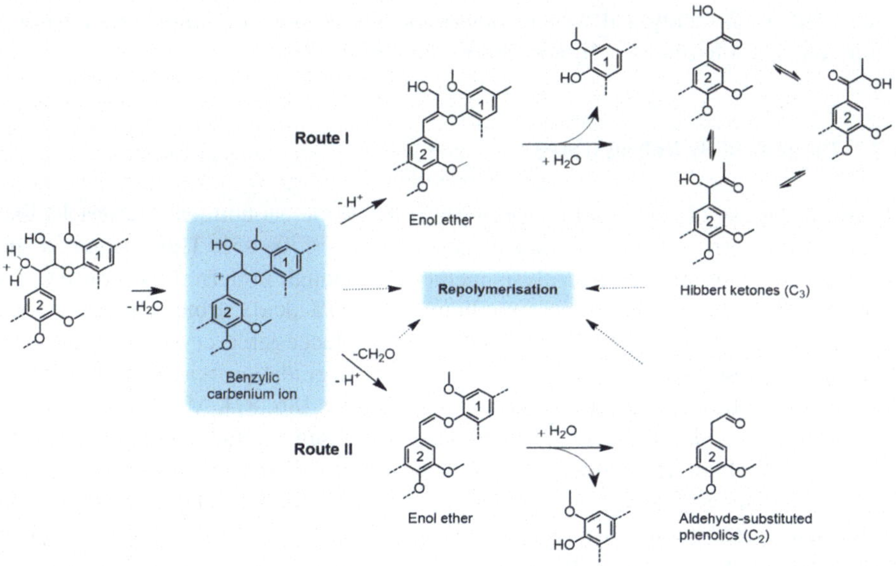

Fig. 4 Fate of the β-O-4 ether bond during acid-catalyzed fractionation of lignocellulosic biomass, illustrating a multitude of lignin intermediates, reactive towards repolymerization

instance the electron-rich *meta*-positions on the aromatic ring [78, 79]. Note that the electron density on the aromatic ring is pH-dependent, shifting from the *ortho*-position at high pH to the *meta*-position at low pH [78]. Moreover, a complex network of repolymerization reactions occurs between carbenium ions, formaldehyde, Hibbert ketones, and C_2-aldehyde substituted phenolics [73, 76, 80]. This ultimately results in a condensed lignin structure, in which the extent of ether bond cleavage and C–C bond formation is related to the severity and characteristics of the applied fractionation process.

3.2.2 Concentrated Acid Hydrolysis (CAH)

CAH is a traditional fractionation method in which cellulose and hemicellulose are efficiently hydrolyzed, reaching yields of over 90% towards water-soluble oligosaccharides [81, 82]. The use of concentrated aqueous acidic solutions (e.g., 42–66 wt% HCl) allows the physicochemical disruption and (partial) depolymerization of insoluble crystalline cellulose fibrils at ambient temperature. However, a post-hydrolysis step under diluted conditions (< 5 wt% acid) and increased temperature is often applied to further depolymerize the soluble carbohydrates towards monomeric pentose and hexose sugars [70, 81, 82]. Overall, the native lignin structure is exposed to strongly acidic conditions, causing nearly complete removal of β-O-4 bonds and a high extent of repolymerization reactions, as discussed within the context of Fig. 4. As a result, lignin is primarily retained as an insoluble highly degraded residue, apart from a small fraction of acid-soluble lignin monomers and short oligomers [83]. The solid lignin residue is easily isolated from the solubilized carbohydrate fraction via filtration. Due to its high fractionation efficiency, this approach is also often used as an analytical tool to determine the amount of lignin (e.g., Klason lignin) and carbohydrates in lignocellulosic materials [83, 84].

3.2.3 Dilute Acid Hydrolysis (DAH)

However, the use of high acid concentrations has some economical drawbacks like severe metal corrosion and inefficient acid recovery [70, 85]. This instigated the development of dilute acid hydrolysis methods. Although preferred by industry, the much lower acidity of the process medium (< 10 wt% acid) strongly increases the required temperature (> 373 K) to hydrolyze the inaccessible crystalline cellulose domains. This in turn leads to unwanted thermal degradation reactions, like sugar dehydration to furans and repolymerization to humins [86, 87]. Moreover, acid-catalyzed reactions (Fig. 4) produce a degraded solid lignin residue, contaminated with insoluble carbohydrate degradation products, often called pseudo-lignin [87–89]. High rates of sugar degradation compared to (hemi) cellulose depolymerization can be tackled by shortening the residence time, which triggered the development of flow-through reactor setups [70]. This also has a positive impact on the preservation of the original lignin structure, as will be discussed in Sect. 5.1.

3.3 (Aqueous) Organosolv Fractionation

Organic solvents have been introduced in biomass fractionation strategies as a way to increase the solubility of lignin in the processing liquor, hereby minimizing lignin redeposition on the cellulose pulp, as observed during mild acid lignocellulose pretreatment [90, 91] and ensuring a more efficient isolation of lignin in a broader range of process conditions [92–95]. Note the analogy with base-catalyzed methods that use lignin ionization (i.e., deprotonation) as a tool for lignin solubilization. A wide range of solvents, including alcohols [96–98], organic acids [99, 100], cyclic ethers [101, 102], short polyols [103], and ketones [104, 105], have been applied, either in combination with water and/or acids (mostly HCl, H_2SO_4, or H_3PO_4) or in their pure form. Organosolv lignins typically possess a high purity and low molecular weight, and they are rich in phenolic hydroxyl groups, which make them soluble in a range of organic solvents [93].

Depending on the process characteristics, the final lignin structure can be extensively degraded or can show a partial preservation of the original functionalities, like the β-O-4 bond [106]. If organic solvents are used in combination with acids or hot water, acid-catalyzed cleavage of hemicellulose, LCCs, and lignin ether bonds results in a soluble mixture of lignin- and hemicellulose-derived products, next to a solid cellulose pulp. Afterwards, lignin is readily isolated from the liquid product mixture via precipitation [92–95]. An alternative strategy is the use of pure organic solvents without the addition of water or acids. Under such pH-neutral conditions, the hydrolysis of both hemicellulose and cellulose is strongly impeded, while an extensive removal of lignin from the lignocellulose matrix can still be achieved, albeit at higher reaction temperatures. Solvolysis has been appointed to cause (partial) cleavage of LCCs and lignin ether bonds [107–110], yet additional investigations are required to clarify the underlying mechanisms [33, 98–100].

4 Are Traditional Lignin Streams a Promising Feedstock for the Production of Chemicals?

Lignin valorization towards chemicals constitutes an alluring biorefinery strategy. This is evidenced by the vast amount of lignin depolymerization studies, starting from 'traditional' lignin streams that are isolated by the lignocellulose fractionation methods discussed in Sect. 3. The production of lignin-derived chemicals, however, requires a multidisciplinary effort, not solely focusing on lignin depolymerization, but aligning all relevant variables between the initial feedstock choice and the target chemicals. Essentially, two complementary research domains can be defined: (1) fractionation and (2) depolymerization, wherein the isolated lignin forms the common element. To optimize the entire valorization chain, the output from both domains (i.e., isolated lignin from fractionation and depolymerized lignin from depolymerization) should be used as feedback mechanisms (see Fig. 5).

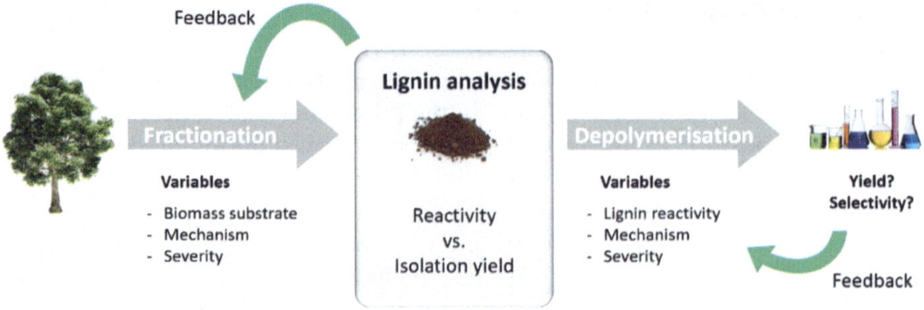

Fig. 5 Production of chemicals from isolated lignin requires reactive lignin isolates as well as effective depolymerization methods, both dependent on a multitude of variables. Robust analytical tools are imperative to allow a sensible interpretation of the product outcome from fractionation and depolymerization processes

4.1 Considerations on the Isolation and Analysis of Lignin

Obviously, the production of chemicals through lignin depolymerization is greatly influenced by the properties of the lignin substrate. Analysis of isolated lignins therefore embodies an important feedback mechanism that enables the optimization of existing fractionation methods. Aryl-ethers, mainly represented by β-O-4 bonds, are the dominant inter-unit linkages in native lignin, but their abundance decreases significantly during most fractionation processes. Unfortunately, a reduced β-O-4 content in the isolated lignin is typically linked to an impaired lignin reactivity and a lower phenolic monomer yield upon lignin depolymerization [12, 106, 111–113]. Recent studies, however, also showed that the correlation between the β-O-4 content and the lignin reactivity is not fully conclusive, and that other factors like, for instance, the presence of impurities [e.g., sulfur, (earth) alkali metals, biomass degradation products] can influence the depolymerization efficiency as well [106, 111, 112]. Nevertheless, the relative amount of β-O-4 bonds, determined by 2D HSQC NMR [114, 115] or thioacidolysis [116], remains a useful and frequently applied measure to assess the reactivity of isolated lignins. Note that potential differences in the applied analytical protocols should always be considered when comparing the results of independent studies [117]. To avoid any ambiguity, designated characterization studies, using a fixed set of analytical techniques to compare the reactivity of isolated lignins, are indispensable.

General statements on the reactivity of lignin, obtained from different fractionation methods (e.g., kraft lignin vs. organosolv lignin), should be treated with caution, since the extent of β-O-4 cleavage can vary significantly within the boundaries of a specific isolation method. Apart from the fractionation mechanism, the process severity and the lignocellulosic feedstock also greatly affect the reactivity of the isolated lignin. This was demonstrated by Barta et al. in a characterization study of 22 organosolv lignins, displaying a wide variation in β-O-4 content, ranging from 2 to 62% [106]. Consequently, a more balanced discussion is required, encompassing all variables that contribute to the obtained lignin properties instead of only referring to the general term of the fractionation method.

A final consideration is the difficulty to reconcile a high lignin reactivity with a high isolated lignin yield. Higher isolation yields usually require an increased process severity, which in turn has a negative impact on the lignin reactivity. Therefore, the isolation yield cannot be overlooked when striving for a maximal lignin reactivity, as it represents a crucial factor to optimize the yield to chemicals based on the original total amount lignin in the biomass feedstock [12, 118].

4.2 Depolymerization of Traditional Lignin

The large amount of existing depolymerization processes can be categorized into five main groups, based on their dominant cleavage mechanism (i.e., reductive, oxidative, acid/base, solvolytic, and thermal depolymerization). Each strategy leads to a characteristic product outcome and will be discussed below. It should, however, be stressed that comparing the depolymerization efficiency of different processes is never straightforward. A large variety of traditional lignin streams are available (see Sect. 3 and 4.1) and different procedures exist to analyze the reaction products after lignin depolymerization. Therefore, results can be described by multiple variables, while several definitions can be used for the same variable [117]. Nevertheless, for the production of chemicals from lignin, a specifically relevant measure is the monomer yield, defined as the wt% of lignin monomers, relative to the original weight of the lignin fraction in the biomass substrate. In addition, the monomer fraction ideally comprises a limited number of monomeric compounds, thus decreasing the product complexity and facilitating down-stream processing. For these reasons, both the monomer yield and the product selectivity within the monomer fraction will be discussed for each depolymerization strategy.

4.2.1 Reductive Depolymerization

Common to all reductive depolymerization processes is the use of a redox catalyst in the presence of a reducing agent. This combination allows the cleavage of arylether linkages through hydrogenolysis, thereby effectuating lignin depolymerization [119–121]. Other chemical reactions that frequently occur are the removal of oxygen via hydrodeoxygenation (HDO) and the saturation of C=C double bonds, C=O double bonds, or aromatic rings through hydrogenation. On the other hand, C–C bond cleavage does normally not take place in reductive processes. Hydrogen is typically applied as the reductant, although the use of other reducing agents such as hydrosilanes has been demonstrated as well [122]. When hydrogen is supplied in the form of H_2 gas, the process is typically called *hydroprocessing. Catalytic transfer hydrogenation* is an alternative option, which is based on hydrogen-donating solvents. Both hydroprocessing and catalytic transfer hydrogenation have been performed with noble metal (e.g., Pt/Al_2O_3 [29, 111, 112], Ru/C [123, 124]) and base metal (e.g., NiMo [125, 126], $CuMgAlO_x$ [127, 128]) catalysts.

In the case of hydroprocessing, a further distinction can be made between three process variants, each yielding characteristic lignin products. *Mild hydroprocessing* is usually performed at temperatures below 593 K in the presence of a solvent (e.g., water, alcohols, cyclic ethers, or solvent mixtures) [29, 111, 112, 129–131]. Under these conditions, the methoxy groups on the aromatic rings are mostly preserved, resulting in a relatively narrow product distribution, mainly containing methoxyphenolic monomers with a propanol- or alkyl-substituent at the *para*-position. Hydroprocessing at temperatures above 593 K can be denoted as harsh hydroprocessing and is characterized by a broader product spectrum, including mainly demethoxylated phenolic species, next to deoxygenated aromatics, alkanes, catechols, and methoxy-phenols [125, 126, 132–134]. Noticeably, harsh hydroprocessing can be performed in the absence of a solvent. Both mild and harsh hydroprocesses typically result in monomer yields below 20 wt% when starting from (partly) degraded traditional lignins, although higher yields have been reported. A third variant, called *bifunctional hydroprocessing* introduces an acidic catalyst in addition to the reductive system. This enables additional reactions such as hydrolysis and dehydration, resulting in a more extensive reduction of the lignin substrate towards alkanes at temperatures below 593 K. To obtain a high alkane selectivity, sufficient acid strength and a good interaction between the acid and metal sites are both key [135–137]. Yields of up to 50 wt% and more towards C_6–C_{18} alkanes have been reached, although these values represent both monomer and dimer products.

In liquid phase reforming, H_2 gas is replaced by hydrogen-donating solvents, such as isopropanol, formic acid, tetralin, but also common alcohols like methanol and ethanol [29, 124, 127, 128, 138]. Advantages of this approach are a lower reactor pressure and reduced safety concerns. Note that the oxidation of the solvent as a result of hydrogen donation (e.g., isopropanol to acetone) implies an additional reduction-step, likely with H_2 gas, to recuperate the solvent. The fate of oxidized solvent molecules is, nonetheless, scarcely described and therefore needs more attention in follow-up studies. A broad product spectrum is obtained, similar to that of harsh hydroprocesses, and monomer yields of up to 30 wt% are frequently observed. Hensen et al. reported even higher yields, reaching up to 60–86 wt%, which was explained by the specific role of ethanol, functioning as capping agent and formaldehyde scavenger [127, 128]. This impedes lignin repolymerization and increases the mass of the lignin products due to extensive alkylation with ethanol.

4.2.2 Oxidative Depolymerization

During oxidative lignin depolymerization, an oxidant (e.g., O_2, H_2O_2, nitrobenzene, metal oxides) is used to convert the lignin polymer into oxygenated low molecular weight compounds [139–141]. In contrast to reductive processing, where primarily ether bond are cleaved, oxidants also enable the scission of C–C bonds. Most oxidation studies describe a selective C–C cleavage of the aliphatic side-chain, resulting in the formation of methoxy-phenolic aldehydes, ketones, and acids. Alternatively, cleavage of the aromatic ring with formation of aliphatic (di)carboxylic acids can be targeted as well [139–143].

The most well-known example of oxidative lignin valorization through C–C cleavage of the aliphatic side-chain is the production of the flavoring component vanillin (4-hydroxy-3-methoxybenzaldehyde) from sulphite liquor (see sulphite pulping in Sect. 3.1) [144, 145]. The process has been commercially practiced for over 80 years and was once the major route to vanillin in the world. Competition with vanillin synthesis from fossil resources, next to environmental concerns and the decreasing market volume of the sulphite pulping process, have now lowered the market share of lignin-derived vanillin to about 15% of the global production, leaving the Norwegian company Borregaard as sole remaining producer [144, 145]. The Borregaard process is performed in a concentrated aqueous NaOH solution and uses O_2 (air) as the preferred oxidizing agent, as it is inexpensive, readily available, and converted to water upon reaction.

Recent studies on aerobic oxidative lignin depolymerization also often apply aqueous NaOH solutions. Nevertheless, other bases (e.g., KOH [146]), acids (e.g., acetic acid [147], H_2SO [41] [46, 149], $H_3PMo_{12}O_{40}$ [148, 149]), and solvents such as methanol [150] have been used as well. Although not essential, a large number of homogeneous and heterogeneous catalysts has been investigated to improve the attainable monomer yield [139–141]. The dominant use of alkaline media for aerobic oxidative lignin depolymerization is related to the fast ionization of phenolic groups at high pH. Though the exact cleavage mechanism remains a subject of discussion, the formation of phenolate ions and their subsequent oxidation towards phenoxy-radicals is generally regarded as a crucial step to initiate C–C bond cleavage [140, 142, 151, 152]. Temperatures of 393-463 K with a (partial) oxygen pressure of 2–15 bar are commonly applied, with the primary products being *p*-hydroxybenzaldehyde, vanillin, and syringaldehyde, next to their acid- and aceto-analogues. These monomer products are prone to secondary (repolymerization) reactions under the applied oxidative conditions, resulting in phenolic monomer yields of around 10% [153, 154]. Interestingly, short reaction times at high temperatures, using a flow-through setup, can be applied to minimize product degradation and increase the phenolic monomer yield [155].

An opposite strategy is to target the further oxidative conversion of unstable phenolic products. Cleavage of the aromatic ring can be favored by increasing the reaction time, the reaction temperature, the O_2 partial pressure, or by applying a stronger oxidant such as H_2O_2. This results in a selective production of aliphatic (di)carboxylic acids, mainly represented by formic, acetic, oxalic, malonic, and/or succinic acid. The ratio in which each acid is formed strongly depends on the process characteristics. A variety of temperatures (333–498 K), pH conditions (acid [156, 157], neutral [158], base [143, 156]), and oxidants (O_2 [116] and H_2O [156–158]) have been applied, resulting in a wide range of total monomer yields from 11 to 56 wt%.

4.2.3 Acid- or Base-Catalyzed Depolymerization

As mentioned previously, acid or base catalysts are often added during reductive or oxidative depolymerization processes as a way to further improve the yield or selectivity of the monomeric products. However, a vast amount of catalytic studies also demonstrated the applicability of homogeneous and heterogeneous acids or bases as catalysts for the depolymerization of lignin in a redox-neutral environment. Typically, the absence of a redox driving force is counteracted by stronger catalysts and/or higher process temperatures (typically > 513 K) [159–165]. No hydrogenation or cleavage of aromatic rings is observed, resulting primarily in a set of methoxy-phenolic products without a side-chain group or with an aldehyde-, aceto-, acid-, alkene-, or alkane-functionality at the *para*-position. Both in acid and base-catalyzed depolymerization, a shift in the product selectivity towards catechols and alkyl-catechols has been observed at increasing temperatures [161–164]. Most studies report monomer yields below 20 wt%, although a few acid-catalyzed depolymerization studies have reported higher phenolic monomer yields between 20 and 60 wt% [164, 165].

As discussed in Sect. 3, lignin is prone to repolymerization reactions in the presence of acid or base catalysts. The repolymerization products have a low solubility in water and are easily converted to chars [160]. A possible solution is the use of (aqueous) organic solvents instead of pure water in order to improve the solubility of the lignin products in the reaction medium. For instance, Hensen et al. showed a doubling in monomer yield (from 6 to 13 wt%) when switching from water to ethanol as the reaction solvent during depolymerization of wheat straw soda lignin with a Lewis acid Cu-acetate catalyst [160]. Superior monomer yields with organic solvents, compared to reactions in water, were also demonstrated for the acid-catalyzed depolymerization of softwood kraft lignin with metal chlorides [164], as well as for the base-catalyzed depolymerization of pine ethanosolv lignin with MgO [159].

4.2.4 Solvolytic Depolymerization

Without oxidants, reductants, and acid/base additives, solvolytic depolymerization processes benefit from their simplicity regarding downstream product isolation and the redundancy of catalyst recovery. Water, organic solvents, and mixtures thereof have been used as reaction media at temperatures in the range of 523–723 K [160, 164, 166–170]. A fairly broad product distribution is obtained, consisting of (methoxy-)phenols with oxygenated and unsaturated side chains for reactions at the lower end of the temperature spectrum, while shifting towards unsubstituted and alkylated (methoxy-)phenols and catechols at higher temperatures or with higher water-to-solvent ratios [166–168]. As solvolytic depolymerization processes often function as the blank run in catalytic studies, a comparison is easily made. Typically, non-catalyzed processes reach inferior monomer yields compared to their corresponding catalytic process, and usually do not surpass 10 wt%. However, higher monomer yields are sometimes reported, for instance reaching 22 wt% from softwood kraft lignin in water at 537 K [169] or 24 wt% from an organosolv lignin (extracted from corncob residue) in tetrahydrofuran at 573 K [170].

4.2.5 Thermal Depolymerization Through Fast-Pyrolysis

In fast-pyrolysis (FP), lignin (or biomass) is rapidly heated to high temperatures of 673–1073 K in the absence of oxygen [171–173]. Under these conditions, vapors are formed next to a solid carbon-enriched char. After removal of the vapors from the heating zone, they are partially condensed into a liquid oil while the remaining fraction is collected in the form of gasses. When targeting the production of chemicals from lignin, a maximization of the liquid fraction is wanted, as it contains the lignin-derived monomers. Higher oil- and monomer yields are normally obtained at temperatures between

673 and 923 K for short vapor residence times (in the order of seconds) [171]. The yields can be further increased by the addition of a catalyst, at which the process is thus called catalytic fast-pyrolysis (CFP). The catalyst serves a dual role, as it (1) stabilizes reactive intermediates to avoid excessive charring, and (2) increases the product selectivity [172]. The latter is especially clear in the case of acid zeolites, which can be used to drastically improve the selectivity towards deoxygenated aromatics like benzene, toluene, and xylene [174–179]. As demonstrated for a range of H-ZSM-5 catalysts, the lower the Si/Al ratio, the higher the acid density and the higher the yield of deoxygenated aromatics [176]. Moreover, Triantafyllidis *et al.* investigated the influence of porosity and external surface area of H-ZSM-5 on the selectivity towards monophenolics or deoxygenated aromatics [180]. Next to H-ZSM-5, other zeolites (sometimes loaded with metals) and oxides have also been applied in CFP [181–184]. During non-catalytic fast-pyrolysis or when the applied catalyst has a low acidity, the product selectivity shifts from deoxygenated products to methoxy-phenolic and catechol products, substituted with oxygenated, unsaturated, or saturated side chains. Overall, most studies on FP and CFP report large numbers of monomer products with total monomer yields below 10%, although higher selectivities and yields of up to 40% have been sporadically observed with CFP [176, 180, 185].

4.3 Closing Remarks on Traditional Lignin Valorization

Traditional lignin valorization is defined here as a two-step lignin isolation-depolymerization approach in which no special attention is given to the fate of the lignin structure during isolation (i.e., fractionation). This approach comes forth from the current dominance of carbohydrate-oriented biorefineries in which a maximal valorization of the carbohydrates is the primary objective. A large side-stream of isolated lignin is obtained this way, which can serve as a substrate for depolymerization towards chemicals. However, as discussed in Sect. 3, the native lignin structure is prone to repolymerization and degradation reactions during the fractionation process, hereby impeding downstream depolymerization. Although promising results have been reported (see Sect. 4), most studies clearly demonstrated the recalcitrant nature of traditional lignins, reaching monomer yields below 20 wt%, often in combination with a broad spectrum of lignin monomer products. In order to address the issue of lignin condensation, alternative strategies for lignin valorization have emerged [186], either (1) targeting the isolation of native-like lignin, in which the original reactivity is maximally preserved or (2) enabling lignin depolymerization to stable monomeric products during lignocellulose processing (see Fig. 6). Both strategies will be discussed in separate sections (Sects. 5, 6), while providing an overview of the most promising examples and their main advantages and disadvantages.

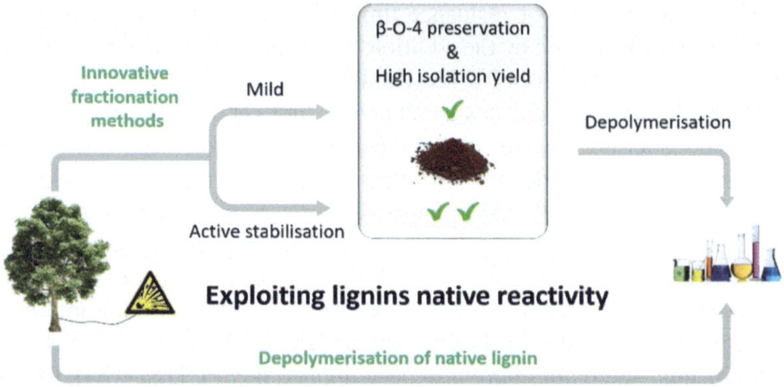

Fig. 6 Lignin valorization to chemicals should exploit the reactive structure of native lignin. Two major routes exist, either targeting lignin isolation with minimal structural alterations (Sect. 5) or directly depolymerizing native lignin to stable monomeric products during lignocellulose processing (Sect. 6)

5 Innovative Lignin Streams: Preserving β-O-4 Bonds During Fractionation

Compared to most isolated lignins, the native lignin structure is more susceptible to depolymerization as a result of its higher ether-bond content and reduced degree of C–C crosslinking. A valid strategy therefore is to isolate lignin from lignocellulosic feedstocks with minimal alteration of its original chemical structural features. This way, the obtained lignin isolate forms a more reactive substrate for further conversions to chemicals. Despite lignin's strong tendency towards condensation, several fractionation methods have achieved the isolation of native-like lignin, especially with respect to the preservation of β-O-4 bonds. To favor lignin isolation over lignin condensation, most of these innovative methods apply mild process conditions, for example by improving the lignin solvating power of the reaction medium. Alternatively, active stabilization mechanisms were introduced recently, allowing a higher process severity and higher isolation yields of native-like lignin.

5.1 Mild Fractionation: Passive Approach

5.1.1 Ammonia

The use of liquid ammonia as a swelling agent for cellulose and as a suitable solvent for lignin solubilization has been known for over 60 years [187, 188]. Several ammonia-based processes are particularly relevant in the context of mild fractionation on herbaceous crops, where high degrees of structural preservation have been achieved in the isolated lignin. Note, however, that a good preservation of β-O-4 bonds often goes hand in hand with low to intermediate lignin extraction yields (circa 50%) [111, 189–191]. Moreover, small quantities of nitrogen are incorporated in the lignin products in the form of amides [189, 192].

Ammonia fiber expansion or AFEX is probably the most recognized ammonia-based method and starts by contacting wet biomass with ammonia. Temperatures close to the critical temperature of ammonia (405 K) are typically applied, resulting in high autogenous pressures (> 100 bar) [193–195]. Alongside other reactions, the lignin polymer is partially solubilized after hydrolysis and ammonolysis of LCCs and to a minor extent lignin ether bonds [194, 196]. Following an explosive pressure release, ammonia is briskly evaporated, opening up the biomass structure while redistributing lignin [194]. As such, AFEX is not a fractionation method, but simply induces physicochemical alterations of the lignocellulose matrix, hereby enhancing its susceptibility towards lignin extraction. By using organic or alkaline extraction solvents, 50-65% of the native lignin in herbaceous crops can be extracted, mainly in the form of oligomeric fragments with a well-preserved amount of β-O-4 bonds [111, 190, 194].

A second method is called *anhydrous ammonia pretreatment* (AAP) and differs from AFEX by the required low moisture content of the lignocellulosic feedstock and the absence of an explosive pressure release [191, 192, 197]. Without this pressure drop, the lignin-containing ammonia solution can be easily separated from the solid carbohydrate fraction. This process is also often referred to as extractive ammonia pretreatment (EAP). However, relatively low extraction yields of circa 45% have been obtained from corn stover [191, 197]. Alternatively, a controlled evaporation of ammonia, followed by an alkaline extraction results in higher isolation yields, up to 65% [192]. The low moisture content of the substrate is required to allow an intense interaction between crystalline cellulose domains and ammonia. This alters the crystalline structure of cellulose from its native C_I form into the C_{III} polymorph, making it more susceptible towards enzymatic hydrolysis [198].

Finally, higher lignin extraction yields, up to 85% from corn stover, have been obtained by performing a flow-through lignin extraction from lignocellulose with aqueous ammonia [199–201]. This process is called *ammonia recycled percolation* (ARP), as it continuously recovers and reuses the ammonia through evaporation. ARP has also been performed on hardwoods, reaching extraction yields of up to 63% [189, 202]. After precipitation from the reaction mixture, a rather low isolated lignin yield of 31% was obtained by Jackson et al., yet showing a good retention of β-O-4 bonds [111, 189].

5.1.2 Ionic Liquids

Ionic liquids (IL) are known for their excellent solubilizing capacity, which can be tailored to the envisioned application by a careful selection of the anion and cation [67, 203–205]. For the effective isolation of native-like lignin from lignocellulose, the *ionosolv process* is a promising method. In analogy to traditional (aqueous) organosolv fractionation, a selective solubilization of lignin and hemicellulose is targeted, while retaining cellulose as a solid product. However, the use of ILs allows to work at lower reaction temperatures (< 433 K) and at atmospheric pressure, which in turn reduces structural alterations in the lignin polymer. As a result, ionosolv lignins retain a significant amount of the original β-O-4 bonds, although the extent of β-O-4 bond cleavage is strongly dependent on the IL and the process conditions [112, 206, 207]. Despite their interesting physico-chemical properties, several challenges need to be addressed before an IL can be labeled as a competitive green solvent. For instance, the use of halogenated anions (e.g., PF^{6-}, BF^{4-}) and organic cations (e.g., substituted imidazolium) in many common ILs, results in a high cost of components [203, 206]. This in turn requires a long-term stability of the IL under the applied reaction conditions of the biorefinery process, as well as an efficient isolation procedure, to allow multiple recycling of the IL. Moreover, a high purity is required as low amounts of IL degradation products or biomass components can severely influence the properties and the dissolution performance of the IL [205]. Finally, the development of environmentally friendly synthesis routes (e.g., deep eutectic solvents [208]) and a better understanding of the toxicity and biodegradability of ILs are important criteria to advance their applicability in the biorefinery [203, 205, 206, 209].

5.1.3 Flow-Through Reactor Setups for Acid Hydrolysis

Compared to batch-mode dilute acid hydrolysis (see DAH in Sect. 3.2), a flow-through reactor setup (FT-DAH) is much more suited to combine high temperatures and short residence times. This way, reactive carbohydrate and lignin intermediates can be removed from the heating zone before degradation occurs [156, 210, 211]. In contrast to the solid lignin residue obtained upon batch mode processing, FT-DAH mainly produces soluble and less degraded oligomers, which can be recovered from the product solution via precipitation. Successful depolymerization of the isolated lignin precipitate clearly indicates a partial preservation of β-O-4 bonds [156, 211].

Two related flow-through processes are dilute acid pretreatment (FT-DAP) and hot water pretreatment (FT-HWP), which apply milder process conditions to target a selective solubilization of hemicellulose and lignin, while retaining cellulose as a solid product [212, 213]. FT-DAP uses lower acid concentrations or temperatures compared to FT-DAH, while FT-HWP represents an autocatalytic process that is based on the acidic properties of hot water in combination with organic acids (e.g., acetic acid), originating from the lignocellulosic substrate [72, 73, 212, 213]. The relatively mild process conditions in both methods result in highly oxygenated low molecular weight lignin products (mainly oligomers), which are proposed to possess a significant part of the original β-O-4 interunit linkages [156, 190, 211], yet additional investigations are required. The high solubility of the lignin-derived product fraction in the hydrolysate solution, however, hampers an effective precipitation and isolation of the lignin fraction [213].

5.1.4 γ-Valerolactone

As mentioned in Sect. 3.3 on organosolv fractionation, the extent of β-O-4 bond cleavage varies significantly with different solvents and process conditions [106]. A remarkably mild organosolv fractionation was reported by Dumesic et al., using an 80/20 wt% mixture of γ-valerolactone (GVL) and water with a low concentration of H_2SO_4 (1.5 wt%) [123]. The use of the polar aprotic solvent GVL promotes the disruption of the lignocellulose matrix [214–218] and allows an effective solubilization of lignin and hemicellulose [123]. The yield of extracted lignin from corn stover reached 75% at a reaction temperature of 423 K. After precipitation, the isolated lignin was analyzed by 2D HSQC NMR, showing great structural similarities with native lignin [123]. Moreover, the preservation of β-O-4 bonds, observed with NMR, was further corroborated by high monomer yields of up to 48%, obtained upon bifunctional hydroprocessing of the lignin precipitate with Ru/C, H_3PO_4, and H_2 [123].

5.1.5 Mechanical Pretreatment

Applying mechanical stress to open up the lignocellulose structure is another strategy to isolate lignin with minimal structural alterations compared to native lignin [219–222]. However, these procedures are lengthy and usually reach limited lignin isolation yields. As a result, this approach is primarily relevant in analytical lignin studies where a maximal preservation of the native lignin structure is the main requirement and no restrictions exist with regard to process costs and efficiency. *Milled wood lignin* (MWL) for instance, is obtained by ball-milling a lignocellulosic substrate for several days, followed by lignin extraction with an aqueous organic solvent [220–222]. The yield of the isolated native-like lignin is generally below 35% [223]. Instead of direct lignin extraction, mechanical pretreatments can also be followed by an enzymatic hydrolysis step, using cellulolytic enzymes. Here, the carbohydrates are converted into soluble mono- and oligosaccharides, while a solid residue is obtained, containing a mixture of lignin, residual carbohydrates, proteins and ashes. Lignin can be extracted from this solid residue and subsequently precipitated, after which it is denominated as cellulolytic enzyme lignin (CEL) [223–225]. Again, a high preservation of β-O-4 bonds and a rather low lignin yield ($<35\%$) is obtained [223, 226]. Lignin isolation can be improved by performing an additional acid hydrolysis step on the solid residue obtained upon enzymatic hydrolysis [226, 227]. This results in the cleavage of lignin-carbohydrate complexes and enables a more extensive solubilization of lignin. After precipitation, a 65% yield of *enzymatic mild acidolysis lignin* (EMAL) can be obtained [223, 226].

5.2 Chemical Stabilization During Fractionation: Active Approach

A general observation with respect to mild fractionation processes is the undesirable trade-off between the lignin isolation yield and the preservation of β-O-4 bonds in the isolated lignin. Although the methods discussed in Sect. 5.1 enable the production of reactive lignins, isolation yields are limited due to the mild process conditions. On the other hand, attempts to improve the isolation yield by increasing the process severity are often linked with increased lignin degradation [106, 203]. A promising advancement here is the use of active stabilization mechanisms that prevent the formation of reactive intermediates during fractionation. Ideally, β-O-4 bonds can be preserved under more severe process conditions, in turn leading to higher lignin isolation yields.

Luterbacher et al. was the first to demonstrate the success of this approach [79]. Related to organosolv processing, the method uses an acidified aqueous dioxane reaction mixture to extract lignin from the lignocellulosic feedstock (i.e., beech wood). With this method, high lignin extraction yields of up to 80% can be obtained at low temperatures of 353–373 K [79]. The innovative aspect lies in the addition of formaldehyde (FA), which undergoes acid-catalyzed condensation with the α- and γ-OH side-chain groups, hereby forming 1,3-dioxane structures (see Fig. 7) [79]. This acetal is relatively stable under the applied conditions and prevents the formation of reactive carbenium ions, the key intermediate in acid-catalyzed depolymerization and repolymerization reactions. A second stabilization mechanism is the blocking of reactive *meta*-positions [78] on the aromatic ring, through electrophilic substitution with FA. This way, *m*-hydroxymethyl groups are formed, though without the formation of recalcitrant methylene bridges [228].

The excellent structural integrity of the FA-stabilized lignin was demonstrated by a subsequent reductive depolymerization with a Ru/C catalyst at 523 K in methanol. A lignin monomer yield of 47%, based on the total lignin in beech wood, was realized, which is around the theoretical maximum [79]. In contrast, when fractionation was performed in the absence of FA, the subsequent depolymerization of the isolated lignin only resulted in a 7% monomer yield. Furthermore, a decent product selectivity towards ethyl- and propyl-substituted guaiacol and syringol can be obtained. After reductive depolymerization, the hydroxymethyl groups at the *meta*-position are retained in the form of methyl groups on the aromatic ring, while FA incorporated in the 1,3-dioxane structure is converted to methane. Note that this strategy implies the consumption of formaldehyde and also lowers the product selectivity by the introduction of additional methyl side chains. Interestingly, aromatic alkylation can be excluded by utilizing other aldehydes than FA, such as acetaldehyde and propionaldehyde [229].

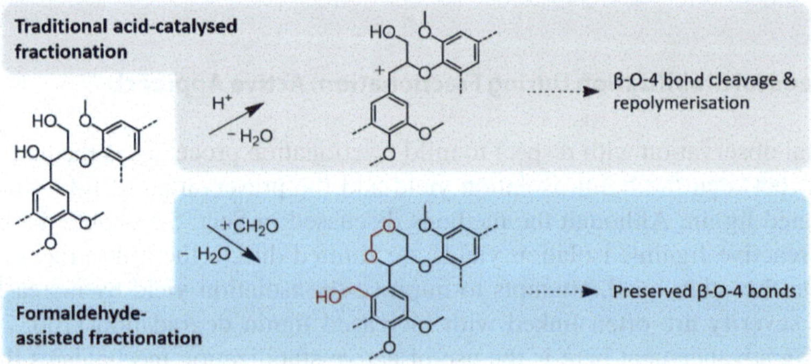

Fig. 7 Lignin stabilization during formaldehyde-assisted fractionation. The formation of 1,3-dioxane structures prevents the formation of reactive carbenium ions (see also Sect. 3.2) while hydroxymethyl groups block reactive *meta*-positions [79]

6 Depolymerization of Native Lignin to Stable Monomers

Instead of targeting the high-yield isolation of native-like lignin (Sect. 5), a complementary valorization approach is to instantly convert native lignin to stable lignin monomers during lignocellulose processing. The key aspect is to provide a process environment in which lignin depolymerization can be performed without the detrimental consequences of lignin repolymerization. Reductive depolymerization-stabilization of native lignin has been investigated most frequently, though other methods exist and are applicable as well.

6.1 Reductive Catalytic Fractionation (RCF)

Hydrogenolysis of native lignin through redox catalysis under reductive conditions was first conveyed in literature by Hibbert, Pepper, and coworkers around 70 years ago [230–233]. Their research efforts were primarily focused on the analytically challenging elucidation of the lignin structures in several lignocellulosic species (e.g., maple, spruce, wheat straw), but also demonstrated the selective production of phenolic monomers with a preserved structure, resembling the original lignin building blocks. These observations were later explained by solvolytic reactions and a reductive catalytic depolymerization-stabilization of lignin (intermediates) [107, 234–236]. Nevertheless, the potential of this approach for the production of lignin-derived chemicals in a lignocellulosic biorefinery was recognized only recently. This *lignin-first* type of biorefinery strategy is referred to here as "reductive catalytic fractionation (RCF)", but is also known under the name "catalytic upstream biorefining (CUB)" [237–239].

In 2008, Yan et al. was one of the first to report near theoretical monomer yields (up to 46%) by processing birch sawdust in a dioxane:water (50:50 vol%) mixture in the presence of a Pt/C catalyst and 4 MPa H_2-gas at 473 K for 4 h [240]. Similar results were obtained by Song et al. (2013), who acquired phenolic monomer yields of around 50% from birch wood in pure methanol [234]. Interestingly, the solvent methanol served as the hydrogen donor for reductive depolymerization-stabilization. An inexpensive base-metal Ni/C catalyst was applied, demonstrating an 89% selectivity towards propyl-substituted guaiacol and syringol (see Fig. 8). Instigated by such promising results, the scientific field of RCF processing has strongly expanded the last few years [241], leading to several fundamental and technological advancements.

Understanding the relation between the applied feedstock and the product outcome is a vital aspect in any biorefinery. Studies have illustrated the positive correlation between the amount of syringyl units (S-units) and the β-O-4 content within the original lignin structure of different feedstocks [42, 43]. This in turn has a direct impact on the attainable lignin monomer yield from RCF. As a result, RCF of hardwoods (up to 80% S-content) generates the highest yields of phenolic monomers (40–55%), in contrast to the monomer yields from softwoods (>95% G-content), which usually do not surpass 30% [46, 234, 240–244]. RCF of herbaceous crops

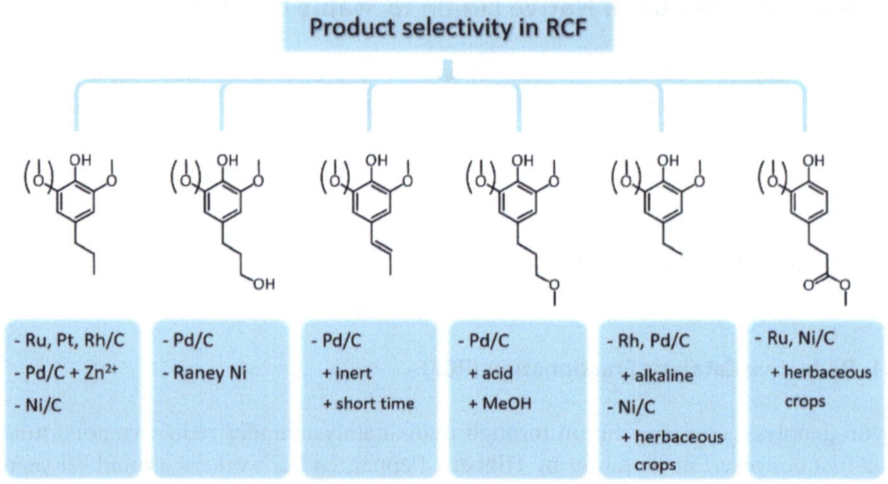

Fig. 8 Overview of the phenolic monomer products that can be selectively obtained from reductive catalytic fractionation (RCF) under specified catalytic conditions (and feedstock)

(5% H, 65% G, 30% S) typically leads to intermediate monomer yields of 25–40%, though yields of over 60% have been reported [46, 237, 245]. Consequently, most RCF studies have been performed on hardwoods. Notwithstanding, the very high selectivity to guaiacyl compounds from softwoods or the unique feature of side-chain ester or acid groups from herbaceous crops (see Fig. 8) justify their use for RCF as well.

Next to the feedstock, the solvent also plays a decisive role during RCF. First of all, the fractionation efficiency (i.e., delignification versus carbohydrate pulp retention) is strongly dependent on the polarity and the lignin solvating power of the solvent (or solvent mixture) [221, 232, 233]. The solvent should hence be carefully selected and optimized in relation to other process parameters, like the reaction temperature and time [234, 246, 247]. Interestingly, acid additives (e.g., H_3PO_4 [238, 240], $Yb(OTf)_3$ [248, 249]) can be used to further tune the fractionation efficiency while reducing the process severity. Moreover, several solvents, such as isopropanol, methanol, and ethanol, have been applied as a hydrogen donor in RCF, thus eliminating the direct need for external H_2 [234, 244]. Alternatively, other components (e.g., formic acid), released during the fractionation process, have been appointed as effective hydrogen donors [242, 250]. Finally, an active participation of the solvent in the (partial) cleavage of lignin carbohydrate complexes and/or β-O-4 bonds has been suggested [234, 242, 243, 245]. It was shown that under specific conditions (e.g., MeOH at 523 K and 12 MPa) lignin solubilization and depolymerization occur predominantly via solvolytic (instead of catalytic) mechanisms [107]. In this case, catalytic action is solely required to stabilize lignin intermediates in solution through hydrogenation. Consequently, no physical contact between the heterogeneous catalyst and the solid lignocellulosic substrate is required during RCF, as demonstrated by the successful use of a catalyst basket [107]. This also facilitates the previously challenging separation between the solid catalyst and the carbohydrate pulp after RCF, and opens up new opportunities in future reactor design (e.g., Carberry reactors, flow-through fixed-bed catalytic reactors) and scale-up initiatives. Interestingly, the above findings were corroborated recently by two independent research groups, successfully demonstrating RCF in a flow-through reactor. A fixed bed containing lignocellulose feedstock was placed in front of a fixed catalyst bed. At elevated temperature, a flow of methanol extracts lignin intermediates from the biomass bed, which are subsequently forced through the catalyst bed. Liquid sampling before and after the catalyst bed therefore enabled the isolation of solvolytic and catalytic processes [235, 236, 251].

A final key feature in the RCF process is the catalytic system. Apart from effectuating high lignin monomer yields through reductive stabilization, the catalyst also functions as a tool to tune the product selectivity, especially with regard to the functionalities on the *para*-substituted side chain (see Fig. 8). As mentioned previously, a high selectivity to propyl-substituted phenolic monomers can be obtained with a Ni/C catalyst and methanol as the hydrogen-donating solvent. Similar results were obtained with Ru/C, Pt/C, and Rh/C catalysts or with a synergistic Pd/C and Zn^{2+} catalytic system, all in the presence of pressurized H_2 [46, 240, 243]. These catalytic systems are capable of removing γ-OH groups through hydrogenolysis or tandem dehydration/hydrogenation. Other catalysts, such as Pd/C and Raney Ni, selectively retain the γ-OH functionality and reach high selectivities to propanol-substituted monomers [244, 252, 253]. Alternatively, Galkin et al. reported a high selectivity to propenyl-substituted guaiacol and syringol under inert atmosphere, using an H_2-activated Pd/C catalyst at a short reaction time of 1 h [242]. Moreover, Pd/C in combination with $Al(OTf)_3$ acid results in methoxypropyl-substituted phenolic monomers, formed by the acid-catalyzed etherification of γ-OH groups with methanol, which is used as the solvent [248]. Finally, also ethyl-substituted monomers (C_2) can be selectively obtained from RCF, for instance through decarboxylation/hydrogenation of acid or ester side chains, present in the monomers from RCF with herbaceous crops (see also previous paragraph on the influence of the lignocellulosic substrate) [237]. Another route is to perform RCF in alkaline media, leading to the base-catalyzed release of formaldehyde from quinone methide intermediates (see Fig. 3). The resulting enol-ether is then further converted to ethyl-substituted monomers through subsequent β-O-4 bond cleavage and hydrogenation [233, 238].

6.2 Reductive One-Pot Processing

Reductive processing of lignocellulosic biomass does not necessarily cause fractionation into a lignin oil and a carbohydrate pulp. Recently, so-called one-pot processes have been disclosed that target the transformation of the complete lignocellulose matrix into soluble low molecular weight products. These processes are characterized by a higher process severity in terms of solvent, acidity, and/or temperature compared to RCF. In short, reductive one-pot processes and RCF processes both benefit in a similar way from the reactivity of native lignin as well as the reductive stabilization of lignin intermediates, but differ with regard to the fate of the carbohydrate fraction.

Li et al. investigated the one-pot conversion of birch wood with a Ni-W_2C/AC catalyst in water at 508 K and 6 MPa H_2. Lignin monomer yields of over 35% were combined with a 70% yield of sugar-derived diols, mainly represented by ethylene glycol and propylene glycol [254]. The same process with methanol as the solvent (i.e., typical RCF conditions) gave a phenolic monomer yield of 47% (around the theoretical maximum) without carbohydrate conversion to diols. Note that in this specific process, the acidity of water at high temperature is required to realize effective carbohydrate hydrolysis as a first step to C_2 –C_4 diols [254]. However, (pure) organic solvents have been successfully applied as well for one-pot processing. Ford et al. applied supercritical methanol (573-593 K at 160-220 bar) in the presence of a copper-doped porous metal oxide to produce substituted cyclohexyl alcohols and ethers from lignin and acyclic C_2-C_6 alcohols and ethers from the carbohydrate fraction [255, 256]. Alternatively, specific acid-redox catalytic systems (e.g., TSA + ht-Ru/C [257, 258], Pt/$NbOPO_4$ [259]) achieve the production of gasoline-range alkanes via extensive depolymerization and hydrodeoxygenation of both lignin and (hemi)cellulose components at relatively mild temperatures of 463-493 K. Interestingly, the alkane product mixture is likely suitable for fuel applications without additional separation steps. In other cases, the cost-efficiency of a one-pot process should always be considered against the costs, related to a potentially more difficult product separation.

6.3 Oxidative Catalytic Fractionation (OCF)

As discussed in Sect. 4.2 on the oxidative depolymerization of isolated lignin, oxidants are known to enable the cleavage of C–C bonds. Nevertheless, most oxidation studies yield low amounts of aromatic monomers (typically $< 15\%$) from isolated lignin [153, 154]. Similar to RCF, oxidative catalytic fractionation (OCF) is performed directly on lignocellulosic substrates, hereby exploiting the reactivity of native lignin to improve the attainable monomer yield. The potential of this approach has already been indicated in the 1930s, when nitrobenzene and later CuO were used as oxidants to depolymerize lignin for compositional analysis [260, 261]. Both methods (i.e., nitrobenzene and CuO oxidation) were compared by Pepper et al., reaching monomer yields of circa 20% from spruce wood and 40% from aspen wood upon reaction at 453 K for 2 h in a concentrated 2 M NaOH solution [261]. According to the guaiacyl/syringyl ratio of the lignocellulosic feedstock, either vanillin (spruce) or syringaldehyde (aspen) was the major monomer product. More recently, Tarabanko et al. successfully applied O_2 as an inexpensive and environmentally benign oxidant during birch processing, using a homogeneous $Cu(OH)_2$ catalyst in a 2.5 M NaOH solution at 443 K for 25 min [262]. A combined monomer yield of 43% was obtained for the aldehyde-substituted monomers vanillin (13%) and syringaldehyde (30%) [262, 263]. Noticeably, in the absence of $Cu(OH)_2$, the selective production of vanillin and syringaldehyde decreased by half. Next to the soluble lignin fraction, a cellulose-enriched carbohydrate pulp was produced, amenable for downstream processing [263]. Although promising, the number of OCF studies is still scarce and additional research will be required to support and further develop this strategy. For instance, the fate of the carbohydrate fraction during OCF has to be further investigated and the required high NaOH concentration is ideally lowered. Also, the implementation of heterogeneous catalysts, as demonstrated earlier for the oxidative depolymerization of lignin model compounds and isolated lignins [153, 154, 264], could benefit the techno-economical aspect of the OCF process.

6.4 Other Opportunities Towards Higher Lignin Monomer Yields from Native Lignin

Several other methods have demonstrated the advantage of processing lignocellulosic feedstock instead of isolated lignin for the selective and high yield production of lignin-derived chemicals.

6.4.1 Thermal (Fast-Pyrolysis)

A mechanistic pyrolysis study performed by Evans et al. demonstrated that native lignin is much less refractory as commonly believed [265]. A selective initial formation of coniferyl alcohol (CA) and sinapyl alcohol (SA) was observed when processing whole biomass samples, as opposed to identical pyrolysis reactions on several isolated lignins [265]. Consequently, under optimized conditions, an improved production of CA- and SA-derived chemicals has been observed after pyrolysis on lignocellulosic substrates [265–267]. For example, Kuroda et al. reported a threefold increase in the monomer yield after fast-pyrolysis of Japanese softwoods: cedar (23%), cypress (21%), and red pine (18%) for 4 s at 773 K, in comparison with the same reaction on the organosolv lignin analogues (yielding 8, 7, and 6%, respectively) [266]. Main monomer products were guaiacol, 4-methylguaicol, 4-vinylguaiacol, eugenol, and isoeugenol. Nevertheless, due its multicomponent composition, thermal processing of whole biomass will generate a more complex pyrolysis oil, compared to that from isolated lignin, especially in the absence of a catalyst.

6.4.2 Solvolytic

Organosolv fractionation typically introduces structural alterations in the lignin polymer (see Sect. 3.3), although milder process variants exist that preserve most β-O-4 bonds in the isolated lignin (see Sect. 5.1). On the other hand, organosolv fractionation at high temperatures and short reaction times can also be applied as a method to obtain high lignin monomer yields directly from lignocellulose. Remarkably, methanol:water (90:10) at 543 K inflicts extensive lignin depolymerization after only 3 min, reaching a high monomer yield of 44% from beech wood, with a high selectivity towards coniferyl- and sinapyl alcohol, as well as their γ-methyl ether analogues [108, 109]. The short reaction time is a key aspect to minimize repolymerization reactions and was enabled by immersing a small 5-ml reactor in a molten tin bath. The high heating and cooling rate forms a challenging aspect for large-scale batch processes and likely requires a transition to flow-through reactor setups.

6.4.3 Base-Catalyzed

Aqueous alkaline pretreatment, a milder version of the traditional soda-pulping process (Sect. 3.1), typically uses less-concentrated NaOH solutions (0.1–1 M instead of 2 M) at temperatures of 373–433 K for 0.5 h [268–270]. These mild process conditions are not suitable for efficient ether bond cleavages, but enable a selective conversion of alkali-labile ester linkages, abundantly present in herbaceous crops (e.g., corn stover, switch grass) [55, 195]. This approach is therefore limited to herbaceous feedstocks, from which high phenolic monomer yields of up to 27% have been achieved [268]. Major monomer products are *p*-coumaric acid, ferulic acid, and vanillic acid.

6.4.4 Acid-Catalyzed

Finally, also acid-catalyzed lignocellulose fractionation can be reconciled with high lignin monomer yields. This, however, requires the stabilization of reactive ketone- and aldehyde-substituted intermediates, typically formed under acidic conditions (see Fig. 4). Several mechanisms can be envisioned and are illustrated in Fig. 9.

Barta et al. applied diols to stabilize reactive carbonyl groups through the formation of cyclic acetals (Fig. 9), thereby preventing acid-catalyzed repolymerization [76, 106]. When ethylene glycol was used in the acid-catalyzed depolymerization (ACD) of walnut dioxasolv lignin, a threefold increase of the lignin monomer yield

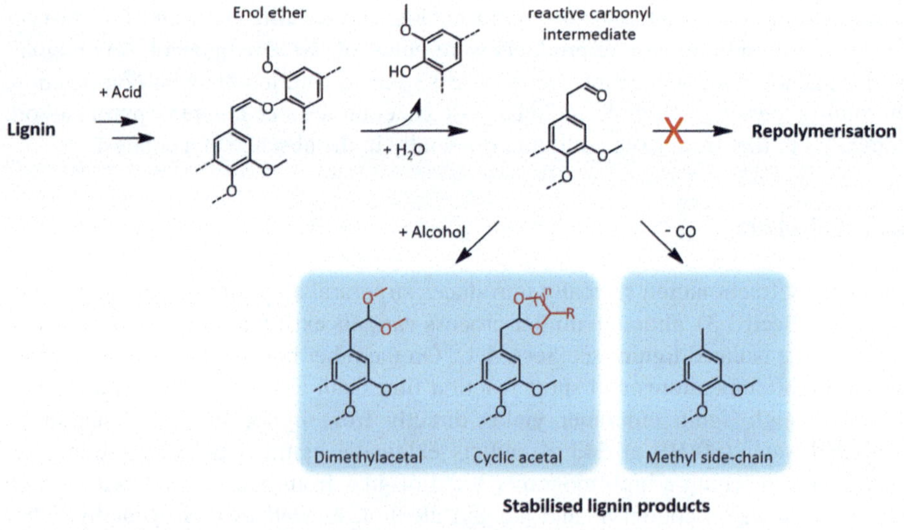

Fig. 9 Lignin stabilization strategies during acid-catalyzed lignin depolymerization/lignocellulose fractionation, converting reactive carbonyl intermediates into acid-stable acetals or methyl groups

was obtained, compared to ACD in the absence of a diol [76]. A successful implementation of this method, directly on native lignin (i.e., during lignocellulose fractionation) is expected to further improve the achieved monomer yields, but remains to be demonstrated. Note that the underlying chemistry of the cyclic acetal formation is related to the formaldehyde-assisted fractionation, discussed in Sect. 5.2, although in that case the carbonyl group is provided by formaldehyde, which reacts with diol-groups in the lignin side chain. Watanabe et al. applied a similar stabilization approach as Barta et al., though relying on acetal formation with methanol instead of ethylene glycol [271]. During acid-catalyzed fractionation of Japanese cedar in toluene, stable monomeric dimethyl-acetals were obtained this way. Finally, an alternative strategy is to use decarbonylation as a way to remove reactive carbonyl groups. Both homogeneous [76] and heterogeneous [272] catalysts have been applied to convert reactive C_2-aldehyde substituted intermediates to stable *p*-methyl substituted products. This approach has not yet been demonstrated on native lignin.

7 Perspective on Lignin-Derived Chemicals

Based on the platform of lignin-derived compounds that can be produced via previously described lignocellulose conversion processes, a myriad of end products can be envisioned. These end products can either be obtained through (1) direct utilization/purification of the lignin monomers, or (2) through downstream catalytic upgrading. Furthermore, a distinction is made between drop-in chemicals with an existing market platform, and new chemicals with potentially beneficial properties as a result of their unique chemical functionality. Here, selected examples are provided to illustrate some general principles and strategies. It is also important to note that most of these technologies are still in an early stage of development and require significant further investigation to assess and improve their industrial viability.

7.1 Direct Utilization

Obviously, the direct utilization of lignin monomers for chemicals is limited by the product outcome of the lignin depolymerization process. Important criteria that should be assessed are the applicability and the (potential) market value of the targeted product, as well as the yield at which this product can be obtained from lignocellulose. This approach seems especially suitable for chemicals with a relatively high market value to compensate for lower product yields and higher costs, associated with the isolation of pure compounds from potentially complex product mixtures. For instance, vanillin is a high-value flavoring and fragrant component, which is produced at a commercial scale via oxidative depolymerization of sulphite liquor (see Sect. 4.1), despite a rather low product yield (~ 5%) and competing fossil-based routes [144, 145]. Nevertheless, many biorefinery schemes lead to substituted phenolics with no or limited existing applications. A future challenge therefore exists in identifying opportunities for new applications, which exploit the distinguishing functionality (e.g., methoxy groups, side chain) of these lignin monomers.

For example, several advantages were observed, using either 4-*n*-propylguaiacol or 4-*n*-propylsyrignol in the synthesis of bisguaiacols (BG) or bissyringols (BS), respectively, as biobased alternatives to bisphenol A from phenol [273]. 4-*n*-propylguaiacol or 4-*n*-propylsyringol were isolated in high yield and selectivity upon RCF of softwoods or hardwoods, respectively (see Sect. 6.1). Compared to phenol, both lignin-derived molecules facilitate BG/BS synthesis and purification, reduce the estrogen potency, and introduce advantageous properties in BG polycarbonates, BG cyanate ester resins, and BS polyesters [273, 274]. As such, lignin-derived products potentially carry a competitive advantage, which can open the market to new and valuable chemicals.

7.2 Chemocatalytic Upgrading

A second strategy is the catalytic upgrading of lignin-derived compounds, for which both biocatalytic and chemocatalytic approaches exist. Within this context, a variety of chemocatalytic processes have been investigated, though this section is mainly focused on the complete or partial defunctionalization of lignin monomers through hydrodeoxygenation (HDO). An interesting feature of HDO is that complex phenolic mixtures, obtained from lignocellulose processing, can be funneled into a limited number of targeted products. This reduces the complexity of the product mixture and allows a higher yield of individual compounds, while lowering the cost associated with product isolation. As a result of funneling, a high market value (as in the case of vanillin) might not always be required for lignin-derived chemicals to become economically viable. Depending on the process characteristics, chemicals such as (1) alkanes, (2) aromatic hydrocarbons, (3) phenols, or (4) cyclohexanols can be selectively obtained.

Complete defunctionalization of substituted phenolics to alkanes is usually achieved with bifunctional acid-redox catalysis. Various *p*-substituted-*o*-methoxylated phenolics as well as isolated lignin streams (see also bifunctional hydroprocessing in Sect. 4) have been selectively converted to cyclic alkanes, e.g., (alkylated) cyclohexanes [7, 136, 240, 275–279]. The resulting alkane (mixture) could be directly applicable as a mid-range fuel additive, hereby eliminating the need for isolation or purification of individual compounds. Alternatively, HDO can lead to the selective production of aromatics, e.g., (alkylated) benzenes, when conditions are applied that disfavor hydrogenation. Therefore, gas-phase reactions at high temperature and under low H_2-pressure have been frequently disclosed, using either traditional sulfided CoMo- and NiMo catalysts or non-sulfided catalysts such as MoO_3 or PtCo/C [42, 280–282].

For the production of phenol or *p*-substituted phenols, the preservation of the phenolic OH-group should be combined with a selective removal of the methoxy group. Again, both traditional sulfided CoMo and NiMo catalysts as well as non-sulfided catalysts (e.g., Pd/C, Fe/C, MoC_x/C) have been successfully applied, though most studies are limited to the conversion of (alkylated) guaiacols [283–285]. Noticeably, Rinaldi et al. demonstrated the production of alkylated phenols, next to aromatic hydrocarbons, starting from a complex pyrolysis oil using a MoC_x/C catalyst

[284]. Alkylphenols (e.g., 4-ethyl- or 4-propyl phenol), could be further dealkylated towards phenol and the corresponding olefin, viz., ethylene or propylene, as disclosed by Sels and coworkers [286]. A related two-step upgrading strategy is the selective demethoxylation of vanillic and syringic acid towards *p*-hydroxybenzoic acid, i.e., a *p*-substituted phenol, which in turn can be carboxylated to terephthalic acid [287].

Cyclohexanols, the fourth major product class, can be directly obtained through demethoxylation and ring hydrogenation of lignin-derived phenolics. Several catalysts, including Ni/CeO_2, $Ru–MnO_x/C$, and Raney Ni have demonstrated high selectivities to (alkylated) cyclohexanols from a variety of substituted phenolics [288–292]. Note that selective removal of *o*-methoxygroups can generate methanol, which for instance could be used as an integrated solvent in the lignocellulosic biorefinery [291]. Moreover, (alkylated) cyclohexanols are readily dehydrogenated to the corresponding ketone, which can be further converted to (alkylated) caprolactone, caprolactam or adipic acid [289, 293]. The non-alkylated analogues are primarily used as building blocks in the production of polyamide (e.g., nylon) and polyesters materials. Interestingly, the alkyl-chain might introduce new and potentially advantageous properties into the polymer, thus creating added value, compared to their fossil-based unsubstituted analogues.

7.3 Biocatalytic Upgrading

The biocatalytic upgrading of lignin-derived products is inspired by aromatic catabolic pathways, studied primarily in aerobic micro-organisms [294–296]. In principle, a wide range of aromatic compounds, including *p*-substituted-*o*-methoxylated phenolics, can be enzymatically converted towards a small set of central metabolic intermediates, such as catechol, protocatechuate, and gallate [296–298]. Subsequent opening of the aromatic ring leads these intermediates into the central carbon metabolism. Based on the elucidated enzymatic reaction pathways, research has focused on the metabolic engineering of microbial strains in order to attain a highly selective production of value-added chemicals. For example, using alkaline pretreatment liquor (see Sect. 6.4) as the aromatic substrate, polyhydroxyalkanoates (PHAs) were produced through catabolic reactions in *Pseudomonas putida* KT2440 [270]. PHAs are valuable polymers that can be used as such in plastics and adhesives, or can be depolymerized and converted to (chiral) chemical precursors or methyl-ester-based fuels [299]. Alternatively, *P. putida* could be engineered towards a selective production of muconic acid via oxidative cleavage of catechol. Amongst others, muconic acid can be further transformed into adipic acid or terephthalic acid, both important polymer building blocks [300, 301]. Also, other bacteria have been studied. For instance, *Rhodococcus opacus* was grown on a minimal medium containing organosolv lignin (see Sect. 3.3), which led to the production of triacylglycerols, a valuable substrate within the field of oleochemistry (e.g., lubricants) or for the production of biodiesel [302]. Note, that in this study, *R. opacus* effectuates both lignin depolymerization as well as further metabolic conversion of released phenolic monomers. Although lignin valorization through biocatalysis constitutes a promising strategy, future research is likely required to further improve the attainable titers, reactions rates, and yields, as well as the robustness of the biocatalysts towards inhibitors (e.g., toxic aromatic compounds).

8 Conclusions

Today, lignin valorization is being recognized as a crucial, yet underexploited component in the lignocellulosic biorefinery. This growing awareness has triggered the development of new technologies, enabling a multitude of future lignin-based applications. One promising opportunity in particular is the selective depolymerization of lignin towards chemicals. Although challenged by lignin's recalcitrant nature, several successful (catalytic) strategies have emerged over the last decade, which were highlighted in this book chapter.

The difficulty of producing chemicals from lignin can be ascribed to the high reactivity of native lignin and its tendency to undergo repolymerization and degradation reactions during lignocellulose fractionation. However, traditional (industrial) fractionation methods have focused primarily on carbohydrate valorization and process conditions have therefore been optimized with little concern about lignin structural alterations. Consequently, commercial lignins embody a challenging feedstock for depolymerization processes, typically leading to low monomer yields and/or a low product selectivity.

In response, mild fractionation methods have been proposed, isolating lignin with a better preserved structure and an increased amenability towards depolymerization. A high retention of native β-O-4 ether bonds can be successfully obtained, while the formation of new recalcitrant C–C bonds through repolymerization reactions was largely avoided. Unfortunately, the applied mild conditions usually do not enable high lignin isolation yields. However, to ensure a full utilization of the lignin fraction present in the lignocellulosic feedstock, the structural preservation of lignin and its isolation in high yields should be combined. An alluring strategy is the active stabilization of β-O-4 bonds, as recently demonstrated with formaldehyde. This stabilization approach allows a higher process severity, leading to a more complete extraction of lignin, while retaining a high content of readily cleavable β-O-4 bonds.

An alternative for the isolation of native-like lignins is the instant depolymerization-stabilization of native lignin towards stable lignin monomers during lignocellulose processing. This approach benefits from depolymerizing lignin in its most reactive form, but requires a fast and efficient stabilization of reactive lignin intermediates in order to minimize lignin repolymerization and maximize the envisioned production of chemicals. Effective stabilization strategies for instance include (1) reductive stabilization, (2) quenching of carbonyl groups through acetal formation, (3) decarbonylation, or (4) a fast physical removal of reactive species from the heating zone.

In conclusion, a selective and high-yield production of lignin-derived platform chemicals can be attained via several strategies as long as three criteria are fulfilled: (1) lignin or lignin-derived products are isolated in high yields, (2) the intrinsic reactivity of native lignin towards depolymerization is fully exploited, and (3) a stabilization mechanism is applied to minimize lignin repolymerization.

Finally, lignin-derived platform chemicals could serve as the starting point for a broad range of existing and new bio-based applications. In particular, opportunities exist in the catalytic conversion of these platform compounds towards base chemicals and fuels, as well as specialty chemicals, pharmaceuticals, and materials.

Acknowledgements This work was performed in the framework of SBO projects ARBOREF & BIOWOOD, EOS project BIOFACT, and Interreg project BIO-HArT. S.V.d.B., T.R., T.V., and W.S. acknowledge the Research Foundation—Flanders (FWO Vlaanderen) for a (post)doctoral fellowship. S.-F.K. acknowledges funding through IWT-SBO project ARBOREF. G.V.d.B acknowledges funding through FISCH-ICON project MAIA.

References

1. Ragauskas AJ, Williams CK, Davison BH, Britovsek G, Cairney J, Eckert CA, Frederick WJ, Hallett JP, Leak DJ, Liotta CL, Mielenz JR, Murphy R, Templer R, Tschaplinski T (2006) Science 311:484–489
2. Hoffert MI, Caldeira K, Benford G, Criswell DR, Green C, Herzog H, Jain AK, Kheshgi HS, Lackner KS, Lewis JS, Lightfoot HD, Manheimer W, Mankins JC, Mauel ME, Perkins LJ, Schlesinger ME, Volk T, Wigley TML (2002) Science 298:981–987
3. McCormick K, Kautto N (2013) Sustainability 5:2589
4. Dale BE, Kim S (2008) In Biorefineries-industrial processes and products. Wiley, 41–66
5. Kamm B, Gruber PR, Kamm M (eds) (2006) Biorefineries-industrial processes and products. Wiley-VCH, Weinheim
6. Wyman CE (ed) (2013) Aqueous pretreatment of plant biomass for biological and chemical conversion to fuels and chemicals. Wiley, Chichester
7. Deneyer A, Renders T, Van Aelst J, Van den Bosch S, Gabriëls D, Sels BF (2015) Curr Opin Chem Biol 29:40–48
8. Chu S, Majumdar A (2012) Nature 488:294–303
9. Bozell JJ, Holladay JE, Johnson D, White JF (2007) Report PNNL 16983, vol 2, pp 1–79
10. Tuck CO, Pérez E, Horváth IT, Sheldon RA, Poliakoff M (2012) Science 337:695–699
11. Ragauskas AJ, Beckham GT, Biddy MJ, Chandra R, Chen F, Davis MF, Davison BH, Dixon RA, Gilna P, Keller M, Langan P, Naskar AK, Saddler JN, Tschaplinski TJ, Tuskan GA, Wyman CE (2014) Science 344:709
12. Schutyser W, Renders T, Van den Bosch S, Koelewijn SF, Beckham GT, Sels BF (2018) Chem Soc Rev 47:852–908
13. Gosselink RJA, de Jong E, Guran B, Abacherli A (2004) Ind Crops Prod 20:121–129
14. Rinaldi R, Jastrzebski R, Clough MT, Ralph J, Kennema M, Bruijnincx PCA, Weckhuysen BM (2016) Angew Chem Int Ed 55:8164–8215
15. Upton BM, Kasko AM (2016) Chem Rev 116:2275–2306
16. Esposito D, Antonietti M (2015) Chem Soc Rev 44:5821–5835
17. Sen S, Patil S, Argyropoulos DS (2015) Green Chem 17:4862–4887
18. Alonso DM, Hakim SH, Zhou S, Won W, Hosseinaei O, Tao J, Garcia-Negron V, Motagamwala AH, Mellmer MA, Huang K, Houtman CJ, Labbé N, Harper DP, Maravelias C, Runge T, Dumesic JA (2017) Sci Adv 3:1–7
19. Huang Y, Duan Y, Qiu S, Wang M, Ju C, Cao H, Fang Y, Tan T (2017) Sustain Energy Fuels 2:637–647
20. Huber GW, Iborra S, Corma A (2006) Chem Rev 106:4044–4098
21. Bidlack JE, Dashek WV (2016) Plant cells and their organelles. Wiley, Chichester, pp 209–238
22. Clark JH, Deswarte FEI (2008) Introduction to chemicals from biomass. Wiley, Chichester, pp 1–20
23. Kudakasseril Kurian J, Raveendran Nair G, Hussain A, Vijaya Raghavan GS (2013) Renew Sustain Energy Rev 25:205–219

24. Himmel ME, Ding SY, Johnson DK, Adney WS, Nimlos MR, Brady JW, Foust TD (2007) Science 315:804–807
25. DeMartini JD, Pattathil S, Miller JS, Li H, Hahn MG, Wyman CE (2013) Energy Environ Sci 6:898–909
26. Davison BH, Parks J, Davis MF, Donohoe BS (2013) Aqueous pretreatment of plant biomass for biological and chemical conversion to fuels and chemicals. Wiley, Chichester, pp 23–38
27. Kumar P, Barrett DM, Delwiche MJ, Stroeve P (2009) Ind Eng Chem Res 48:3713–3729
28. Saini JK, Saini R, Tewari L (2015) 3 Biotech 5:337–353
29. Zakzeski J, Jongerius AL, Bruijnincx PCA, Weckhuysen BM (2012) Chemsuschem 5:1602–1609
30. Klemm D, Heublein B, Fink H-P, Bohn A (2005) Angew Chem Int Ed 44:3358–3393
31. Heinze T (2016) Cellulose chemistry and properties: fibers, nanocelluloses and advanced materials. Rojas OJ (ed) Springer International Publishing, Cham, 2016, pp 1–52
32. Schädel C, Blöchl A, Richter A, Hoch G (2010) Plant Physiol Biochem 48:1–8
33. Girio FM, Fonseca C, Carvalheiro F, Duarte LC, Marques S, Bogel-Lukasik R (2010) Bioresour Technol 101:4775–4800
34. Huang F (2014) Materials for biofuels, World Scientific, pp 1–26
35. U.S Department of Energy Genome Programs image gallery, http://genomics.energy.gov (18/4/2017)
36. Schutyser W, Renders T, Van den Bossche G, Van den Bosch S, Koelewijn S-F, Ennaert T, Sels BF (2017) Nanotechnology in catalysis. Wiley, Chichester, pp 537–584
37. Zakzeski J, Bruijnincx PCA, Jongerius AL, Weckhuysen BM (2010) Chem Rev 110:3552–3599
38. Chundawat SPS, Beckham GT, Himmel ME, Dale BE (2011) Ann Rev Chem Biomol Eng 2:121–145
39. Yuan T-Q, Sun S-N, Xu F, Sun R-C (2011) J Agric Food Chem 59:10604–10614
40. Ralph J, Lundquist K, Brunow G, Lu F, Kim H, Schatz PF, Marita JM, Hatfield RD, Ralph SA, Christensen JH, Boerjan W (2004) Phytochem Rev 3:29–60
41. Vanholme R, Demedts B, Morreel K, Ralph J, Boerjan W (2010) Plant Physiol 153:895–905
42. Laskar DD, Yang B, Wang H, Lee J (2013) Biofuels. Bioprod Biorefining 7:602–626
43. Mottiar Y, Vanholme R, Boerjan W, Ralph J, Mansfield SD (2016) Curr Opin Biotechnol 37:190–200
44. Liu E, Das L, Zhao B, Crocker M, Shi J (2017) BioEnergy Res
45. Brunow G (2008) Biorefineries-Industrial Processes and Products. Wiley, Chichester, pp 151–163
46. Van den Bosch S, Schutyser W, Vanholme R, Driessen T, Koelewijn SF, Renders T, De Meester B, Huijgen WJJ, Dehaen W, Courtin CM, Lagrain B, Boerjan W, Sels BF (2015) Energy Environ Sci 8:1748–1763
47. del Rio JC, Rencoret J, Prinsen P, Martinez AT, Ralph J, Gutierrez A (2012) J Agric Food Chem 60:5922–5935
48. Kamm B, Kamm M, Gruber PR, Kromus S (2008) Biorefineries-industrial processes and products. Wiley, Chichester, pp 1–40
49. Mikkola J-P, Sklavounos E, King AWT, Virtanen P (2016) Ionic liquids in the biorefinery concept: challenges and perspectives. The Royal Society of Chemistry, pp 1–37
50. Balat M, Balat H (2009) Appl Energy 86:2273–2282
51. Sánchez ÓJ, Cardona CA (2008) Biores Technol 99:5270–5295
52. Li M, Pu Y, Ragauskas AJ (2016) Front Chem 4:45
53. Zhao X, Zhang L, Liu D (2012) Biofuels. Bioprod Biorefining 6:465–482
54. Lora JH (2016) Quality living through chemurgy and green chemistry. Lau PCK, Springer, pp 221–261
55. Sjöström E (1993) Wood chemistry, 2nd edn. Academic Press, San Diego, pp 114–164
56. Gierer J (1985) Wood Sci Technol 19:289–312
57. Gierer J, Norén I, Wännström S (1987) Journal 41:79
58. Gellerstedt G (2009) Pulping chemistry and technology, pp 91–120
59. Chakar FS, Ragauskas AJ (2004) Ind Crops Prod 20:131–141
60. Gellerstedt G (2015) Ind Crops Prod 77:845–854
61. Ragnar M, Henriksson G, Lindström ME, Wimby M, Blechschmidt J, Heinemann S (2014) Ullmann's encyclopedia of industrial chemistry. Wiley, Chichester
62. Calvo-Flores FG, Dobado JA, Isac-García J, Martín-MartíNez FJ (2015) Lignin and lignans as renewable raw materials. Wiley, Chichester, pp 113–144
63. Azadi P, Inderwildi OR, Farnood R, King DA (2013) Renew Sustain Energy Rev 21:506–523

64. Sixta H (1998) Lenzinger Berichte, 18–27
65. Lora J (2008) Monomers, polymers and composites from renewable resources. Gandini MNB (ed) Elsevier, Amsterdam, pp 225–241
66. Aro T, Fatehi P (2017) Chemsuschem 10:1861–1877
67. Gillet S, Aguedo M, Petitjean L, Morais ARC, da Costa Lopes AM, Lukasik RM, Anastas P (2017) Green Chem
68. Anderson S, Dimmel D, Izsak P (2003) J Wood Chem Technol 23:141–159
69. Vishtal AG, Kraslawski A (2011) BioResources 6:3547–3568
70. Luterbacher JS, Martin Alonso D, Dumesic JA (2014) Green Chem 16:4816–4838
71. Katzen R, Schell DJ (2008) Biorefineries-industrial processes and products. Wiley, pp 129–138
72. Pu Y, Hu F, Huang F, Davison BH, Ragauskas AJ (2013) Biotechnol Biofuels 6:15
73. Pu Y, Hu F, Huang F, Ragauskas AJ (2015) BioEnergy Res 8:992–1003
74. Sturgeon MR, Kim S, Lawrence K, Paton RS, Chmely SC, Nimlos M, Foust TD, Beckham GT (2013) ACS Sustain Chem Eng 2:472–485
75. Adler E (1977) Wood Sci Technol 11:169–218
76. Deuss PJ, Scott M, Tran F, Westwood NJ, de Vries JG, Barta K (2015) J Am Chem Soc 137:7456–7467
77. Kulka M, Fisher HE, Baker SB, Hibbert H (1944) J Am Chem Soc 66:39–41
78. Shuai L, Saha B (2017) Green Chem 19:3752–3758
79. Shuai L, Amiri MT, Questell-Santiago YM, Héroguel F, Li Y, Kim H, Meilan R, Chapple C, Ralph J, Luterbacher JS (2016) Science 354:329–333
80. Imai T, Yokoyama T, Matsumoto Y (2011) J Wood Sci 57:219–225
81. Saeman JF, Bubl JL, Harris EE (1945) Ind Eng Chem Anal Ed 17:35–37
82. Bergius F (1937) Ind Eng Chem 29:247–253
83. C. W. Dence, in *Methods in Lignin Chemistry*, eds. S. Lin and C. Dence, Springer Berlin Heidelberg, 1992, ch. 3, pp 33–61
84. Courtin CM, Van den Broeck H, Delcour JA (2000) J Chromatogr A 866:97–104
85. Zheng Y, Zhao J, Xu F, Li Y (2014) Prog Energy Combust Sci 42:35–53
86. Saeman JF (1945) Ind Eng Chem 37:43–52
87. Hu F, Ragauskas A (2014) RSC Adv 4:4317–4323
88. Hu F, Jung S (2012) Ragauskas, Bioresour Technol 117
89. van Zandvoort I, Koers EJ, Weingarth M, Bruijnincx PCA, Baldus M, Weckhuysen BM (2015) Green Chem 17:4383–4392
90. Nitsos CK, Matis KA, Triantafyllidis KS (2013) Chem Sus Chem 6:110–122
91. Nitsos CK, Choli-Papadopoulou T, Matis KA, Triantafyllidis KS (2016) ACS Sustain Chem Eng 4:4529–4544
92. Sannigrahi P, Ragauskas AJ (2013) Aqueous pretreatment of plant biomass for biological and chemical conversion to fuels and chemicals, Wiley, pp 201–222
93. Zhao X, Cheng K, Liu D (2009) Appl Microbiol Biotechnol 82:815–827
94. Li M-F, Yang S, Sun R-C (2016) Biores Technol 200:971–980
95. Zhang Z, Harrison MD, Rackemann DW, Doherty WOS, O'Hara IM (2016) Green Chem 18:360–381
96. Huijgen WJJ, Telysheva G, Arshanitsa A, Gosselink RJA, de Wild PJ (2014) Ind Crops Prod 59:85–95
97. Wildschut J, Smit AT, Reith JH, Huijgen WJJ (2013) Biores Technol 135:58–66
98. Hallac BB, Sannigrahi P, Pu Y, Ray M, Murphy RJ, Ragauskas AJ (2010) Ind Eng Chem Res 24:2723–2732
99. Snelders J, Dornez E, Benjelloun-Mlayah B, Huijgen WJJ, de Wild PJ, Gosselink RJA, Gerritsma J, Courtin CM (2014) Biores Technol 156:275–282
100. Abdelkafi F, Ammar H, Rousseau B, Tessier M, El Gharbi R, Fradet A (2011) Biomacromol 12:3895–3902
101. Grande PM, Viell J, Theyssen N, Marquardt W, Dominguez de Maria P, Leitner W (2015) Green Chem 17:3533–3539
102. Quesada-Medina J, López-Cremades FJ, Olivares-Carrillo P (2010) Biores Technol 101:8252–8260
103. Alriols MG, Tejado A, Blanco M, Mondragon I, Labidi J (2009) Chem Eng J 148:106–114
104. Katahira R, Mittal A, McKinney K, Ciesielski PN, Donohoe BS, Black SK, Johnson DK, Biddy MJ, Beckham GT (2014) ACS Sustain Chem Eng 2:1364–1376
105. Huijgen WJJ, Reith JH, den Uil H (2010) Ind Eng Chem Res 49:10132–10140

106. Deuss PJ, Lancefield CS, Narani A, de Vries JG, Westwood NJ, Barta K (2017) Green Chem 19:2774–2782
107. Van den Bosch S, Renders T, Kennis S, Koelewijn SF, Van den Bossche G, Vangeel T, Deneyer A, Depuydt D, Courtin CM, Thevelein JM, Schutyser W, Sels BF (2017) Green Chem 19:3313–3326
108. Minami E, Saka S (2003) J Wood Sci 49:0073–0078
109. Minami E, Saka S (2005) J Wood Sci 51:395–400
110. Minami E, Kawamoto H, Saka S (2003) J Wood Sci 49:158–165
111. Bouxin FP, McVeigh A, Tran F, Westwood NJ, Jarvis MC, Jackson SD (2015) Green Chem 17:1235–1242
112. Lancefield CS, Rashid GMM, Bouxin F, Wasak A, Tu W-C, Hallett J, Zein S, Rodríguez J, Jackson SD, Westwood NJ, Bugg TDH (2016) ACS Sustain Chem Eng 4:6921–6930
113. Si X, Lu F, Chen J, Lu R, Huang Q, Jiang H, Taarning E, Xu J (2017) Green Chem
114. Ralph SR, Ralph J, Landucci LL. NMR database of lignin and cell wall model compounds. http://ars.usda.gov/Services/docs.htm?docid=10491
115. Ralph J, Landucci LL (2011) Lignin and lignans: advances in chemistry. Heitner C, Dimmel DR, Schmidt JA. CRC Press, pp 137–244
116. Rolando C, Monties B, Lapierre C (1992) Methods in Lignin Chemistry. Lin SY, Dence CW (eds) Springer, Berlin, pp 334–349
117. Galkin M, Di Francesco D, Edlund U, Samec JSM (2017) Faraday discussions
118. Lancefield CS, Panovic I, Deuss PJ, Barta K, Westwood NJ (2017) Green Chem 19:202–214
119. Zaheer M, Hermannsdörfer J, Kretschmer WP, Motz G, Kempe R (2014) Chemcatchem 6:91–95
120. Galkin MV, Sawadjoon S, Rohde V, Dawange M, Samec JSM (2014) ChemCatChem 6:179–184
121. Parsell TH, Owen BC, Klein I, Jarrell TM, Marcum CL, Haupert LJ, Amundson LM, Kenttamaa HI, Ribeiro F, Miller JT, Abu-Omar MM (2013) Chem Sci 4:806–813
122. Feghali E, Carrot G, Thuery P, Genre C, Cantat T (2015) Energy Environ Sci 8:2734–2743
123. Luterbacher JS, Azarpira A, Motagamwala AH, Lu F, Ralph J, Dumesic JA (2015) Energy Environ Sci 8:2657–2663
124. Kloekhorst A, Shen Y, Yie Y, Fang M, Heeres HJ (2015) Biomass Bioenerg 80:147–161
125. Kumar CR, Anand N, Kloekhorst A, Cannilla C, Bonura G, Frusteri F, Barta K, Heeres HJ (2015) Green Chem 17:4921–4930
126. Oasmaa A, Alén R, Meier D (1993) Biores Technol 45:189–194
127. Huang X, Korányi TI, Boot MD, Hensen EJM (2014) Chemsuschem 7:2276–2288
128. Huang X, Koranyi TI, Boot MD, Hensen EJM (2015) Green Chem 17:4941–4950
129. Torr KM, van de Pas DJ, Cazeils E, Suckling ID (2011) Biores Technol 102:7608–7611
130. Barta K, Warner GR, Beach ES, Anastas PT (2014) Green Chem 16:191–196
131. Molinari V, Clavel G, Graglia M, Antonietti M, Esposito D (2016) ACS Catal 6:1663–1670
132. Ratcliff MA, Johnson DK, Posey FL, Chum HL (1988) Appl Biochem Biotechnol 17:151–160
133. Kloekhorst A, Heeres HJ (2015) ACS Sustain Chem Eng 3:1905–1914
134. Kloekhorst A, Heeres HJ (2016) Catal Sci Technol 6:7053–7067
135. Kasakov S, Shi H, Camaioni DM, Zhao C, Barath E, Jentys A, Lercher JA (2015) Green Chem 17:5079–5090
136. Wang H, Ruan H, Feng M, Qin Y, Job H, Luo L, Wang C, Engelhard MH, Kuhn E, Chen X, Tucker MP, Yang B (2017) Chemsuschem 10:1846–1856
137. Zhang J, Teo J, Chen X, Asakura H, Tanaka T, Teramura K, Yan N (2014) ACS Catal 4:1574–1583
138. Ma R, Hao W, Ma X, Tian Y, Li Y (2014) Angew Chem Int Ed 53:7310–7315
139. Li C, Zhao X, Wang A, Huber GW, Zhang T (2015) Chem Rev 115:11559–11624
140. Ma R, Xu Y, Zhang X (2015) Chemsuschem 8:24–51
141. Ma R, Guo M, Zhang X (2017) Catal Today
142. Gierer J (1986) Wood Sci Technol 20:1–33
143. Demesa AG, Laari A, Turunen I, Sillanpää M (2015) Chem Eng Technol 38:2270–2278
144. Fache M, Boutevin B, Caillol S (2016) ACS Sustain Chem Eng 4:35–46
145. Rodrigues Pinto PC, Borges da Silva EA, Rodrigues AE (2012) Biomass conversion: the interface of biotechnology, chemistry and materials science. Baskar C, Baskar S, Dhillon RS (eds) Springer Berlin, pp 381–420
146. Kagawa S, Rokugawa M (1971) Jpn Tappi J 25:506–511
147. Partenheimer W (2009) Adv Synth Catal 351:456–466
148. Voitl T, Rudolf von Rohr P (2008) ChemSusChem 1:763–769
149. Voitl T, Rohr PRV (2010) Ind Eng Chem Res 49:520–525

150. Deng W, Zhang H, Wu X, Li R, Zhang Q, Wang Y (2015) Green Chem 17:5009–5018
151. Tarabanko VE, Hendogina YV, Petuhov DV, Pervishina EP (2000) React Kinet Catal Lett 69:361–368
152. Tarabanko VE, Petukhov DV (2003) Chem Sustain Dev 11:655–667
153. Lange H, Decina S, Crestini C (2013) Eur Polymer J 49:1151–1173
154. Das L, Kolar P, Sharma-Shivappa R (2012) Biofuels 3:155–166
155. Werhan H, Assmann N, Rudolf von Rohr P (2013) Chem Eng Process 73:29–37
156. Xiang Q, Lee YY (2000) Appl Biochem Biotechnol 84:153–162
157. Ma R, Guo M, Zhang X (2014) Chemsuschem 7:412–415
158. Hasegawa I, Inoue Y, Muranaka Y, Yasukawa T, Mae K (2011) Energy Fuels 25:791–796
159. Long J, Zhang Q, Wang T, Zhang X, Xu Y, Ma L (2014) Biores Technol 154:10–17
160. Güvenatam B, Heeres EHJ, Pidko EA, Hensen EJM (2016) Catal Today 259:460–466
161. Hepditch MM, Thring RW (2000) Can J Chem Eng 78:226–231
162. Katahira R, Mittal A, McKinney K, Chen X, Tucker MP, Johnson DK, Beckham GT (2016) ACS Sustain Chem Eng 4:1474–1486
163. Beauchet R, Monteil-Rivera F, Lavoie JM (2012) Biores Technol 121:328–334
164. Zhang X, Zhang Q, Long J, Xu Y, Wang T, Ma L, Li Y (2014) BioResources 9:3347–3360
165. Deepa AK, Dhepe PL (2014) RSC Adv 4:12625–12629
166. Lee H-S, Jae J, Ha J-M, Suh DJ (2016) Biores Technol 203:142–149
167. Ye Y, Zhang Y, Fan J, Chang J (2012) Ind Eng Chem Res 51:103–110
168. Ye Y, Fan J, Chang J (2012) J Anal Appl Pyrol 94:190–195
169. Onwudili JA, Williams PT (2014) Green Chem 16:4740–4748
170. Jiang Z, He T, Li J, Hu C (2014) Green Chem 16:4257–4265
171. Bridgwater AV (2012) Biomass Bioenerg 38:68–94
172. Liu C, Wang H, Karim AM, Sun J, Wang Y (2014) Chem Soc Rev 43:7594–7623
173. Mu W, Ben H, Ragauskas A, Deng Y (2013) BioEnergy Res 6:1183–1204
174. Yu Y, Li X, Su L, Zhang Y, Wang Y, Zhang H (2012) Appl Catal A 447:115–123
175. Mihalcik DJ, Mullen CA, Boateng AA (2011) J Anal Appl Pyrol 92:224–232
176. Ma Z, Troussard E, van Bokhoven JA (2012) Appl Catal A 423:130–136
177. Bond JQ, Upadhye AA, Olcay H, Tompsett GA, Jae J, Xing R, Alonso DM, Wang D, Zhang T, Kumar R, Foster A, Sen SM, Maravelias CT, Malina R, Barrett SRH, Lobo R, Wyman CE, Dumesic JA, Huber GW (2014) Energy Environ Sci 7:1500–1523
178. Zhou G, Jensen PA, Le DM, Knudsen NO, Jensen AD (2016) Green Chem 18:1965–1975
179. Thilakaratne R, Tessonnier J-P, Brown RC (2016) Green Chem 18:2231–2239
180. Lazaridis PA, Fotopoulos AP, Karakoulia SA, Triantafyllidis KS (2018) Front Chem 6:1–21
181. Zhao Y, Deng L, Liao B, Fu Y, Guo Q-X (2010) Energy Fuels 24:5735–5740
182. Jackson MA, Compton DL, Boateng AA (2009) J Anal Appl Pyrol 85:226–230
183. Ma Z, Custodis V, van Bokhoven JA (2014) Catal Sci Technol 4:766–772
184. Ennaert T, Van Aelst J, Dijkmans J, De Clercq R, Schutyser W, Dusselier M, Verboekend D, Sels BF (2016) Chem Soc Rev 45:584–611
185. Zhang M, Resende FLP, Moutsoglou A (2014) Fuel 116:358–369
186. Renders T, Van den Bosch S, Koelewijn SF, Schutyser W, Sels BF (2017) Energy Environ Sci 10:1551–1557
187. Barry AJ, Peterson FC, King AJ (1936) J Am Chem Soc 58:333–337
188. Yan MM, Purves CB (1956) Can J Chem 34:1582–1590
189. Bouxin FP, David Jackson S, Jarvis MC (2014) Biores Technol 162:236–242
190. Singh S, Cheng G, Sathitsuksanoh N, Wu D, Varanasi P, George A, Balan V, Gao X, Kumar R, Dale BE, Wyman CE, Simmons BA (2015) Front Energy Res 2
191. da Costa Sousa L, Foston M, Bokade V, Azarpira A, Lu F, Ragauskas AJ, Ralph J, Dale B, Balan V (2016) Green Chem 18:4205–4215
192. Mittal A, Katahira R, Donohoe BS, Pattathil S, Kandemkavil S, Reed ML, Biddy MJ, Beckham GT (2017) ACS Sustain Chem Eng 5:2544–2561
193. Balan V, Bals B, Chundawat SP, Marshall D, Dale BE (2009) Methods Mol Biol 581:61–77
194. Chundawat SPS, Donohoe BS, da Costa Sousa L, Elder T, Agarwal UP, Lu F, Ralph J, Himmel ME, Balan V, Dale BE (2011) Energy Environ Sci 4:973–984
195. Kim JS, Lee YY, Kim TH (2016) Bioresour Technol 199:42–48

196. Chundawat SPS, Bals B, Campbell T, Sousa L, Gao D, Jin M, Eranki P, Garlock R, Teymouri F, Balan V, Dale BE (2013) Aqueous pretreatment of plant biomass for biological and chemical conversion to fuels and chemicals. Wiley, Chichester, pp 169–200
197. da Costa Sousa L, Jin M, Chundawat SPS, Bokade V, Tang X, Azarpira A, Lu F, Avci U, Humpula J, Uppugundla N, Gunawan C, Pattathil S, Cheh AM, Kothari N, Kumar R, Ralph J, Hahn MG, Wyman CE, Singh S, Simmons BA, Dale BE, Balan V (2016) Energy Environ Sci 9:1215–1223
198. Chundawat SPS, Bellesia G, Uppugundla N, da Costa Sousa L, Gao D, Cheh AM, Agarwal UP, Bianchetti CM, Phillips GN, Langan P, Balan V, Gnanakaran S, Dale BE (2011) J Am Chem Soc 133:11163–11174
199. Kim TH, Lee YY (2005) Biores Technol 96:2007–2013
200. Kim TH, Kim JS, Sunwoo C, Lee YY (2003) Biores Technol 90:39–47
201. Yang B, Wyman CE (2008) Biofuels Bioprod Biorefining-Biofpr 2:26–40
202. Yoon HH, Wu ZW, Lee YY (1995) Appl Biochem Biotechnol 51:5–19
203. Brandt A, Grasvik J, Hallett JP, Welton T (2013) Green Chem 15:550–583
204. George A, Tran K, Morgan TJ, Benke PI, Berrueco C, Lorente E, Wu BC, Keasling JD, Simmons BA, Holmes BM (2011) Green Chem 13:3375–3385
205. Badgujar KC, Bhanage BM (2015) Biores Technol 178:2–18
206. Brandt-Talbot A, Gschwend FJV, Fennell PS, Lammens TM, Tan B, Weale J, Hallett JP (2017) Green Chem 19:3078–3102
207. Weigand L, Mostame S, Brandt-Talbot A, Welton T, Hallett JP (2017) Faraday Discuss 18
208. Zhang Q, De Oliveira Vigier K, Royer S, Jerome F (2012) Chem Soc Rev 41:7108–7146
209. Petkovic M, Seddon KR, Rebelo LPN, Silva C (2011) Pereira. Chem Soc Rev 40:1383–1403
210. Yan L, Zhang L, Yang B (2014) Biotechnol Biofuels 7:76
211. Wang H, Ben H, Ruan H, Zhang L, Pu Y, Feng M, Ragauskas AJ, Yang B (2017) ACS Sustain Chem Eng 5:1824–1830
212. Liu C, Wyman CE (2003) Ind Eng Chem Res 42:5409–5416
213. Bhagia S, Li H, Gao X, Kumar R, Wyman CE (2016) Biotechnol Biofuels 9:245
214. Luterbacher JS, Rand JM, Alonso DM, Han J, Youngquist JT, Maravelias CT, Pfleger BF, Dumesic JA (2014) Science 343:277–280
215. Alonso DM, Wettstein SG, Mellmer MA, Gurbuz EI, Dumesic JA (2013) Energy Environ Sci 6:76–80
216. Mellmer MA, Sener C, Gallo JMR, Luterbacher JS, Alonso DM, Dumesic JA (2014) Angew Chem Int Ed 53:11872–11875
217. Shuai L, Luterbacher J (2016) Chemsuschem 9:133–155
218. Shuai L, Questell-Santiago YM, Luterbacher JS (2016) Green Chem 18:937–943
219. Barakat A, Mayer-Laigle C, Solhy A, Arancon RAD, de Vries H, Luque R (2014) RSC Advances 4:48109–48127
220. Rencoret J, Marques G, Gutiérrez A, Nieto L, Jiménez-Barbero J, Martínez ÁT, del Río JC (2009) Ind Crops Prod 30:137–143
221. El Hage R, Brosse N, Chrusciel L, Sanchez C, Sannigrahi P, Ragauskas A (2009) Polym Degrad Stab 94:1632–1638
222. Guerra A, Mendonça R, Ferraz A, Lu F, Ralph J (2004) Appl Environ Microbiol 70:4073–4078
223. Tolbert A, Akinosho H, Khunsupat R, Naskar AK, Ragauskas AJ (2014) Biofuels Bioprod Biorefining 8:836–856
224. Hu Z, Yeh T-F, Chang HM, Matsumoto Y, Kadla John F (2006) Journal 60:389
225. Chang HM, Cowling Ellis B, Brown W (1975) Journal 29:153
226. Guerra A, Filpponen I, Lucia LA, Argyropoulos DS (2006) J Agric Food Chem 54:9696–9705
227. Guerra A, Filpponen I, Lucia LA, Saquing C, Baumberger S, Argyropoulos DS (2006) J Agric Food Chem 54:5939–5947
228. van der Klashorst GH (1989) Lignin. American Chemical Society, vol. 397, ch. 26, pp 346–360
229. Wu L, Talebi AM (2018) Angewandte Chemie International Edition 57:1356–1360
230. Bower JR, Cooke LM, Hibbert H (1943) J Am Chem Soc 65:1192–1195
231. Brewer CP, Cooke LM, Hibbert H (1948) J Am Chem Soc 70:57–59
232. Pepper JM, Brounstein CJ, Shearer DA (1951) J Am Chem Soc 73:3316–3319
233. Pepper JM, Supathna P (1978) Can J Chem 56:899–902
234. Song Q, Wang F, Cai JY, Wang YH, Zhang JJ, Yu WQ, Xu J (2013) Energy Environ Sci 6:994–1007

235. Anderson EM, Stone ML, Katahira R, Reed M, Beckham GT, Román-Leshkov Y (2017) Joule 1:613–622
236. Kumaniaev I, Subbotina E, Savmarker J, Larhed M, Galkin MV, Samec JSM (2017) Green Chem 19:5767–5771
237. Anderson EM, Katahira R, Reed M, Resch MG, Karp EM, Beckham GT, Román-Leshkov Y (2016) ACS Sustain Chem Eng
238. Renders T, Schutyser W, Van den Bosch S, Koelewijn S-F, Vangeel T, Courtin CM, Sels BF (2016) Acs Catal 6:2055–2066
239. Ferrini P, Rezende CA, Rinaldi R (2016) Chemsuschem 9:3171–3180
240. Yan N, Zhao C, Dyson PJ, Wang C, Liu LT, Kou Y (2008) Chemsuschem 1:626–629
241. Galkin MV, Samec JSM (2016) Chemsuschem 9:1544–1558
242. Galkin MV, Samec JSM (2014) Chemsuschem 7:2154–2158
243. Parsell T, Yohe S, Degenstein J, Jarrell T, Klein I, Gencer E, Hewetson B, Hurt M, Kim JI, Choudhari H, Saha B, Meilan R, Mosier N, Ribeiro F, Delgass WN, Chapple C, Kenttamaa HI, Agrawal R, Abu-Omar MM (2015) Green Chem 17:1492–1499
244. Ferrini P, Rinaldi R (2014) Angewandte Chemie-International Edition 53:8634–8639
245. Luo H, Klein IM, Jiang Y, Zhu H, Liu B, Kenttämaa HI, Abu-Omar MM (2016) ACS Sustain Chem Eng 4:2316–2322
246. Schutyser W, Van den Bosch S, Renders T, De Boe T, Koelewijn SF, Dewaele A, Ennaert T, Verkinderen O, Goderis B, Courtin CM, Sels BF (2015) Green Chem 17:5035–5045
247. Renders T, Van den Bosch S, Vangeel T, Ennaert T, Koelewijn S-F, Van den Bossche G, Courtin CM, Schutyser W, Sels BF (2016) ACS Sustain Chem Eng 4:6894–6904
248. Huang X, Morales Gonzalez OM, Zhu J, Koranyi TI, Boot MD, Hensen EJM (2017) Green Chem 19:175–187
249. Huang X, Zhu J, Korányi TI, Boot MD, Hensen EJM (2016) Chemsuschem
250. Galkin MV, Smit AT, Subbotina E, Artemenko KA, Bergquist J, Huijgen WJJ, Samec JSM (2016) Chemsuschem 9:3280–3287
251. Anderson EM, Stone ML, Hülsey MJ, Beckham GT, Román-Leshkov Y (2018) ACS Sustain Chem Eng 6:7951–7959
252. Van den Bosch S, Schutyser W, Koelewijn S-F, Renders T, Courtin CM, Sels BF (2015) Chem Commun
253. Pepper JM, Lee YW (1969) Can J Chem 47:723–727
254. Li C, Zheng M, Wang A, Zhang T (2012) Energy Environ Sci 5:6383–6390
255. Matson TD, Barta K, Iretskii AV, Ford PC (2011) J Am Chem Soc 133:14090–14097
256. Barta K, Ford PC (2014) Acc Chem Res 47:1503–1512
257. Op de Beeck B, Dusselier M, Geboers J, Holsbeek J, Morre E, Oswald S, Giebeler L, Sels BF (2015) Energy Environ Sci 8:230–240
258. Deneyer A, Ennaert T, Cavents G, Dijkmans J, Vanneste J, Courtin CM, Dusselier M, Sels BF (2016) Green Chem 18:5594–5606
259. Xia Q, Chen Z, Shao Y, Gong X, Wang H, Liu X, Parker SF, Han X, Yang S, Wang Y (2016) 7:11162
260. Chen C-L (1992) Methods in lignin chemistry. Lin SY, Dence CW (eds) Springer, Berlin, pp 301–321
261. Pepper JM, Casselman BW, Karapally JC (1967) Can J Chem 45:3009–3012
262. Koropachinskaya N, Tarabanko V, Chernyak M (2003) Chem Plant Raw Mater (Russia), 9–14
263. Kuznetsov BN, Kuznetsova SA, Danilov VG, Tarabanko VE (2005) Chem Sustain Dev 13:531–539
264. Behling R, Valange S, Chatel G (2016) Green Chem 18:1839–1854
265. Evans RJ, Milne TA, Soltys MN (1986) J Anal Appl Pyrol 9:207–236
266. Kuroda K-I, Inoue Y, Sakai K (1990) J Anal Appl Pyrol 18:59–69
267. Kotake T, Kawamoto H, Saka S (2015) J Anal Appl Pyrol 113:57–64
268. Karp EM, Nimlos CT, Deutch S, Salvachua D, Cywar RM, Beckham GT (2016) Green Chem 18:4750–4760
269. Karp EM, Resch MG, Donohoe BS, Ciesielski PN, O'Brien MH, Nill JE, Mittal A, Biddy MJ, Beckham GT (2015) ACS Sustain Chem Eng 3:1479–1491
270. Linger JG, Vardon DR, Guarnieri MT, Karp EM, Hunsinger GB, Franden MA, Johnson CW, Chupka G, Strathmann TJ, Pienkos PT, Beckham GT (2014) Proc Natl Acad Sci 111:12013–12018
271. Kaiho A, Kogo M, Sakai R, Saito K, Watanabe T (2015) Green Chem 17:2780–2783

272. Jastrzebski R, Constant S, Lancefield CS, Westwood NJ, Weckhuysen BM, Bruijnincx PCA (2016) Chemsuschem 9:2074–2079
273. Koelewijn S-F, Van den Bosch S, Renders T, Schutyser W, Lagrain B, Smet M, Thomas J, Dehaen W, Van Puyvelde P, Witters H, Sels BF (2017) Green Chem 19:2561–2570
274. Koelewijn SF, Cooreman C, Renders T, Andecochea Saiz C, Van den Bosch S, Schutyser W, De Leger W, Smet M, Van Puyvelde P, Witters H, Van der Bruggen B, Sels BF (2018) Green Chem 20:1050–1058
275. Zhao C, Kou Y, Lemonidou AA, Li X, Lercher JA (2009) Angewandte Chemie-International Edition 48:3987–3990
276. Yan N, Yuan Y, Dykeman R, Kou Y, Dyson PJ (2010) Angew Chem Int Ed 49:5549–5553
277. Zhang W, Chen J, Liu R, Wang S, Chen L, Li K (2014) ACS Sustain Chem Eng 2:683–691
278. Luska KL, Migowski P, El-Sayed S, Leitner W (2015) Angew Chem Int Ed 54:15750–15755
279. Wang H, Wang H, Kuhn E, Tucker MP, Yang B (2017) ChemSusChem
280. Yohe SL, Choudhari HJ, Mehta DD, Dietrich PJ, Detwiler MD, Akatay CM, Stach EA, Miller JT, Delgass WN, Agrawal R, Ribeiro FH (2016) J Catal 344:535–552
281. Prasomsri T, Shetty M, Murugappan K, Roman-Leshkov Y (2014) Energy Environ Sci 7:2660–2669
282. Wang H, Male J, Wang Y (2013) ACS Catal 3:1047–1070
283. Sun J, Karim AM, Zhang H, Kovarik L, Li XS, Hensley AJ, McEwen J-S, Wang Y (2013) J Catal 306:47–57
284. Cao Z, Engelhardt J, Dierks M, Clough MT, Wang G-H, Heracleous E, Lappas A, Rinaldi R, Schüth F (2017) Angew Chem Int Ed 56:2334–2339
285. Joshi N, Lawal A (2013) Ind Eng Chem Res 52:4049–4058
286. Verboekend D, Liao Y, Schutyser W, Sels BF (2016) Green Chem 18:297–306
287. Bai Z, Phuan WC, Ding J, Heng TH, Luo J, Zhu Y (2016) ACS Catal 6:6141–6145
288. Schutyser W, Van den Bossche G, Raaffels A, Van den Bosch S, Koelewijn S-F, Renders T, Sels BF (2016) ACS Sustain Chem Eng 4:5336–5346
289. Schutyser W, Van den Bosch S, Dijkmans J, Turner S, Meledina M, Van Tendeloo G, Debecker DP, Sels BF (2015) Chemsuschem 8:1805–1818
290. Wang XY, Rinaldi R (2012) Energy Environ Sci 5:8244–8260
291. Nakagawa Y, Ishikawa M, Tamura M, Tomishige K (2014) Green Chem 16:2197–2203
292. Ishikawa M, Tamura M, Nakagawa Y, Tomishige K (2016) Appl Catal B 182:193–203
293. Dijkmans J, Schutyser W, Dusselier M, Sels BF (2016) Chem Commun 52:6712–6715
294. Jiménez JI, Miñambres B, García JL, Díaz E (2002) Environ Microbiol 4:824–841
295. Barbe V, Vallenet D, Fonknechten N, Kreimeyer A, Oztas S, Labarre L, Cruveiller S, Robert C, Duprat S, Wincker P, Ornston LN, Weissenbach J, Marliere P, Cohen GN, Medigue C (2004) Nucleic Acids Res 32:5766–5779
296. Masai E, Kamimura N, Kasai D, Oguchi A, Ankai A, Fukui S, Takahashi M, Yashiro I, Sasaki H, Harada T, Nakamura S, Katano Y, Narita-Yamada S, Nakazawa H, Hara H, Katayama Y, Fukuda M, Yamazaki S, Fujita N (2012) J Bacteriol 194:534–535
297. Bugg TDH, Ahmad M, Hardiman EM, Rahmanpour R (2011) Nat Prod Rep 28:1883–1896
298. Masai E, Katayama Y, Fukuda M (2007) Biosci Biotechnol Biochem 71:1–15
299. Chen GQ (2009) Chem Soc Rev 38:2434–2446
300. Vardon DR, Franden MA, Johnson CW, Karp EM, Guarnieri MT, Linger JG, Salm MJ, Strathmann TJ, Beckham GT (2015) Energy Environ Sci 8:617–628
301. Xie N-Z, Liang H, Huang R-B, Xu P (2014) Biotechnol Adv 32:615–622
302. Kosa M, Ragauskas AJ (2013) Green Chem 15:2070–2074

Topics in Current Chemistry (2019) 377:26
https://doi.org/10.1007/s41061-019-0251-6

REVIEW

Lignin Depolymerization to BTXs

Luis Serrano[1] · Juan Antonio Cecilia[2] · Cristina García-Sancho[2] · Araceli García[3]

Received: 27 May 2019 / Accepted: 31 August 2019 / Published online: 16 September 2019

Abstract
Lignin, one of the main components of lignocellulosic biomass, is the largest renewable source of aromatics on the planet and presents an extraordinary opportunity for being used in the production of bio-based products. It can be transformed for the substitution of aromatic chemical-derived petrol as BTXs. The wide range of applications that it can be obtained from BTXs building blocks makes the selective depolymerization of lignin a great scientific challenge. This review emphasizes the different strategies for the fragmentation of lignin to monomers or aromatics hydrocarbons. Thus, a by-product traditionally discarded or used for energy generation, it could be valorized into high added-value products.

Keywords Lignin · Depolymerization · BTXs · Catalysis

1 Introduction

Currently, the growing world energy demand and the depletion of fossil fuel reserves together with the environmental concerns lead to the development of alternative and renewable resources [1, 2]. In recent years, biomass has emerged as an

---

Chapter 7 was originally published as Serrano, L., Cecilia, J. A., García-Sancho, C. & García, A. Topics in Current Chemistry (2019) 377: 26. https://doi.org/10.1007/s41061-019-0251-6.

✉ Luis Serrano
iq3secal@uco.es

✉ Araceli García
qo2ganua@uco.es

1 Departamento de Química Inorgánica e Ingeniería Química, Universidad de Córdoba, Campus de Rabanales, Edificio Marie Curie, 14071 Córdoba, Spain

2 Departamento de Química Inorgánica, Cristalografía y Mineralogía (Unidad Asociada al ICP-CSIC), Facultad de Ciencias, Universidad de Málaga, Campus de Teatinos, 29071 Málaga, Spain

3 NANOVAL FQM-383 Research Group, Departamento de Química Orgánica, Universidad de Córdoba, 14071 Córdoba, Spain

attractive renewable source for the production of biofuels and valuable chemicals. Lignocellulose is the most promising type of biomass due to its abundance, with an estimated annual production of about 2×10^{11} tons. It is inexpensive and widely available, mainly from non-food lignocellulose such as agricultural and forest wastes [3–5]. Lignocellulosic biomass is mainly composed of polymeric carbohydrates (cellulose and hemicellulose) and aromatic polymer lignin with smaller amounts of pectin, proteins, and inorganic salts [6, 7]. Recently, the fractionation of lignocellulose to its main components (cellulose, hemicellulose, and lignin) has been widely studied; thus, these major constituents should be fully valorized in order to create economically feasible biorefineries. Lignin is the most underutilized component of lignocellulosic biomass and is generally employed for heat and power production by combustion [8]. However, the higher carbon content and the lower oxygen content of lignin-derived compounds allowed obtaining more thermally stable and higher energy-density fuels than other fuels derived from other biomass constituents [5]. Therefore, the valorization of lignin, the largest renewable source of aromatics, is required to reach the economic viability and environmental sustainability of a biorefinery [9–11].

Lignin, one of the three major components of lignocellulosic biomass, provides the structural reinforcement and resilience to plant tissues and plays critical roles in water and nutrient transport [7, 12]. Lignin is a three-dimensional amorphous hydrophobic polymer consisting of methoxylated phenylpropane (Fig. 1) [13].

Therefore, the lignin is mainly composed of three basic phenol derivatives or monolignols: *p*-coumaryl alcohol, coniferyl alcohol, and sinapyl alcohol. These monolignols are incorporated to lignin aromatic in the form of phenyl propanoids, namely *p*-hydroxyphenyl, guaiacyl, and syringyl units, respectively [10, 14, 15]. The

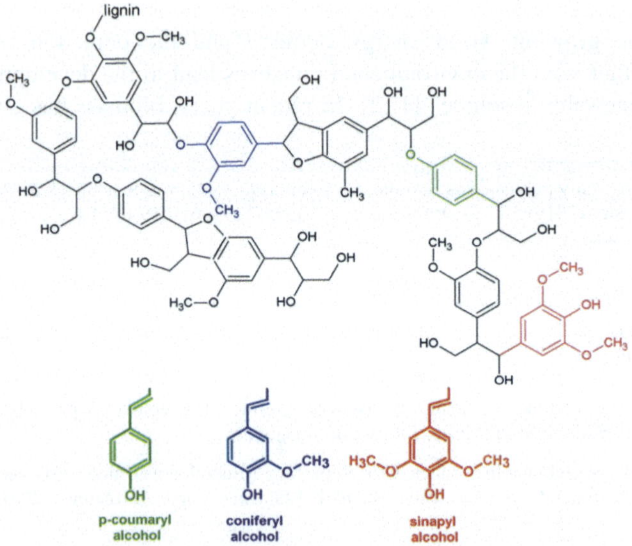

Fig. 1 Representative fragment of lignin structure and its main monolignols

relative distribution of these three subunits in the lignin macromolecules differs for different plan species. Thus, softwood lignin is exclusively composed of guaiacyl units, hardwood lignin is generally formed of guaiacyl and syringyl units, whereas grass lignins also contain *p*-hydroxyphenyl units in addition to the guaiacyl and syringyl units [15, 16]. Other compounds such as *p*-coumarate, ferulate, hydroxycinnamates, *p*-hydroxybenzoate, tricin, or acetate can also be found in lignin [11]. These differences in lignin composition complicate its valorization into chemicals and fuels, so lignin has been traditionally burnt to generate heat and power. However, lignin exhibits a great potential as a low-cost raw material in various sectors, such as cement, carbon fibers, polymers/resins, binders, foams, syngas, etc. [16, 17]. Likewise, lignin is the largest renewable source of aromatic compounds in such a way that it can be used as a starting material for the direct preparation of aromatic chemicals currently petroleum-derived such as BTXs (benzene, toluene, and xylene) [17–20].

BTXs are versatile building blocks employed currently for the petrochemical industry with numerous applications such as fuel additives in order to enhance octane number in gasoline, solvents, or even as feedstocks for the synthesis of plastics, including polyesters and polyurethanes, detergents, dyes, synthetic fibers, alkanes, etc. [19–22]. Nowadays, BTXs are mainly synthesized from fossil sources by the catalytic reforming of naphtha in petroleum refineries [23]. Considering the diminishing of petroleum reserves and the growing demand of BTXs combined with environmental problems provoked by fossil fuels, the use of lignin as a feedstock for sustainable and renewable BTX production has been considered of great interest in recent years [24–26]. The synthesis of BTXs requires the fragmentation of lignin in their respective monomers and subsequent selective removal of oxygen atoms by hydrodeoxygenation (HDO), dehydration, decarboxylation, etc. [17]. Most depolymerization-valorization processes studied for lignin are based on two main routes: a multi-step process involving converting biomass into platform chemicals and subsequent catalytic upgrading or the catalytic fast pyrolysis in one single reactor [16, 27–29]. The present review compiles the development of different processes for lignin depolymerization into aromatic hydrocarbons in order to obtain an efficient and sustainable supply of BTXs.

2 State of the Art

The huge interest of the scientific community in lignin and its depolymerization has led to many of research works. According to the SciFinder search tool, the first published document dealing with lignin depolymerization dates back to 1928, as a conference paper [30] about the production of phenol-derived products from the treatment of humic acids and lignin with different solvents. Since then, the scientific concern about the lignin challenge has grown significantly, resulting in more than 2000 documents related to lignin depolymerization (almost 200 patents, 300 conference papers, and 1600 journal articles).

As detailed in Fig. 2, more than 75% of this scientific production has been carried out during the last 10 years, showing a clear chronological change in the lignin

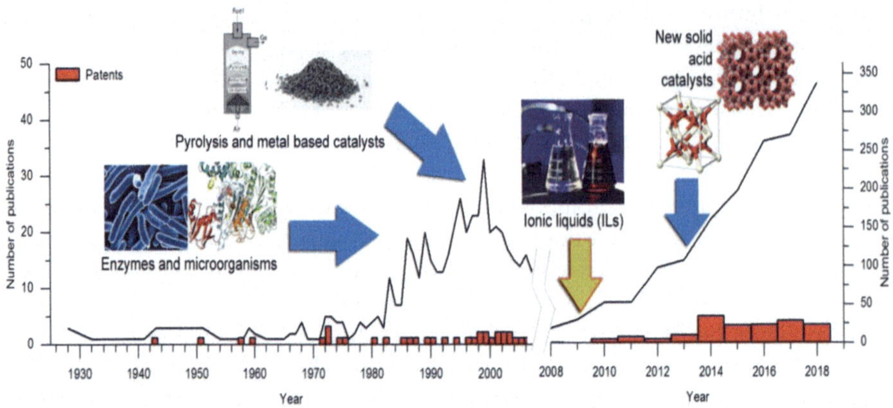

Fig. 2 Chronological study of fuels and BTX production by lignin depolymerization. (source: SciFinder®, data recorded on December 2018)

depolymerization strategies. Up to 1980, most of the works found in the literature applied simple, severe, low-efficiency procedures based on thermal, enzymatic, or alkaline treatment of lignin. In 1989, the scientific trend was heading towards the simultaneous use of oxidative reagents based on copper (II) or cerium (IV) ammonium nitrates [30]. Later, this approach was enhanced by the combination of pyrolysis and acid-catalyzed reaction pathways [31], multistep depolymerization techniques, the use of steam explosion for the improvement of hydrolytic depolymerization of lignin [32], or intensified treatments by using new reaction technologies [33]. The emergence of ionic liquids (ILs) as depolymerization hydrogen-donor solvents for phenol derivatives obtaining from lignin occurred in 1999 [34]. Further investigations have used these ILs together with amines or transition metals [35–37] as an improved synergetic alternative for obtaining lignin-derived fuels and phenol compounds.

In the last years, relevant works have been published proposing multi-step catalytic valorization of lignin to BTXs and fuels [38] using cubic CeO_2 nanoparticles [39] and zeolites such as HZSM-5, MCM-22, or HY [40]. More recent publications on lignin depolymerization propose the use of solid acid catalysts assisted by microwave heating [41] or of *tert*-butyl-hydroperoxide as oxidant agent, initiated by titanium grafted into the lignin structure [42].

The improved behavior in the research on BTXs obtaining from lignin has conducted the scientific community towards the industrial exploitation of this unique natural source of phenolic compounds. Thus, the first patent regarding lignin depolymerization [43] was published in 1943 about obtaining reactive lignin from the lignin produced by the saccharification of wood, addressing the issue of lignin treatment for obtaining an active substrate for further chemical transformations.

After that, different studies were patented, improving the yield, efficiency, and selectivity of the applied depolymerization method. Thus, during the 1970s, most of the industrial exploitation of lignin depolymerization consisted of the use of solvents, ozone, or high temperature [44]. However, new technologies and pathways were considered at the beginning of the 21st century for the production of fuels from lignin, such as basic catalysts or metal complexes [45, 46] followed by a selective hydrocracking step [47, 48]. More recently, the trend has advanced towards the use of enzymes [49, 50], ionic liquids [51], or novel solid nanocatalysts [52].

3 Plant Lignin Structure

The catalytic transformation of lignin into platform chemicals requires the necessity of a strict knowledge about lignin structure. Numerous studies have demonstrated the differences in the content and structure of lignin between types of plants, botanical species, and even morphological parts of the plant [53, 54]. Lignin acts as a glue between cellulose fibers and hemicelluloses, and also gives mechanical support to plant tissue and fibers, as well as physical strength and rigidity to the cell system.

With respect to the plant mass, lignin represents by weight about 30% in softwood, 20–25% in hardwood, and 10–15% in grasses [14, 55, 56]. Table 1 presents the variation of major linkages found in softwood, hardwood, and grass species. It is particularly remarkable that hardwood contains higher methoxyl content than softwood species [54, 57].

The main influence on the lignin reactivity consists of the content of these major linkages with the functional groups present in the structure. Moreover, the lignin functional groups constitute the key for its transformation into high added-value products [60]. Methoxyl, phenolic and aliphatic hydroxyl, benzyl alcohol, noncyclic benzyl ether, and carbonyl groups are the main ones. Thereby, typical values could be 1.09 for aliphatic hydroxyl, 0.26 for phenolic hydroxyl, and 0.2 for total carbonyl per C9 units [61].

4 Extraction and Isolation of Lignin. Repolymerization Phenomenon

The complex lignin structure is modified depending, apart from the original lignocellulosic biomass, also on the isolation method carried out for its recovery.

With respect to the extraction process, the main source of lignin is the pulp and paper industry (around 50 million tons/year) and the wastes obtained in bioethanol generation [14, 19, 26, 62]. These processes constitute a fractionation technology where the dissolution of lignin is involved, leaving cellulose and some part of the hemicelluloses as insoluble fraction. This group includes classic pulping processes as kraft, sulfite, organosolv, alkaline, and alternative processes such as the use

Table 1 Major linkages found in hardwood and softwood species [58, 59]

Linkages (%)	β–*O*–4	5–5	α–*O*–4	4–*O*–5	β-5	β-1	β–β
Softwood	45–50	9.5–11	6–8	3.5–4	9–12	7	2
Hardwood	60	4.5	6-8	6.5	6	7	3
Grasses	40	–	–	–	4	1	6

of ionic liquids (ILs). One second group with processes as enzymatic and Klason belongs to the methods where the cellulose and hemicelluloses are solubilized, leaving the lignin as a solid fraction [54].

Kraft process is the most common method used in a pulp and paper mill. This treatment, using a combination of aqueous sodium hydroxide and sodium sulfide, provokes the dissolution of lignin, modifying its structure by cleavage of most α-aryl ether and β-aryl ether bonds [13, 56, 63]. Kraft lignin usually presents high phenolic content, lower molecular weight than original lignin, hydrophobic character, and low sulfur content. Conversely, it partially contains hemicelluloses and repolymerization products in its structure. The works of Crestini et al. should be noted [64, 65] about the structural elucidation of kraft lignin. NMR techniques are presented as the most powerful tool for analyzing the heterogeneity and polydispersity on kraft lignin samples.

Sulfite is also a relatively common process in the pulp and paper industry. Lignocellulosic biomass is delignified by means of HSO^{3-} and SO^{3-} ions generating lignosulphonates as a by-product. Lignosulphonates present a considerable amount of sulfonate groups in their structure, have high solubility in water, some polar organics and amines, and a higher molecular weight than kraft lignin. This kind of lignin is currently commercially produced and applied in a wide variety of products as animal feed, construction materials, a source of vanillin, plasticizer, oil additive, surfactant, and adhesives, etc. [66, 67].

Organosolv lignin is generated from treatment of lignocellulosic biomass with organic solvents and water under a mild process (140–220 °C) with the support of acids as catalysts [68]. Ethanol and methanol fractionation processes are amply used without added acid and this process is considered a green process because it uses mild conditions with easily recoverable solvents by distillation. The obtained lignins are considered as high-purity lignins in comparison with other methods, with high solubility in polar organic solvents [66] and low molecular weight (500–5000 g/mol) [69, 70].

Soda or alkaline isolation is similar to the kraft process without the presence of sulphur in the medium. It comes from the traditional pulping processes using soda or soda-anthraquinone in the cooking liquor to improve the stability of carbohydrates [71]. This process breaks the β–*O*–4 bonds, resulting in fractions of molecular weight between 1000 and 3000 Da [56].

The use of ionic liquids has acquired special relevance recently due to the green character of these solvents. Ionic liquids present great capacity of biomass dissolution [72] but their high cost reduces their application on an industrial scale. Ionic liquids can be designed introducing interesting catalyst properties to be used as dissolution and depolymerization media in a one-step reaction [73].

Thus, the extraction process is a key point in the valorization of lignin. The extraction method strongly affects the lignin structure and it will mark the future strategies for its depolymerization. Economic and sustainable treatments of the lignocellulosic biomass that result with high-purity lignin samples could be the future for the development of integrated biorefineries.

However, a great difference should be noted between the results of depolymerization reactions using technical lignin samples and lignin model compounds. Many studies have been published about the depolymerization of lignin model compounds. This number is largely reduced using technical lignin samples. The fail comes because the lignin model compounds do not achieve to represent the complexity of the lignin structure [74].

As it has been described before, the lignin structure presents an enormous dependence on its origin (plant types, botanical species, and even between trees and morphological parts of the tree [53]) and isolation method (soda, organosolv, sulfite, kraft, enzymatic or ionic liquids, among others [75]). The lignin obtained after the isolation process is known as technical lignin, which differs from the native lignin mainly in the abundance of different C–C and C–O linkages [76]. The diversity of monolignols units and the multiplicity of bond types (Table 2 and Fig. 1) explain the complexity of lignin depolymerization due to the heterogeneity in the molecule.

The real challenge for the scientific community is the effective depolymerization of technical lignins because of the occurrence of the repolymerization phenomenon. Repolymerization reactions consisted in the reactions between instable lignin fragments and original lignin producing unconverted lignin with higher molecular weight than the rough lignin during hydrothermal depolymerization processes [81–83]. This phenomenon presents a critical role and it is difficult to find with the use of lignin model compounds. This condensation reaction takes place via a carbenium ion, generated at the α-carbon atom. The ether bond in the intermediate can be cleaved (depolymerized) or attacked by an adjacent aromatic ring forming a stable carbon–carbon bond (polymerized) [84]. The key for the solution could be the protection of phenolic OH groups to suppress the repolymerization and some interesting studies are being conducted on this sense [85, 86].

5 Strategies for Lignin Conversion

It is well reported in the literature that the depolymerization of lignin requires a hydrothermal process with sub- or supercritical conditions of pressure and temperature using acid, base, or oxidant reagents since lignin displays a stable three-dimensional structure [87].

Table 2 Most frequent linkage types found by some authors in technical lignin samples

Linkage	Erikson et al. (1973) [77]	Nimz (1974) [78]	Alder (1977) [79]	Glasser and Glasser (1981) [80]
β–*O*–4	49–51	65	48	55
α–*O*–4	6–8	–	–	–
β-5	9–15	6	9–12	16
β-1	2	15	7	9
5-5	9.5	2.3	9.5–11	9
4–*O*–5	3.5	1.5	3.5–4	3
β–β	2	5.5	2	2

5.1 Basic and Acid-Catalyzed Depolymerization

Base-catalyzed depolymerization (BCD) reaction requires an excess of the strong base, such as NaOH, KOH, or K_2CO_3 in water, alcohol, dioxane or THF as solvents. This process takes place at a temperature of about 240–340 °C and high pressure, which favors the formation of side products (Table 3). The temperature reaction must be limited, since the use of higher temperatures is not encouraged since the molecule tends to get completely degraded to form both coke and gases [8]. BCD diminishes the oxygen content in a 50% as compared to the lignin feed, obtaining as main products anisole, guaiacol, phenyl ether, biphenyl, and benzyl phenyl ether so it is necessary the use of an additional hydrodeoxygenation or hydrocracking step to obtain benzene, toluene and xylenes.

In all cases, phenyl ether β–*O*–4 linkages were effectively broken in the BCD reaction while carbon–carbon linkages were less affected (Fig. 3) [87]. Traditionally, H_2O has been used as solvent in the depolymerization of lignin; however, the use of solvents with lower dielectric moment than water helps to readily dissolve relatively high molecular weight products derived from biomass [88]. However, these alcohols can react with phenyl ethers, forming phenols and alkyl-ethers, and also alkylation of phenols and catechols [59]. Thus, a sustainable alternative is the use of H_2O/alcohol solvents. Other authors have pointed out that the use of aprotic solvents, such as dioxane, tetrahydrofuran, or acetone also attain high performance on the depolymerization of lignin [8]. A drawback of the BCD reaction is related to the re-polymerization of the obtained monomers due to the drastic reaction conditions [59]; however, the use of H_2O/alcohol or acetone was efficient to minimize the polymerization and suppressing char formation [8].

The acid-catalyzed depolymerization (ACD) is known from 1943 when Hewson and Hibbert did a series treatment of maple woodmeal using a different combination of acids and alcohols, including HCl/ethanol and formic acid/ethylene glycol, with the purpose of separating the lignin into water-soluble and water-insoluble components [89]. Recently, wt% of formic acid associated with 77 wt% of ethanol was employed in the reaction with wheat straw lignin by Gasson et al. [90] where the formic acid acted as hydrogen source in the hydrolysis of lignin due to the protonation of the –*O*–4 bond facilitating the depolymerization by this zone (Table 4). In general, ACP requires harsh reaction conditions, which could increase the cost of reaction facility and post-handling. On the other hand, the acid conditions and high temperature can also cause intramolecular condensations of the lignin network [8].

Solvents in supercritical conditions behave differently than in subcritical conditions (Table 5), which supposes interesting advantages as increased species mixing, heat and mass transfer, fast reaction typically at a few minutes level, environmentally benign, good scalability, simple and easy for continuous production. The unique properties at supercritical conditions, i.e., strong dependence of the solubility of a material in a super critical fluid to its density and good contact between oxidants and reactants, make supercritical fluids ideal for separation and extraction of useful products and for oxidation of organic materials.

Table 3 Base-catalyzed lignin depolymerization (BCD)

Lignin	Base catalyst aqueous solution	Reaction conditions		Major products	Yield wt%
		Temperature (°C)	Pressure (MPa)		
Steam explosion hemp lignin	5 wt. % NaOH	300–330	3.5	Guaiacol	0.9–2.8
				Catechol	0.8–3.0
				Vanillin	0.5–0.8
Steam explosion softwood lignin	5 wt% NaOH	300–330	3.5	Guaiacol	1.2–2.1
				Catechol	0.1–3.2
				Vanillin	0.3–0.5
Organosolv lignin	2 wt% NaOH	300	25	Syringol	4.1
				Hydroxyacetophenone	1.6
				Guaiacol	1.1
Kraft lignin	5 wt% NaOH	270–315	13	Pyrocatechol	0.5–4.9
Organosolv olive tree pruning lignin	4 wt% NaOH, KOH, $Ca(OH)_2$, LiOH or K_2CO_3	300	90	Catechol	0.1–2.4

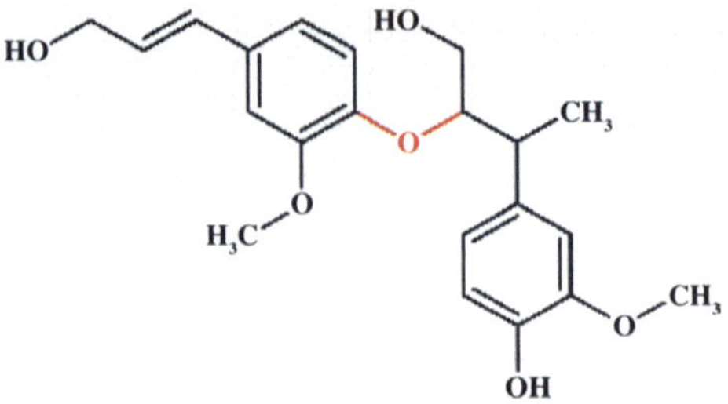

Fig. 3 Scission of the phenyl ether β–*O*–4 linkage

Table 4 Acid-catalyzed lignin depolymerization

Lignin	Acid catalyst aqueous solution	Reaction conditions		Major products	Yield wt%
		Temperature (°C)	Pressure (MPa)		
Wheat straw lignin	10 wt% formic acid 77 wt% ethanol	360	25	Methoxyphenols	1.3
				Catechols	0.5
				Phenols	0.3
Wheat straw lignin	10 wt% formic acid 77 wt% ethanol	380	25	Methoxyphenols	2.0
				Catechols	1.7
				Phenols	1.5
Wheat straw lignin	10 wt% formic acid 81 wt% ethanol	380	25	Methoxyphenols	2.9
				Catechols	1.5
				Phenols	2.0

Table 5 Supercritical conditions of some solvents used in the lignin depolymerization

Solvent	Critical temperature (K)	Critical pressure (MPa)	Critical density (g cm^{-3})
Water	647.1	22.1	0.322
Methanol	512.6	8.09	0.272
Ethanol	513.9	6.14	0.276
Acetone	508.1	4.70	0.278

The hydrolysis of lignin using H_2O in supercritical conditions occurs first by hydrolysis and then by dealkylation, yielding low molecular weight fragments. The yields of monomeric phenols are than expected by the repolymerization of the monomers forming char [91]. The use of alcohol in supercritical conditions as solvents favors a faster depolymerization due to the fact that they act as hydrogen transfer favoring β–*O*–4- linkage breaking; however, the repolymerization cannot be avoided [92]. The use of ethanol as solvent resulted in a much higher yield of monomers than methanol due to the slightly lower dielectric constant. In addition, ethanol can act as a scavenger for formaldehyde formed during the removal of methoxyl groups from lignin, avoiding repolymerization reactions. Ethanol can also serve as a capping agent in the presence of Lewis acid sites to stabilize reaction-intermediates by O-alkylating the phenolic hydroxyl groups and C-alkylating the aromatic rings [59,

93]. Other authors have pointed out that cross-linking reaction can also be prevented by the addition of phenol in the reaction medium [91, 92].

5.2 Oxidative Depolymerization

The use of oxidants and/or oxidizing protocols interesting alternative due to the large number of hydroxyl groups in lignin the pre-oxidation of lignin will weaken the β–*O*–4 bond, then make it more reactive and can be easily converted to monophenols. Most reported protocols involve the use of homogeneous conditions (e.g., alkali or mineral acids) combined with hydrogen peroxide [26] or the combination of homogeneous heteropolyacid catalysts with transition metals (Cu^{2+}, Fe^{3+}, etc.) under high temperatures and pressures [8]. On the other hand, previous research has reported that benzylic ketone can be converted into veratric acid (88% yield) and guaiacol (42% yield) in alkali-H_2O_2 conditions. This fact supposes that the coupling oxidations with C–C bond cleavage are also favored under these reaction conditions [8].

The oxidative depolymerization may not be ideal in some cases since they produce radicals that can lead to partial re-polymerization and form more complex lignin structures [89, 94]. Another drawback is related to the feasibility of product separation and/or isolation (especially for low quantities of generated products).

5.3 Ionic-Liquid Catalyzed Depolymerization

The use of ionic liquids (ILs) for the depolymerization of lignin has shown to be a promising alternative to conventional chemical depolymerization. ILs are formed by inorganic/organic salts, which are liquid at low temperatures (< 100 °C), and possess interesting properties such as thermal and chemical stability, low vapor pressure, and high ability to dissolve a wide variety of compounds [95] (Table 6). It has been reported in the literature that the ionic liquid anions mostly affect the structural integrity of the lignin biopolymer while the ionic liquid cations only have spectator roles [72]. The lignin depolymerization using ionic liquid anions is directly related with its nucleophilicity since the anion can possibly attack highly electron-deficient protonated C–O bonds (Fig. 4). The activity of these ionic liquids follows the next trend: alkylsulfonate > lactates > acetates > chlorides > phosphates [72].

The most important advantage of the ionic liquid is that these compounds can act as acid catalyst and as solvent. Thus, 1–*H*–3-methylimidazolium chloride can promote acidolytic cleavage of β–*O*–4 linkages in lignin via protonation of the ether linkages and subsequent attack by a nucleophile, as H_2O using mild temperature range (110–150 °C) [96]. Another advantage of the use of ionic liquid is associated with their ability to dissolve relatively larger amounts of the starting polymer (500 g lignin per kg ionic liquid), although the conversions are relatively low [96]. In order to increase the depolymerization degree, transition-metal catalysts are employed together with the ionic liquids to favor the depolymerization via oxidation [97]. The use of ionic liquid also has a drawback since their utilization for industrial applications is limited due to the high costs and investment, environmental hazards, and recyclability, etc. [8].

Table 6 Ionic liquids used in the lignin depolymerization

Lignin	Ionic liquid	Reaction conditions: Temperature (°C)	Reaction conditions: Pressure (MPa)	Major products	Yield wt. %
Eugenol	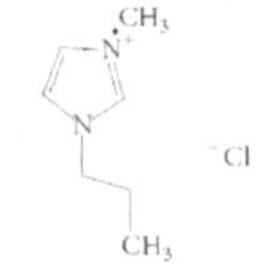	200	-	Guaiacol	7.9
Oak Wood lignin	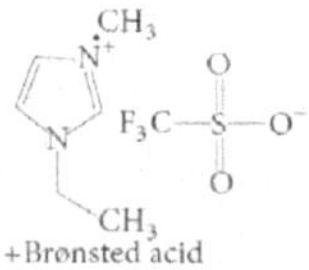 +Brønsted acid	110-150	-	Alkyl aryl ether linkages	-
Organosolv bleech lignin	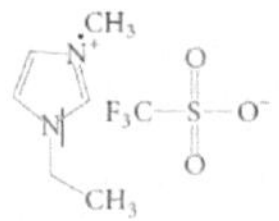 $+Mn(NO_3)_2$	110	8.4	2,6-Dimethoxy-1,4-benzoquinone	11.5
Veratrylglycerol-*β*-guaiacyl ether	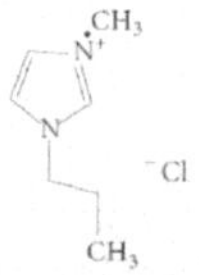	150	-	Guaiacol	71.5
Guaiacylglycerol-*β*-guaiacyl ether	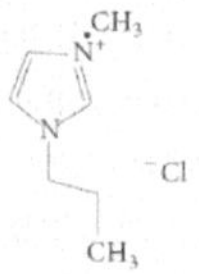	150	-	*β*-O-4 Ether bond cleavage	-
Guaiacylglycerol-*β*-guaiacyl ether	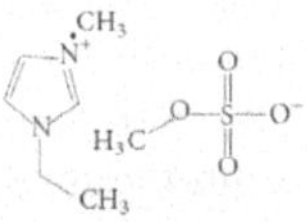	150	-	Guaiacol	75

6 Heterogeneous Catalysis Depolymerization

As was indicated previously, lignin is a heterogeneous polymer with variable composition depending on the raw material and the isolation method. The synthesis of benzene, toluene, and xylenes (BTXs) requires the fragmentation of the polymer in their respective monomers and then, the O-removal of these monomers. Generally, the fragmentation of lignin takes place from a fast-pyrolysis process, obtaining an energy-rich liquid mixture of oxygenated aromatic and aliphatic compounds, denoted as bio-oil. This bio-oil contains a high proportion of oxygen and water, which provides poor chemical and thermal instability in comparison to the chemicals from the petrochemical industry [25, 98]. The use of heterogeneous catalysis can favor the selectivity towards specific products, as BTXs.

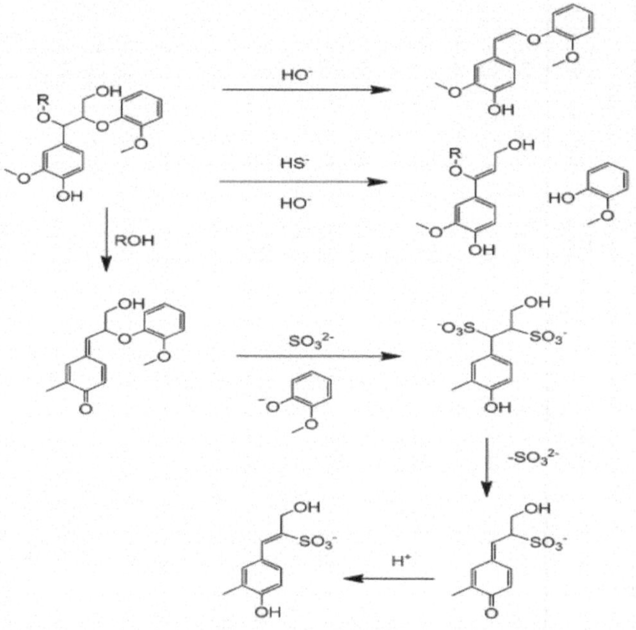

Fig. 4 Mechanism proposed in the depolymerization of lignin using ionic liquids (alkylsulfonate) as active phase

The lignin conversion into valuable chemicals, as BTXs, can be approached in two different ways.

1. Conversion in one step. Lignin is directly converted into the desired products. This methodology is useful to synthesize functionalized products.
2. Conversion in two steps. Firstly, lignin is transformed in a mixture of monomeric and oligomeric oxygen-rich compounds (bio-oil), and then it is deoxygenated to produce the targeted aromatic compounds.

6.1 Deoxygenation from Lignin (One step)

Bio-oil obtained from the fast-pyrolysis can be improved using a solid acid catalyst, since these catalysts increase the yield to deoxygenated products and decrease the char formation, resulting in a product with higher *C*/*O* ratio with higher quality than that obtained from the fast pyrolysis [99]. The yield towards fully deoxygenated compounds, such as BTXs, in a single step is relatively low, obtaining mainly alkylphenol monomers.

As takes place in the petrochemical industry, zeolites are the main catalysts used in the lignin cracking. Among the wide variety of zeolites, several authors have reported that H-ZSM-5 is the most efficient catalyst to obtain deoxygenated liquid fraction and aromatic and naphthenic compounds with lower percentage of coke [100] (Table 7). Catalytic fast pyrolysis, using H-ZSM-5 as catalyst, is thought

Table 7 Depolymerization and deoxygenation of lignin using zeolites as catalysts

Catalyst (Si/Al ratio)	Feed	Reactor	T (°C)	Feed/Cat. ratio	Aromatic yield	Aromatic distribution		
						Benzene	Toluene	Xylenes
H-ZSM-5 (30)	Pine wood sawdust	Fluidized bed	600	0.2	11	23.1	30.0	13.9
ZSM-5	Pine wood	Fluidized bed	600	0.35	13.9	20.8	37.1	19.8
ZSM-5	Pine wood	Fluidized bed	450	0.35	5.9	10.8	32.2	38.0
ZSM-5	Pine wood (36 wt%) + Methanol (64 wt%)	Fluidized bed	450	0.56	21.1	5.8	16.9	62.9
ZSM-5	Pine wood (41 wt%) + *n*-propanol (59 wt%)	Fluidized bed	450	0.58	16.3	11.0	39.3	39.2
ZSM-5	Pine wood (47 wt%) + *n*-butanol (53 wt%)	Fluidized bed	450	0.64	17.2	10.6	38.7	40.2
ZSM-5	Pine wood (45 wt%) + 2-butanol (55 wt%)	Fluidized bed	450	0.64	15.6	10.4	38.6	40.2
ZSM-5(56)	Harwood lignin-acetone (1:2 wt.)	Packed bed	500	5	89.4	8.6	33.1	31.5
ZSM-5(56)	Hardwood lignin-acetone (1:2 wt.)	Packed bed	600	5	87.9	13.6	42.4	22.7
ZSM-5(56)	Hardwood lignin-acetone (1:2 wt.)	Packed bed	600	2.5	74.6	9.3	31.0	25.0
Co-ZSM-5(50) 5 wt% Co	Pine wood lignin	Pyroprobe	650	0.11	39.8	8.0	11.1	9.6
Ni-ZSM-5(50) 5 wt.% Ni	Pine wood lignin	Pyroprobe	650	0.11	41.3	7.4	10.6	10.0
Mo-ZSM-5(50) 5 wt% Mo	Pine wood lignin	Pyroprobe	650	0.11	42.3	6.4	11.5	11.0
Pt-ZSM-5(50) 0.5 wt% Pt	Pine wood lignin	Pyroprobe	650	0.11	46.4	6.7	12.0	11.9
Ga/Meso-MFI (17.5) 1 wt% Ga	Radiata pine sawdust	Packed bed	500	10	–	2.3	7.8	11.8
Ga/Meso-MFI (17.5) 5 wt% Ga	Radiata pine sawdust	Packed bed	500	10	–	0.4	2.1	5.1
La/ZSM-5 (23) 6 wt% La	Wheat straw lignin	Packed bed	600	0.33	–	0.58	1.65	1.29

to result from a series of reactions where non-volatile compounds (lignin) are first cracked to heavy volatile compounds, which are subsequently cracked to volatile alkyl aromatics and ultimately to coke and gas [101]. In this sense, it has been reported feed/catalyst ratio as well as the conversion values in the packed bed reactor reaches higher conversion values in comparison to the fluidized-bed one. However, the packed-bed reactors display several disadvantages related to its high proportion of non-detected products by the formation of a larger amount of carbonaceous deposits on the surface of the H-ZSM-5 catalyst. Other authors have carried out a study with several zeolites, where the yield towards deoxygenated compounds follows the next trend: HZSM-5 > H-beta > H-mordenite > H-ferrierite ~ H–Y [102]. In this way, other authors have pointed out that H–Y, silicalite, and silica-alumina produce more aliphatic than aromatic hydrocarbons, while H-ZSM-5 and H-mordenite lead to higher proportions of aromatic hydrocarbons with a lower percentage of coke [93] so the dimension of the cavities is a key parameter in the activity and selectivity of the cracking and deoxygenation processes. The pyrolysis of Chinese fir and rice straw was evaluated using H-ZSM-5 as catalyst with different Si/Al molar ratio between 550 and 900 °C, obtaining the higher BTXs yields for the H-ZSM-5 with the lowest Si/Al molar ratio and being toluene the main product. The absence of catalyst under the same reactions conditions led to guaiacol-, syringol-, and phenol-type compounds [103]. This fact indicates that the presence of a higher proportion of acid sites and the strength of these sites are key factors in the catalytic behavior of the H-ZSM-5 zeolites. On the other hand, the incorporation of heteroatoms seems to improve the production of aromatic compounds. Thus, Ni, Co, or Pt are involved in cracking, hydrogenolysis, hydrocracking, and dehydrogenation reactions [104], while Ga is involved dehydrocyclization of reaction intermediates [105].

The incorporation of heteroatoms, such as Al or Zn, into mesoporous silica has emerged as an alternative to zeolites to obtain BTXs compounds [19]. Similarly to the H-ZSM-5, the catalytic activity of Al-MCM-48 is directly related to the amount of active sites, while Zr-MCM-48 catalyst displays an inflexion point since the catalyst with highest proportion of acid sites reaches lower BTXs yields.

Several authors have proposed hydrogenolysis reactions as an alternative to the catalytic fast pyrolysis. Several authors have proposed the use of bi-functional catalysts where an active support (solid acid catalysts) is combined with a hydrogenating phase (Ni, Pd, Pt). These catalysts achieve an efficient depolymerization of lignin up to their monomers. However, the formation of BTXs is still limited, so these bi-functional systems are not sustainable at a larger scale since the use of noble metals rises the prize of its implementation [106]. In the same way, other hydrogenating catalysts such as $NiMoS/Al_2O_3$ [107], copper chromite [108], or transition-metal phosphides [109] have been studied in the depolymerization of lignin, obtaining alkyl-phenols, guaiacols, and syringols in all cases.

In any case, the deoxygenation from lignin in one step is relatively low, obtaining a wide range of products. The two-stage process is more sustainable to obtain higher yield values. Thus, the depolymerization of lignin takes place in a first step by fast pyrolysis or catalytic fast pyrolysis and then these monomers are deoxygenated in a second step.

6.2 Deoxygenation from Bio-oil or Target Oxygenated Compounds (two steps)

Several authors have carried out the deoxygenation of bio-oil obtained from fast pyrolysis to obtain deoxygenated aromatic compounds. Ether approaches is the deoxygenation of the light vapors from bio-oil or the use of different model compounds from lignin-derived bio-oils including anisole, guaiacol, vanillin, eugenol, phenol, and cresol.

Zeolite, which exhibited excellent behavior in the depolymerization of lignin, has also shown a good behavior in the deoxygenation of monomers to obtain benzene toluene and xylene by dehydration reactions. It has been reported in the literature that H-ZSM-5 displays the highest aromatic yields, suggesting that the catalytic activity is ascribed to the presence of Brønsted acid sites. However, these catalysts tend to suffer a strong deactivation due to the formation of carbonaceous deposits on these acidic centers.

Most of the catalytic processes to obtain BTXs proceed from hydrodeoxygenation reactions (HDO) in liquid or gas phase. Liquid-phase hydrogenation involves treatment with high pressure (50–100 bar) and temperature (250–500 °C) in a solvent with H_2 or an H donor (e.g., alcohol). The use of a gas phase is a sustainable alternative since the separation of the solvent is much easier and requires softer conditions of pressure and temperature.

Traditionally, hydrotreating catalysts are based on molybdenum or tungsten sulfide and promoted by cobalt or nickel as the active phases with mainly porous γ-alumina as a material support (Table 8). The first studies using these catalysts in the HDO reaction were carried out by Ollis et al., which deoxygenated phenol to benzene and then cyclohexane due to the high H_2 pressure [110]. From these points, several authors have evaluated the catalytic behavior of these CoMo-based catalysts using alumina or silica-alumina as support using phenol, guaiacol, methoxyphenol, eugenol, or anisole as target molecules, obtaining benzene, toluene, and xylenes in all cases. However, these catalysts also display several disadvantages related to its deactivation by the formation of carbonaceous deposits and by the oxidation of the sulfide species by the presence of H_2O, obtained as a by-product. In fact, some authors co-feed H_2S to extend the life of the catalyst.

The use of noble metals, such as Pd, Pt, Ru, or Rh, has emerged as a promising alternative to traditional sulfides [111, 112] due to the low carbon deposition and the softer catalytic conditions used in the noble metal-based catalysts (Table 9). It is well reported in the literature the use of these noble metals to upgrade different model compounds from lignin-derived bio-oils including anisole, guaiacol, vanillin, eugenol, phenol and cresol. These catalysts display a high hydrogenating behavior, which is increased by the use of acid support, so the temperature and pressure must be modulated to avoid hydrogenation of the aromatic ring. In addition, the presence of these acid sites favors the transalkylation reaction, favoring the methylation of the aromatic ring to form toluene and xylenes [112].

Other authors have proposed the use of cheaper transition-metal phosphides, such as Ni [113, 114], Fe [115, 116], NiFe [117], NiMo [118], or NiCu [119] (Table 9). These catalysts are active for short times or low amount of oxygenated compound per unit of time since they tend to undergo a process of deactivation by oxidation of

Table 8 Deoxygenation of lignin-derived compounds using metal sulfides as catalysts

Catalyst	Support	Reactor configuration	*T* (K)	*P* (bar)	*t* (h)	Lignin-derived compound	Major products
CoMo	Al_2O_3	Batch	613–628	70	–	4-methylphenol	Toluene, methylcyclohexane
CoMo	Al_2O_3	Flow reactor	523	15	–	Phenol	Benzene
CoMo	Al_2O_3	Batch	573	50	4.17	4-methylphenol	Toluene
CoMo	Al_2O_3	Batch	598	69	1.68	4-methulguaiacol	Toluene, cresol, methylcyclohexane
CoMo	Al_2O_3	Batch	573	69	4	Eugenol	Propylcyclohexane, propylphenol, propyl-guaiacol, propylphenol
CoMo	Al_2O_3	Batch	523–598	34	6.67–10	Anisole	Phenol, benzene, cyclohexane
CoMo	Al_2O_3	Batch	523	34	20	Guaiacol	Catechol, phenol, benzene, cyclohexane
CoMo	Al_2O_3	Batch	553	70	2.5	Guaiacol	Phenol, catechol
CoMo	Al_2O_3	Fixed bed	573	50	–	Anisole	Phenol, 2-methylphenol, benzene
CoMo	Al_2O_3	Flow reactor	523	15	–	Phenol	Benzene
CoMo	Al_2O_3	Flow reactor	523	15	–	Anisole	Toluene, phenol, benzene, 2-methylphenol
CoMo	Al_2O_3	Batch	573	55	4	Guaiacol, phenol	Benzene, toluene, xylenes

Table 9 Deoxygenation of lignin-derived compounds using transition-metals, phosphides, carbides and nitrides as catalysts

Catalyst	Support	Reactor configuration	T (K)	P (bar)	t (h)	Lignin-derived compound	Major products
Pt	Al_2O_3	Packed bed	573	1.4	–	Anisole	Phenol, 2-methylphenol, benzene
Pt	SiO_2–Al_2O_3	Packed bed	573	1.4	–	4-methylanisole	4-methylphenol, toluene
Pt	SiO_2	Fixed bed	673	1	0–5	Anisole	Benzene
Pt	H-Beta zeolite	Fixed bed	673	1	0–5	Anisole	Benzene, toluene, xylenes
Rh	ZrO_2	Batch	573	80	3	Guaiacol	Benzene
Ru/MoO_3	Activated carbon	Packed bed	623–673	40	0.067	Guaiacol	Benzene, phenol
Ni_2P	SiO_2	Packed bed	573	1	0.33	Guaiacol	Benzene
Ni_2P	SiO_2	Fixed bed	573	15	0–10	Anisole	Phenol, benzene, cyclohexane
NiMoP	SiO_2	Fixed bed	573	15	0–10	Anisole	Phenol, benzene, cyclohexane
MoP	SiO_2	Packed bed	573	1	0.33	Guaiacol	Benzene
MoP	SiO_2	Fixed bed	573	15	0–10	Anisole	Phenol, benzene, cyclohexane
Co_2P	SiO_2	Packed bed	573	1	0.33	Guaiacol	Benzene
Fe_2P	SiO_2	Packed bed	573	1	0.33	Guaiacol	Phenol
WP	SiO_2	Packed bed	573	1	0.33	Guaiacol	Phenol
Mo_2C	–	Fixed bed	553	1	6.66	Anisole	Benzene, toluene
Mo_2N	Activated carbon	Batch	573	50	6	Guaiacol	Phenol
Fe	SiO_2	Fixed bed	623–723	1	4	Guaiacol	Phenol, benzene
Fe	Activated carbon	Fixed bed	623	1	3	Guaiacol	Phenol, benzene, Toluene
NiFe	H-Beta	Fixed bed	523–673	1	1–6	Guaiacol	Phenol, benzene, toluene, xylenes
Ni	SiO_2	Fixed bed	423	1	0–4	Phenol	Cyclohexanone, cyclohexanol, benzene
Ni	SiO_2	Fixed bed	573	50	–	Anisole	Phenol, benzene
Ni	SiO_2	Flow reactor	563–583	3	–	Anisole	Benzene
Ni-Cu	Al_2O_3	Packed bed	573	10	4	Anisole	Benzene, toluene, cyclohexane, methylcyclohexane
Ni-Co	Al-MCM-41	Packed bed	673	1	4	Guaiacol	Cresol, phenol, benzene, toluene
Ni-Mo	SiO_2	Packed bed	683	1	5	Guaiacol, Anisole, Phenol	Benzene, toluene, xylenes
MoO_x	Al_2O_3	Fixed bed	613	40	5	Cresol	Toluene, methylcyclohexanes

the active phase or the formation of coke. On the other hand, MoOx phase has also shown to be active in the HDO reaction by the presence of coordinatively unsaturated sites (CUS) [120].

In order to obtain inexpensive catalysts with high activity and resistance to deactivation processes, new catalytic approaches have been taken by combining transition metals such as Fe, Ni, Co, Mo, W, with N (nitrides) [121], C (carbides) [122], and P (phosphides) [123, 124] (Table 9). The presence of these heteroatoms modifies the electronic density of the metal sites as well as increases the dispersion of the available sites and diminishes the deactivation by coke deposition. In addition, the unreduced P-species can act as acid sites that also favor the transalkylation of the aromatic ring.

7 Enzymatic Depolymerization

Currently, different methods have been developed in order to convert lignin to value-added chemicals including chemical and biological routes. In the case of biological methods, a lower energetic demand is required due to higher selectivities attained at mild reaction conditions in comparison to chemical methods usually conducted at high temperatures and pressures, which leads to high energy costs [59, 125]. Within the biological delignification strategies, the microbial treatment is slower and displays higher interference of side reactions than the enzymatic processes [126]. Thus, biological depolymerization has been extensively studied in recent decades in such a way that it has been demonstrated that several types of oxidative enzymes are secreted by fungi [127]. A common fungus widely studied for lignin degradation is *Phanerochaete chrysosporium*, which produces a lignin peroxidase enzyme that is able to deconstruct it in the presence of nitrogen and carbon sources [128]. Thus, fungi can degrade lignin by secreting enzymes denominated by "ligninases" that can be classified as phenol oxidases (laccase) or heme peroxidases [lignin peroxidase (LiP), manganese peroxidase (MnP) and versatile peroxidase (VP)] (Fig. 5). Therefore, white-rot fungi can secrete one or more of these powerful oxidative enzymes in addition to other compounds required for effective degradation of lignin [129, 130]. Although these enzymes cannot be found in fungi, they can be produced by different bacterial species that are able to grow on lignin [131]. Moreover, these enzymes show wide substrate specificity and utilize radical intermediates for lignin depolymerization [127]. It has recently been demonstrated that some bacteria can also secrete many of these enzymes so it is not limited to the use of fungi [132, 133].

Therefore, the use of enzymes for lignin depolymerization is an alternative route to remove lignin from biomass and produce aromatic compounds. However, very few studies obtained an efficient lignin depolymerization by using enzymes, since most oxidative enzymes are able to depolymerize lignin and polymerize aromatic compounds [134]. As mentioned above, highly specialized ligninases such as laccases or heme peroxidases (LiP, MnP, and VP) can be developed from white-rot enzymes and are potentially involved in degradation of lignin. In general, laccases employ O_2 as electron acceptors while peroxidases use H_2O_2 as a co-substrate [130].

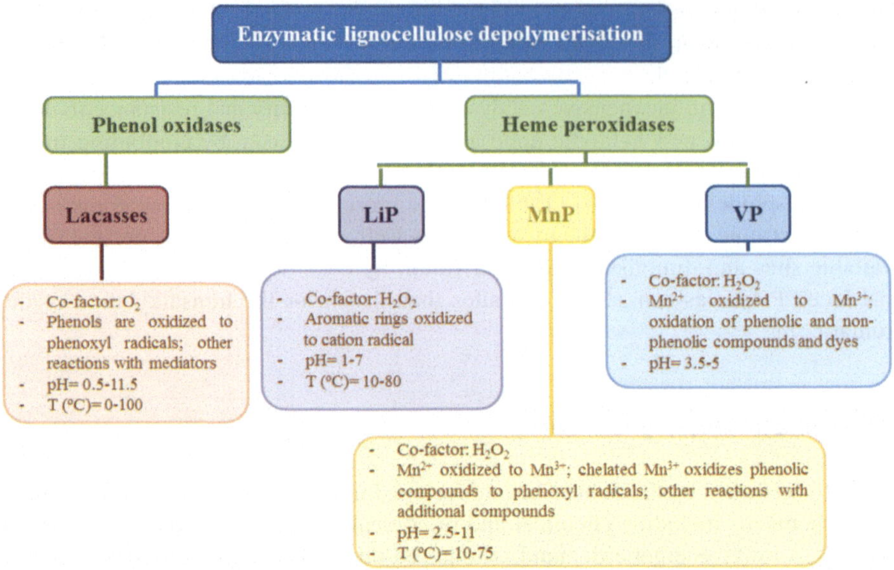

Fig. 5 Schematic diagram of lignolytic enzymes involved in lignin depolymerization

On one hand, laccases are a type of copper-containing enzymes able to oxidize phenolic rings to phenoxy radicals while using molecular oxygen as oxidant. Although these enzymes do not catalyze directly the oxidation of non-phenolic aromatic rings of lignin, they can degrade high-redox-potential substrates by using small chemical oxidants acting as redox mediators, which can penetrate through the lignocellulosic matrix [38]. In recent years, the use of laccases for lignin degradation has gained considerable attention since it has demonstrated to be effective in degrading lignin of pulped fibers in the paper industry and purified lignin or soluble monomers [125, 135, 136]. However, the efficacy of these enzymes for lignin degradation is debated due to the fact that lignin is too large to penetrate the laccase active sites, as an addition of a mediator is necessary [128]. In addition, the laccases oxidize the phenolic groups of lignin, resulting in polymerization in the absence of suitable mediators [137]. Therefore, it is not clear what the role of these enzymes is for lignin degradation, but considering that these enzymes can be produced at higher scale than ligninolytic peroxidases, they must be considered for the lignin depolymerization process in industrial processes [38]. Bourbonnais et al. [138] and Shleev et al. [139] reported the interaction of Kraft lignin with laccase but they did not identify which key factors had influence on the process. Baiocco et al. [140] demonstrated the mechanism of the laccase mediator towards nonphenolic substrates by following either an electron transfer or a radical hydrogen atom transfer. Chen et al. [125] evaluated the synergy of laccase as ligninolytic enzyme with cellulose and hemicellulose amendments on ensiled corn stover. They found that the partial hydrolysis of cellulose and hemicellulose into soluble sugars led to ensilage facilitated laccase penetration into the lignocellulose complex and improved the lignin degradation.

On the other hand, LiP are heme-containing peroxidases in which the heme groups act independently and reduce its substrate in the presence of H_2O_2. These enzymes catalyze a wide range of lignin depolymerization reactions with soluble lignin models compounds. It is known that LiPs can oxidize the aromatic rings of lignin (e.g., guaiacol, vanillyl alcohol, catechol, syringic acid, acetosyringone) via long-range electron transfer [141]. Moreover, the unspecific aromatic ring oxidation produces unstable cation radicals which undergo different non-enzymatic processes that can lead to the release of aromatic aldehydes, demethoxylation with subsequent methanol release, or aromatic ring cleavage prior to the formation of muconate-type structures [141, 142]. Likewise, MnP utilizes H_2O_2 as co-substrate and its catalytic cycle is similar to LiP and involves the production of oxidized intermediates. In the case of MnP, Mn^{2+} is oxidized to Mn^{3+} in such a way that Mn^{3+} is subsequently stabilized by organic acids like oxalic acid or lactic acid. These Mn^{3+} chelates facilitate the detachment of Mn^{3+} from the MnP active site and stimulate the MnP activity by increasing the rate of oxidation since they act as redox mediators oxidizing phenolic lignin structures [143, 144]. Therefore, MnP might play a relevant role in lignin depolymerization. Kapich et al. [145, 146] showed that peroxidized unsaturated fatty acids were able to oxidize non-phenolic β–*O*–4-linked lignin model compounds irrespective of MnP, confirming that direct contact between MnP and lignin was not required but, rather, that the formed peroxyl radicals acted as agents for lignin degradation. In 1999, Camarero et al. found a new peroxidase enzyme combining two major peroxidase properties from LiP and MnP that was called versatile peroxidase, VP [147]. VP is a hybrid of LiP and MnP is capable of oxidizing Mn^{2+} and phenolic compounds as well as non-phenolic aromatic compounds [141].

Dordick et al. reported that report that horseradish peroxidase and milk lactoperoxidase, which were inactive toward lignin in water, could vigorously depolymerize both synthetic and natural lignins in organic media and even native lignocellulose originating from wheat straw in the case of horseradish peroxidase [148]. Salvachúa et al. employed two selected fungal secretomes with high levels of laccases and peroxidases for DMR-EH (deacetylation, mechanical refining and enzymatic hydrolysis) lignin depolymerization [127]. They reduced the lignin average molecular weight by 63 and 75% at pH 7 compared to the value of the control treated at the same conditions by using the secretome obtained from *Pleurotus eryngii*, which exhibited the highest laccase activity. They also demonstrated that the incubation of an aromatic-catabolic microbe, *Pseudomonas putida* KT2440, with the fungal secretome and DMR-EH lignin enhanced lignin depolymerization and partially prevented the repolymerization. This study confirmed that ligninolytic enzymes can be used to partially depolymerize lignin and the presence of an aromatic-catabolic bacterium as a "microbial sink" can improve the extent of enzymatic lignin depolymerization. Likewise, other authors have confirmed that the presence of bacteria favors lignin depolymerization and enhances the enzymatic hydrolysis in the pretreatment [133, 149, 150]. Moreover, there are some reports of additional enzymes that might exhibit the potential to act on lignin. Ortiz-Bermúdez et al. reported that chloroperoxidases might contribute to lignin degradation since they catalyzed the cleavage of the β–*O*–4 bond of a dimeric lignin model compound [151]. More recently, also dye-decolorizing peroxidases (DyP) were reported to degrade lignin

[132, 152–154]; however, not all DyPs are capable of oxidizing non-phenolic lignin model compounds [155].

All of these oxidative enzymes have in common that they cleave the bonds present in lignin in a non-selective fashion using radical-based mechanisms but there are other enzymes that use a non-radical mechanism such as glutathione lyases and β-etherases; however, the depolymerization efficiency by using these enzymes depends on the lignin starting material [131]. Thus, different β-etherases have been tested for lignin depolymerization but the oxidation of the Cα-hydroxyl groups of the β–*O*–4 aryl ether bonds in the lignin polymer is firstly required, so these enzymes need a carbonyl group adjacent to the aryl ether linkage [156]. Reiter et al. assayed an in vitro one-pot enzyme cascade by combining Cα-dehydrogenase (LigD), β-etherase (LigF), and glutathione lyase (LigG) for the depolymerization of kraft-type and organosolv lignins, adding a glutathione reductase [157]. Nevertheless, they obtained low monomeric aromatic yields except for vanillin production from softwood alkali lignin. Picart et al. employed glutathione-dependent β-etherases and glutathione lyases as enzymes for the biocatalytic depolymerization of lignin, successfully carrying out one-pot multi-step processes [158]. Ohta et al. employed two β-etherases, GST4 and GST5, and one glutathione lyase, GST3, for depolymerization of milled-wood lignin and improved aromatic yields [159]. Gall et al. also tested another enzyme cascade by using nantiocomplementary Cα-dehydrogenases, LigD and LigN, enantiocomplementary β-etherases, LigE and LigF, glutathione lyase and glutathione reductase for lignin depolymerization in one pot [160]. They achieved significant amounts of syringyl-hydroxypropanone when they employed a hybrid poplar lignin highly enriched in syringyl units. Likewise, they attained high amounts of guaiacyl-hydroxypropanone and syringyl-hydroxypropanone from a more complex lignin obtained from maize corn stover. However, they conclude that this cascade of enzymes was limited to the depolymerization of smaller lignin oligomers. Recently, Picart et al. tested β-etherase enzymes for depolymerization of beech wood OrganoCat lignin in such a way that this lignin was firstly oxidized by using laccase enzyme in order to generate the Cα carbonyl groups required for β-etherase activity [161]. By the use of β-etherases LigE, they attained the isolation of soluble lignin oil (12.5 wt%) with low molecular weight aromatics. However, the laccase-mediator system also favored the repolymerization of the initial lignin.

Therefore, different enzymes have been evaluated in the literature for lignin depolymerization. However, BTXs have not been detected yet by using these catalytic systems, as it is necessary to complete these studies and find reaction media able to dissolve lignin and being enzyme-compatible capable of attaining high yields of monomeric aromatic compounds.

8 Future Perspectives

The valorization to lignin to obtain valuable chemicals such as benzene, toluene, and xylenes (BTXs) has received increasing interest by the scientific community in recent years. The aim of this review is to highlight the use of the lignin as starting source to obtain aromatic compounds (BTXs), which have traditionally been obtained from fossil fuels such as oil. As is well known, fossil fuel reserves are located in very specific places on Earth. This fact has given rise to fossil fuels are subject to speculation and are trigger of international conflicts. In contrast, the use of biomass in general and lignin in particular provides energetic independence to each region.

Taking into account these premises, this review is focused on highlighting the processes and technologies employed to depolymerize the lignin. The valorization of the lignin can contribute to the development of a future bio-based economy that is more sustainable and environmentally more benign. In spite of the fact that lignin has great potential as sustainable source to produce energy as well as valuable chemicals such as BTXs, the processes and technologies still need to be developed.

As indicated above, the lignocellulosic biomass is formed by cellulose, hemicellulose, and lignin. Both cellulose and hemicellulose are formed by monosaccharide units, which are linked through β (1$\rightarrow$4)-glycosidic bonds so it can be considered a homogeneous source for obtaining valuable chemicals. However, lignin presents a more heterogenous and variable composition depending on the source of biomass employed as feedstock. This fact implies the use of different reaction parameters, such as temperature, which gives rise to catalytic processes with low selectivity due to the wide range of products formed. Another challenge associated with the inherent complexity of the lignin structure is given by the catalysts used in the depolymerization processes. Thus, most of the catalysts provide the catalytic surface for different bond cleavage to happen via either protonation/deprotonation, nucleophilic attack, or hydrogenolysis, so predicting the reaction products and the mechanism involved in these depolymerization processes is a challenge for the upcoming years. In this sense, the scientific community is trying to replace homogeneous catalysts by solid catalysts due to their easier separation and recyclability. In addition, heterogenous catalysts are less corrosive, which implies a reduction of costs, and are more respectful to the environment. The complexity of lignin structure implies that each case must be approached following a specific protocol. This fact is relatively well defined in the case of the homogeneous catalysis; instead, protocols for heterogeneous catalysts have barely been studied. In addition, the mass transfer from lignin feedstock to the surface of catalysts is a limiting step to reach high-yield values [82, 83, 162, 163].

Among various depolymerization methodologies, thermal treatment with high hydrogen pressures (hydrotreating) is emerging to improve lignin depolymerization yields [82, 162]. In these systems, besides the catalyst, the solvent, pressure, temperature, and especially the raw material of lignin must be considered. In this field, several advances have already been reported; however, the scientific community must deepen this topic to obtain sustainable and more economical catalytic systems that present high yields towards aromatic compounds in such a way that these processes can be integrated in a biorefinery.

Considering the heterogeneity of the lignin and the complexity to separate the obtained products [9, 26, 87], another challenge in the lignin depolymerization is focused on an efficient purification, which should be environmentally sustainable and not energy intensive, and diversification towards chemicals and fuels as end products from lignin. In this sense, the extraction under supercritical conditions has reached excellent results in the purification step [164].

On the other hand, one of the main drawbacks of lignin depolymerization is related to the high content of the oxygenated species obtained after the treatment. These oxygenated species display a poor energy potential so their removal by hydrodeoxygenation (HDO) reactions is necessary to obtain aromatic monomers, such as BTXs. The methodology proposed in previous sections exposes that the depolymerization and the HDO step can take place at the same stage although oxygenated products are still predominant in the selectivity pattern. Taking into account this fact, one of the main objectives of future research is related to the development of catalytic systems that provide high efficiency in both the depolymerization and HDO steps [82, 162, 165, 166].

In order to obtain BTXs, a critical issue in the lignin depolymerization is related to the cleavage of the C–O bond without modifying the aromatic core. In this sense, the hydrogenating character of the catalysts plays an important role in the selectivity pattern. Thus, noble metal-based catalysts tend to hydrogenate the aromatic ring under severe pressure and temperature conditions, obtaining cycloalkanes [167]. A cheaper alternative is the use of catalysts based on transition metals, such as Ni, which has the ability to break the C–C and C–O bonds without modifying the structure of the aromatic ring [82, 162] and that in turn it must be resistant to water molecules formed as a by-product in the hydrogenolysis of the C–O bond. In this sense, process engineering and design towards mild depolymerization conditions, using mild conditions of temperature and pressure and process intensification, such as continuous flow reactions, can result in catalytic systems where parameters are optimized depending on the raw material of lignin to control C–C and C–O bond cleavage and hydrogenation processes [26].

The lignin valorization may only effectively reach the market if a cheap, efficient, and green protocol is developed for lignin depolymerization. This protocol must follow the next premises:

- Development of an efficient deconstruction method.
- Effective and competitive processes for the separation, purification, and isolation of the obtained chemicals.
- The use of mild conditions that avoid the formation of collateral reactions.
- Conducting the tests on a larger scale and then implement it in a biorefinery plant.

Acknowledgements The authors would like to thank the Spanish Ministry of Economy, Industry and Competitiveness (postdoctoral contract Juan de la Cierva Incorporacion IJCI-2015-23168 and Ramon y Cajal contract RYC-2015-17109) for financial support during this work. J.A.C. and C.G.S. thank Malaga University for financial support.

References

1. Ragauskas AJ, Williams CK, Davison BH, Britovsek G, Cairney J, Eckert CA, Frederick WJ, Hallett JP, Leak DJ, Liotta CL (2006) Science 311:484–489
2. Luque R, Herrero-Davila L, Campelo JM, Clark JH, Hidalgo JM, Luna D, Marinas JM, Romero AA (2008) Energy Environ Sci 1:542–564
3. Wang T, Nolte M, Shanks B (2014) Green Chem 16:548–572
4. Demirbas A (2001) Energy Convers Manage 42:1357–1378
5. Cheng F, Brewer CE (2017) Renew Sust Energ Rev 72:673–722
6. Taherzadeh M, Karimi K (2008) Int J Mol Sci 9:1621–1651
7. Brandt A, Grasvik J, Hallett J, Welton T (2013) Green Chem 15:550–583
8. Xu C, Arancon RAD, Labidi J, Luque R (2014) Chem Soc Rev 43:7485–7500
9. Ragauskas AJ, Beckham GT, Biddy MJ, Chandra R, Chen F, Davis MF, Davison BH, Dixon RA, Gilna P, Keller M (2014) Science 344:1246843
10. Sun Z, Fridrich B, de Santi A, Elangovan S, Barta K (2018) Chem Rev 118:614–678
11. Schutyser W, Renders T, Van den Bosch S, Koelewijn SF, Beckham GT, Sels BF (2018) Chem Soc Rev 47:852–908
12. Boyce CK, Zwieniecki MA, Cody GD, Jacobsen C, Wirick S, Knoll AH, Holbrook NM, Natl P (2004) Acad Sci USA 101:17555–17558
13. Chakar FS, Ragauskas AJ (2004) Ind Crops Prod 20:131–141
14. García Calvo-Flores F, Dobado JA (2010) Chemsuschem 3:1227–1235
15. Menon V, Rao M (2012) Prog Energ Combust Sci 38:522–550
16. Lazaridis PA, Fotopoulos AP, Karakoulia SA, Triantafyllidis KS (2018) Front Chem 6:295
17. Olcese RN, Francois J, Bettahar MM, Petitjean D, Dufour A (2013) Energy Fuels 27:975–984
18. Rinaldi R, Jastrzebski R, Clough MT, Ralph J, Kennema M, Bruijnincx PCA, Weckhuysen BM (2016) Angew Chem Int Ed 55:8164–8215
19. Elfadly AM, Zeid IF, Yehia FZ, Rabie AM, Aboualala MM, Park SE (2016) Int J Biol Macromol 91:278–293
20. Sirous-Rezaei P, Jae J, Ha JM, Ko CH, Kim JM, Jeon JK, Park YK (2018) Green Chem 20:1472–1483
21. Heeres A, Schenk N, Muizebelt I, Blees R, De Waele B, Zeeuw AJ, Meyer N, Carr R, Wilbers E, Heeres HJ (2018) ACS Sustain Chem Eng 6:3472–3480
22. Zheng A, Zhao Z, Chang S, Huang Z, Wu H, Wang X, He F, Li H (2014) J Mol Catal A 383:23–30
23. Sultana A, Fujitani T (2017) Catal Commun 88:26–29
24. Bond JQ, Alonso DM, Wang D, West RM, Dumesic JA (2010) Science 327:1110–1114
25. Huber GW, Iborra S, Corma A (2006) Chem Ver 106:4044–4098
26. Zakzeski J, Bruijnincx PCA, Jongerius AL, Weckhuysen BM (2010) Chem Rev 110:3552–3599
27. Vispute TP, Zhang HY, Sanna A, Xiao R, Huber GW (2010) Science 330:1222–1227
28. Agblevor FA, Beis S, Mante O, Abdoulmoumine N (2010) Ind Eng Chem Res 49:3533–3538
29. Carlson TR, Cheng YT, Jae J, Huber GW (2011) Energy. Environ Sci 4:145–161
30. Schultz TP, Fisher TH, Dershem SM (1989) In: Adhesives from renewable resources, ACS Publications, Washington
31. Britt PF, Buchanan AC, Thomas KB, Lee SK (1995) J Anal Appl Pyrolysis 33:1–19
32. Heitz M, Wu G, Lapointe J, Rubio M (1995) J Wood Chem Technol 15:515–528
33. Miller JE, Evans L, Littlewolf A, Trudell DE (1999) Fuel 78:1363–1366
34. Dorrestijn E, Kranenburg M, Poinsot D, Mulder P (2005) Holzforschung 53:611–616
35. Jia S, Cox BJ, Guo X, Zhang ZC, Ekerdt JG (2010) Holzforschung 64:577–580
36. Zakzeski J, Jongerius AL, Weckhuysen BM (2010) Green Chem 12:1225–1236
37. Stärk K, Taccardi N, Bösmann A, Wasserscheid P (2010) Chemsuschem 3:719–723
38. Gasser CA, Hommes G, Schäffer A, Corvini PFX (2012) Appl Microbiol Biotechnol 95:1115–1134
39. Seong G, Dejhosseini M, Adschiri T (2018) Appl Catal A 550:284–296
40. Fan M, Deng S, Wang T, Li Q (2014) Chin J Chem Phys 27:221–226
41. Zhou M, Sharma BK, Li J, Zhao J, Xu J, Jiang J (2019) Fuel 239:239–244
42. Fernandes MRC, Huang X, Abbenhuis HCL, Hensen EJM (2019) Int J Biol Macromol 123:1044–1051
43. Kast & Ehinger Gmbh (1943) Procédé d'extraction de lignine entrant en réaction, à partir de la lignine de sacchariication du bois, Patent code: BE 448826 19430200, Belgium

44. Mayer C (1975) Obtaining aromatic fuels and intermediate products from fossil natural products by thermal depolymerization and stabilizing pyrolysis, Patent code: DE 2358960 A1 19750528, Germany
45. Shabtai JS, Zmierczak WW, Chornet E (1999) Process for conversion of lignin to reformulated hydrocarbon gasoline, Patent code: WO 9910450 A1 19990304, USA
46. Watanabe T, Messner K, Koller K (1999) Chemical method for lignin depolymerization, Patent code: WO 9909244 A1 19990225, Japan
47. Shabtai JS, Zmierczak WW, Chornet E, Johnson D (2003) Process for converting lignins into a high octane blending component, Patent code: US 20030115792 A1 20030626, USA
48. Shabtai JS, Zmierczak WW, Chornet E (2000) Process for conversion of lignin to reformulated, partially oxygenated gasoline, Patent code: WO 2000011112 A1 20000302, USA
49. Zhang X (2014) Conversion of lignin to phenolic and carboxylate compounds which can be used as antimicrobial agents, for polymer applications, or hydrogenated to hydrocarbon fuels, Patent code: WO 2014200509 A1 20141218, USA
50. Sarkanen S, Wang Y, Nutsubidze NN (2000) Isolation and activities of lignin depolymerase from Trametes cingulata and its use for depolymerizing lignin and delignifying kraft pulp, Patent code: WO 2000073426 A1 20001207, USA
51. Laxmikant DP, Kelkar AA, Matsagar BM, Singh SK (2014) Production of aromatic compounds by depolymerization of lignin, Patent code: WO 2014181360 A1 20141113, India
52. Li W, Wang H, Liu Q, Ma L, Zhang Q, Wei X (2018) Method for catalytically depolymerizing lignin, Patent code: CN 108014782 A 20180511, China
53. Lewin M, Goldstein JS (1991) Wood structure and composition. International fiber science and technology. CRC Press, Boca Raton
54. Sjöström E (1981) Wood chemistry, fundamentals and applications. Academic Press, Millbrae
55. Kamm B, Gruber PR, Kamm M (2006) Biorefineries—industrial processes and products. Wiley-VCH Verlag, Weinheim
56. Li C, Zhao X, Wang A, Huber GW, Zhang T (2015) Chem Rev 115:11559–11624
57. Brunow G (2006) In: Biorefineries-Industrial proccesses and products, Wiley-VCH, Weinheim
58. Wen JL, Xue BL, Xu F, Sun RC, Pinkert A (2013) Ind Crops Prod 42:332–343
59. Pandey MP, Kim CS (2011) Chem Eng Tech 34:29–41
60. Serrano L, Esakkimuthu ES, Marlin N, Brochier-Salon MC, Mortha G, Bertaud F (2018) Energy Fuels 32:5969–5977
61. Sakakibara A, Sano Y (2001) Wood and cellulose chemistry. Marcel Dekker, New York
62. Sannigrahi P, Pu Y, Ragauskas A (2010) Curr Opin Env Sust 2:383–393
63. Lewis NG, Yamamoto E (1990) Annu Rev Plant Physiol Plant Mol Biol 41:455–496
64. Sette M, Wechselberger R, Crestini C (2011) Chem Eur J 17:9529–9535
65. Crestini C, Lange H, Setter M, Argyropolulos DS (2017) Green Chem 19:4104–4121
66. Vishtal A, Kraslawski A (2011) BioResources 6:3547–3568
67. Rahimi A, Ulbrich A, Stahl SS, Coon JJ (2014) Nature 515:249–252
68. Laskar DD, Yang B, Wang H, Lee J (2013) Biofuels Bioprod Biorefin 7:602–626
69. Deuss PJ, Lancefield CS, Narani A, de Vries JG, Westwood NJ, Barta K (2017) Green Chem 19:2774–2782
70. Chung H, Washburn NR (2013) Green Mater 1:137–160
71. Galkin MV, Samec JSM (2016) Chem Sus Chem 9:1544–1558
72. Chatel G, Rogers RD (2014) ACS Sust Chem Eng 2:322–339
73. Long J, Lou W, Wang L, Yin B, Li X (2015) Chem Eng Sci 122:24–33
74. Li C, Zhao X, Wang A, Huber GW, Zhang T (2015) Chem Rev 115:11559–11624
75. Gillet S, Aguedo M, Petitjean L, Morais ARC, da Costa Lopes AM, Łukasik RM, Anastas PT (2017) Green Chem 00:1–3
76. Bouxin FP, McVeigh A, Tran F, Westwood NJ, Jarvis MC, Jackson SD (2015) Green Chem 17:1235–1242
77. Erickson M, Larsson S, Miksche GE (1973) Acta Chem Scand 27:903–904
78. Nimz HH (1974) Agew Chem Int Ed 13:313–321
79. Alder E (1977) Wood Sci Technol 11:169–218
80. Glasser WG, Glasser HR (1981) Pap Puu 63:71–83
81. Miller JE, Evans LR, Littlewolf AE, Trudell DE (2002) Sandia National Laboratories report SAND2002-1317, USA
82. Toledano A, Serrano L, Labidi J, Pineda A, Balu AM, Luque R (2013) Chem Cat Chem 5:977–985

83. Toledano A, Serrano L, Pineda A, Romero AA, Luque R, Labidi J (2014) Appl Catal B-Environ 145:43–55
84. Li J, Henriksson G, Gellerstedt G (2007) Biores Tech 98:3061–3068
85. Roberts VM, Stein V, Reiner T, Lemonidou A, Li X, Lercher JA (2011) Chem Eur J 17:5939–5948
86. Toledano A, Serrano L, Labidi J (2014) Fuel 116:617–624
87. Zhao C, Lercher JA (2013) In: The role of catalysis for the sustainable production of bio-fuels and bio-chemicals, Elsevier Science, Amsterdam
88. Yamazaki J, Minami E, Saka S (2006) J Wood Sci 52:527–532
89. Hewson WB, Hibbert H (1943) J Am Chem Soc 65:1173–1176
90. Gasson JR, Forchheim D, Sutter T, Hornung U, Kruse A, Barth T (2012) Ind Eng Chem Res 51:10595–10606
91. Aida TM, Sato T, Sekiguchi G, Adschiri T, Arai K (2002) Fuel 81:1453–1461
92. Tsujino J, Kawamoto H, Saka S (2003) Wood Sci Technol 37:299–307
93. Adjaye JD, Bakhshi NN (1995) Fuel Process Technol 45:161–183
94. Rahimi A, Azarpira A, Kim H, Ralph J, Stahl SS (2013) J Am Chem Soc 135:6415–6418
95. Pu Y, Jiang N, Ragauskas AJ (2007) J Wood Chem Technol 27:23–33
96. George A, Tran K, Morgan TJ, Benke PI, Berrueco C, Lorente E, Wu BC, Keasling JD, Simmons BA, Holmes BM (2011) Green Chem 13:3375–3385
97. Crestini C, Crucianelli M, Orlandi M, Saladino R (2010) Catal Today 156:8–22
98. Oasmaa A, Czernik S (1999) Energy Fuels 13:914
99. Ma Z, Troussard E, van Bokhoven JA (2012) Appl Catal A 423:130–136
100. Jackson MA, Compton DL, Boateng AA (2009) J Anal Appl Pyrolysis 85:226–230
101. Huber GW, Corma A (2007) Angew Chem Int Ed 46:7184–7201
102. Rezaei PS, Shafaghat H, Daud WMAW (2014) Appl Catal A 469:490–511
103. Shen D, Zhao J, Xiao R (2016) Energ Convers Manage 124:61–72
104. Park HJ, Park KH, Jeon JK, Kim J, Ryoo R, Jeong KE, Park SH, Park YK (2012) Fuel 97:379–384
105. Choi SJ, Park SH, Jeon JK, Lee IG, Ryu C, Suh DJ, Park YK (2013) Renew Energ 54:105–110
106. Shabtai JS, Zmierczak WW, Chornet E (2001) US Patent 6172272
107. Sheu YHE, Anthony RG, Soltes EJ (1988) Fuel Process Technol 19:31–50
108. Harris EE, D'Ianni J, Adkins H (1938) J Am Chem Soc 60:1467–1470
109. Horacek J, Homola F, Kubickova I, Kubicka D (2012) Catal Today 179:191–198
110. Odebunmi EO, Ollis DF (1983) J Catal 80:65–75
111. Sun J, Karim AM, Zhang H, Kovarik L, Li XS, Hensley AJ, McEwen JS, Wang Y (2013) J Catal 306:47–57
112. Zhu X, Lobban LL, Mallinson RG, Resasco DE (2011) J Catal 281:21–29
113. Huuska MK (1986) Polyhedron 5:233–236
114. Yang Y, Ochoa-Hernández C, O'Shea VA, Pizarro P, Coronado JM, Serrano DP (2014) Appl Catal B: Environ 145:91–100
115. Olcese RN, Bettahar M, Petitjean D, Malaman B, Giovanella F, Dufour A (2012) Appl Catal B 115–116:63–73
116. Olcese R, Bettahar MM, Malamanc B, Ghanbajac J, Tibavizco L, Petitjean D, Dufour A (2013) Appl Catal B 129:528–538
117. Xu X, Jiang E, Du Y, Li B (2016) Renew Energ 96:458–468
118. He T, Liu X, Ge Y, Han D, Li J, Wang Z, Wu J (2017) Catal Commun 102:127–130
119. Ardiyanti AR, Khromova SA, Venderbosch RH, Yakovlev VA, Heeres HJ (2012) Appl Catal B 117–118:105–117
120. Whiffen VML, Smith KJ (2010) Energy Fuels 24:4728–4737
121. Ghampson IT, Sepúlveda C, Garcia R, Radovic LR, Fierro JLG, DeSisto WJ, Escalona N (2012) Appl Catal A 439–440:111–124
122. Lee WS, Wang Z, Wu RJ, Bhan A (2014) J Catal 319:44–53
123. Li K, Wang R, Chen J (2011) Energy Fuels 25:854–863
124. Zhao HY, Li D, Bui P, Oyama ST (2011) Appl Catal A 391:305–310
125. Chen Q, Marshall MN, Geib SM, Tien M, Richard TL (2012) Bioresource Technol 117:186–192
126. Roberts SM, Turner NJ, Willetts AJ, Turner MK (1995) Camb Univ Press 2:34–40
127. Salvachúa D, Katahira R, Cleveland NS, Khanna P, Resch MG, Black BA, Purvine SO, Zink EM, Prieto A, Martínez MJ, Martínez AT, Simmons BA, Gladden JM, Beckham GT (2016) Green Chem 18:6046–6062
128. Kirk TK, Farrell RL (1987) Annu Rev Microbiol 41:465–501

129. Salvachúa D, Prieto A, Martínez AT, Martínez MJ (2013) Appl Environ Microbiol 79:4316–4324
130. Dashtban M, Schraft H, Syed TA, Qin W (2010) Int J Biochem Mol Biol 1:36–50
131. Husarcíková J, Voß H, Domínguez de María P, Schallmey A (2018) Appl Microbiol Biotechnol 102:5391–5401
132. Ahmad M, Roberts JN, Hardiman EM, Singh R, Eltis LD (2011) Bugg TD 50:5096–5107
133. Salvachúa D, Karp EM, Nimlos CT, Vardon DR, Beckham GT (2015) Green Chem 17:4951–4967
134. Rico A, Rencoret J, del Río J, Martínez AT, Gutiérrez A (2014) Biotechnol Biofuels 7:6
135. Kolb M, Sieber V, Amann M, Faulstich M, Schieder D (2012) Bioresour Technol 104:298–304
136. Tengerdy RP, Szakacs G (2003) Biochem Eng J 13:169–179
137. Rochefort D, Leech D, Bourbonnais R (2004) Green Chem 6:14–24
138. Bourbonnais R, Paice M, Reid I, Lanthier P (1995) Yaguchi. M Appl Environ Microb 61:1876–1880
139. Shleev S, Persson P, Shumakovich G, Mazhugo Y, Yaropolov A, Ruzgas T (2006) Gorton. Enzyme Microb Tech 39:841–847
140. Baiocco P, Barreca AM, Fabbrini M, Galli C, Gentili P (2003) Org Biomol Chem 1:191–197
141. Ruiz-Dueñas FJ, Martínez AT (2009) Microb Biotechnol 2:164–177
142. Martínez AT, Speranza M, Ruiz-Dueñas FJ, Ferreira P, Camarero S, Guillén F, Martínez MJ, Gutiérrez A, del Río JC (2005) Int Microbiol 8:195–204
143. Wariishi H, Dunford HB, MacDonald ID, Gold MII (1989) J Biol Chem 264:3335–3340
144. Gold MH, Youngs HL, Gelpke MD (2000) Met Ions Biol Syst 37:559–586
145. Kapich A, Hofrichter M, Vares T, Hatakka A (1999) Biochem Biophys Res Commun 259:212–219
146. Kapich AN, Jensen KA, Hammel KE (1999) FEBS Lett 461:115–119
147. Camarero S, Sarkar S, Ruiz-Dueiias FJ, Martinez MAJ, Martinez AT (1999) J Biol Chem 274:10324–10330
148. Dordick JS, Marletta MA, Klibanov AM (1986) Proc Natl Acad Sci USA 83:6255–6257
149. Shi Y, Chai L, Tang C, Yang Z, Zheng Y, Chen Y, Jing Q (2013) Bioprocess Biosyst Eng 36:1957–1965
150. Liu D, Yan X, Zhuo S, Si M, Liu M, Wang S, Ren L, Chai L, Shi Y (2018) Bioresour Technol 257:62–68
151. Ortiz-Bermúdez P, Srebotnik E, Hammel KE (2003) Appl Environ Microbiol 69:5015–5018
152. Brown ME, Barros T, Chang MCY (2012) ACS Chem Biol 7:2074–2081
153. Rahmanpour R, Bugg TDH (2015) Arch Biochem Biophys 574:93–98
154. de Gonzalo G, Colpa DI, Habib MHM, Fraaije MW (2016) J Biotechnol 236:110–119
155. Hofrichter M, Ullrich R, Pecyna M, Liers C, Lundell T (2010) Appl Microbiol Biotechnol 87:871–897
156. Picart P, Müller C, Mottweiler J, Wiermans L, Bolm C, Domínguez de María P, Schallmey A (2014) Chemsuschem 7:3164–3171
157. Reiter J, Strittmatter H, Wiemann LO, Schieder D, Sieber V (2013) Green Chem 15:01373–01381
158. Picart P, Sevenich M, Domínguez de María P, Schallmey A (2015) Green Chem 17:4931–4940
159. Ohta Y, Hasegawa R, Kurosawa K, Maeda AH, Koizumi T, Nishimura H, Okada H, Qu C, Saito K, Watanabe T, Hatada Y (2016) Chemsuschem 10:425–433
160. Gall DL, Kontur WS, Lan W, Kim H, Li Y, Ralph J, Donohue TJ, Noguera DR (2018) Appl Environ Microbiol 84:e02076-17
161. Picart P, Liu H, Grande PM, Anders N, Zhu L, Klankermayer J, Leitner W, Domínguez de María P, Schwaneberg U, Schallmey A (2017) Appl Microbiol Biotechnol 101:6277–6287
162. Toledano A, Serrano L, Pineda A, Balu AM, Labidi J, Luque R (2013) Chemsuschem 6:529–536
163. Song Q, Wang F, Cai J, Wang Y, Zhang J, Yu W, Xu J (2013) Energy Environ Sci 6:994–1007
164. Gosselink RJA, Teunissen W, Van Dam JEG, de Jong E, Gellerstedt G, Scott EL, Sanders JPM (2011) Bioresour Technol 106:173–177
165. Barta K, Warner GR, Beach ES, Anastas PT (2014) Green Chem 16:191–196
166. Jongerius AL, Bruijnincx PCA, Weckhuysen BM (2013) Green Chem 15:3049–3056
167. Güvenatam B, Jurs O, Heeres EHJ, Pidko EA, Hensen EJM (2014) Catal Today 233:83–91

Topics in Current Chemistry (2019) 377:36
https://doi.org/10.1007/s41061-019-0260-5

REVIEW

Heterogeneous Catalyzed Thermochemical Conversion of Lignin Model Compounds: An Overview

Mikel Oregui-Bengoechea[1] · Ion Agirre[1] · Aitziber Iriondo[1] · Alexander Lopez-Urionabarrenechea[1] · Jesus M. Requies[1] · Iker Agirrezabal-Telleria[1] · Kepa Bizkarra[1] · V. Laura Barrio[1] · Jose F. Cambra[1]

Received: 29 May 2019 / Accepted: 18 October 2019 / Published online: 14 November 2019

Abstract
Thermochemical lignin conversion processes can be described as complex reaction networks involving not only de-polymerization and re-polymerization reactions, but also chemical transformations of the depolymerized mono-, di-, and oligomeric compounds. They typically result in a product mixture consisting of a gaseous, liquid (i.e., mono-, di-, and oligomeric products), and solid phase. Consequently, researchers have developed a common strategy to simplify this issue by replacing lignin with simpler, but still representative, lignin model compounds. This strategy is typically applied to the elucidation of reaction mechanisms and the exploration of novel lignin conversion approaches. In this review, we present a general overview of the latest advances in the principal thermochemical processes applied for the conversion of lignin model compounds using heterogeneous catalysts. This review focuses on the most representative lignin conversion methods, i.e., reductive, oxidative, pyrolytic, and hydrolytic processes. An additional subchapter on the reforming of pyrolysis oil model compounds has also been included. Special attention will be given to those research papers using "green" reactants (i.e., H_2 or renewable hydrogen donor molecules in reductive processes or air/O_2 in oxidative processes) and solvents, although less environmentally friendly chemicals will be also considered. Moreover, the scope of the review is limited to those most representative lignin model compounds and to those reaction products that are typically targeted in lignin valorization.

Keywords Lignin · Model compounds · Heterogeneous catalysis · Thermochemical conversion · Reaction mechanism

---

Chapter 8 was originally published as Oregui-Bengoechea, M., Agirre, I., Iriondo, A., Lopez-Urionabarrenechea, A., Requies, J. M., Agirrezabal-Telleria, I., Bizkarra, K., Barrio, V. L. & Cambra, J. F. Topics in Current Chemistry (2019) 377: 36. https://doi.org/10.1007/s41061-019-0260-5.

✉ Mikel Oregui-Bengoechea
mikel.oregui@ehu.eus

Extended author information available on the last page of the article

1 Introduction

Global warming, volatile oil prices, and world political instability suggests the necessity of accelerating the transition from a fossil-based to a more sustainable and "green" economy. While energy can be produced from different renewable sources, other fossil-based chemicals can only be produced from biomass [1]. In particular, lignocellulosic biomass found in both agricultural and forestry residues is an abundant and relatively unexploited resource with enormous potential to replace non-renewable fuels and chemicals [2].

Lignocellulosic biomass is a heterogeneous feedstock comprising typically 40–45 wt% cellulose, 25–35 wt% hemicellulose, 15–30 wt% lignin, and up to 10 wt% inorganic components [3]. Over the last decades, most of the effort has been focused on the development of more efficient processes for the conversion of sugars—from the cellulose and hemicellulose fractions- into fuels and chemicals [4, 5]. In contrast, the efficient conversion of lignin, is still a challenge. Only approximately 2% of the lignin residues are used commercially, with the remaining volumes being burned as low value fuel [6]. Thus, the efficient valorization of the lignocellulose feedstock requires the development of new and efficient strategies for the conversion of lignin.

Lignin is a highly complex polyphenol macromolecule (600–15,000 kDa) formed through the chemical bonding of three basic propyl-phenyl units, the so-called monolignols (Fig. 1): coumaryl (H unit), coniferyl (G unit), and sinapyl alcohol (S unit) [7]. The relative abundance of each monolignol is related to the plant taxonomy: (1) softwood lignin (gymnosperm) contains more G units, (2) hardwood (angiosperm) lignin is mainly a mixture of G and S units, while (3) grass lignin presents a mixture of all three units [8]. Monolignols are predominantly linked by C–O–C ether or C–C bonds (Fig. 2). In native lignin, ether bonds account for over two-thirds of the total linkages [9]. The most abundant linkage is the β-*O*-4 linkage accounting for 43–50% of the bonds in softwood and 50–65% in hardwood [10]. Other mayor lignin linkages are β-5 (phenylcoumaran) and β–β (resinol) bonds. Additional linkages such as α-*O*-4 (α-aryl ether), 4-*O*-5 (diaryl ether), 5–5 and β-1 (spirodienone) are also found at lower concentration (Fig. 2) [7, 11]. The reactivity of lignin is largely determined by the nature and relative abundance of the aforementioned lignin linkages. In addition, the reactivity of the biopolymer is affected, to a lesser degree, by the presence

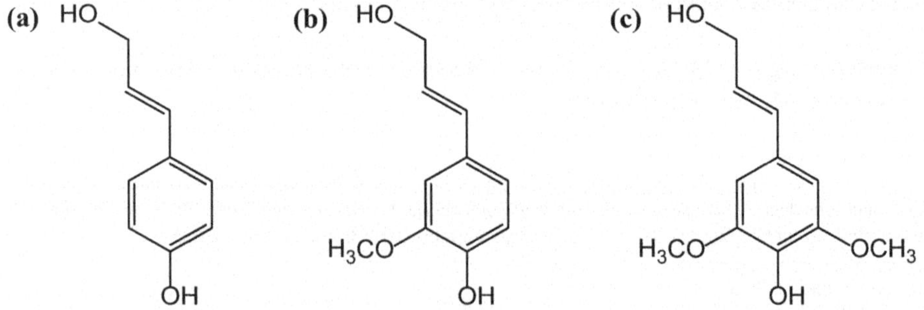

Fig. 1 Monolignols: **a** *p*-coumaryl (H unit), **b** coniferyl (G unit), and **c** sinapyl alcohol (S unit)

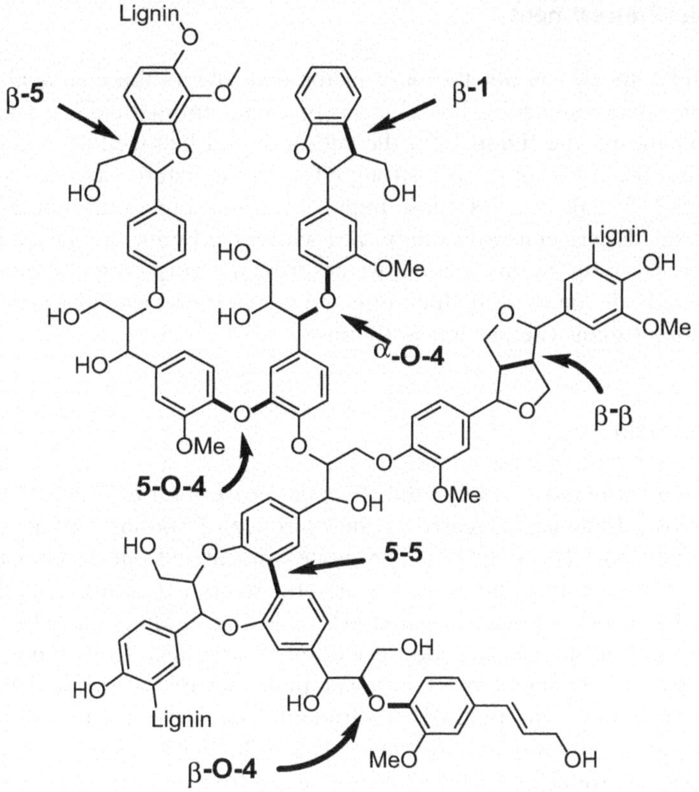

Fig. 2 Representative structure of lignin showing the major ether and C–C bonds of lignin

of functional groups such as the methoxyl, hydroxyl, and carbonyl groups attached to the aromatic rings and ether and C–C linkages.

Lignin is the only direct source of renewable aromatics and its conversion can also lead to the production of fuel and fuel blends [12, 13]. Hence, numerous strategies have been applied for the deconstruction of lignin into components such as phenols, aromatic aldehydes, cyclic hydrocarbons, etc. [9]. Reductive, oxidative, pyrolytic, hydrolytic, and even gasification methods have been applied for the conversion of lignin, but these processes are often inefficient [9]. The highly cross-linked amorphous structure of lignin makes it extremely resistant to thermal and chemical attack [7]. Moreover, lignin thermochemical conversion processes often lead to side reactions that convert a considerable amount of the biopolymer into organic solids or chars, reducing the process selectivity [14]. Consequently, only one large-scale commercial process currently uses lignin as raw material: the production of vanillin by oxidation of lignosulfonates (Borregaard) [7, 10].

1.1 Biomass Pretreatment

The botanical species is not the only factor that affects the chemical structure of the lignin macromolecule. The biomass pretreatment method, i.e., the method used for separating the lignin from the cellulose and hemicellulose components (i.e., biomass fractionation), has a strong effect in the nature and reactivity of the lignin bonds [15, 16]. Besides, those impurities found in natural biomass (minerals, organic acids) are generally solubilized within the lignin streams and can play a role in its reactivity. In this section, we describe the most common biomass pretreatment methods found in the literature and the characteristic properties of their corresponding lignins (i.e., technical lignins).

1.1.1 Kraft Lignin

Kraft pulping is the most widely used chemical process for biomass pretreatment. Lignocellulosic biomass is treated in the presence of sodium sulfide and under alkaline conditions (Na_2S/NaOH). The process is carried out at temperatures of 155–175 °C for several hours yielding a solid fraction (cellulose) and a lignin containing fluid (black liquor) fraction [16, 17]. The lignin is partially depolymerized and thiol groups are introduced within its structure. Kraft lignin is water-insoluble and its molecular mass is lower than that of the original lignin [16, 18]. In kraft pulping, the lignin is traditionally obtained in a two-step process. First, the lignin is precipitated after neutralizing the black liquor. Later the lignin is redissolved in water and acid to overcome conventional filtering and sodium separation problems. An additional technology for extracting high-quality lignin from a kraft pulp mill is the LignoBoost process. In the LignoBoost process, the lignin is obtained by evaporating the black liquor after its neutralization with CO_2 [16, 19].

1.1.2 Alkali-Pulping Lignin

Among alkali-pulping processes, soda pulping is the most common technology to obtain alkali lignin. Soda (NaOH) is used as the main pulping chemical employed. Additives such as anthraquinone might also be added to the reaction mixture in order to decrease carbohydrate degradation. In soda pulping, the lignin recovery consist of a three-step precipitation-maturation-filtration process. This type of lignin is sulfur-free [20].

1.1.3 Sulfite Process Lignin

The sulfite process yields to the so-called lignosulfonates, a class of modified lignin molecules characterized by its water solubility and the presence of sulfonated groups within its structure [15, 16]. The sulfite processes consist on the impregnation of biomass with an aqueous solution of sulfur dioxide at different pHs. Lignosulfonates are widely used in the industry as plasticizers or water reducers [21].

1.1.4 Steam Explosion

Steam explosion does not yield a high-purity technical lignin. This process only makes the cellulose more accessible for a subsequent hydrolysis pretreatment process. Thus, steam explosion is generally coupled to a dilute hydrolysis process in order to obtain a high-purity technical lignin. The steam explosion process typically consists of two-step process. Firstly, biomass reacts with steam at high temperatures (between 180 and 240 °C), high pressures (1–3.5 MPa), and short reaction times. In a second step, the biomass is exploded in the presence of different chemicals; a sudden pressure release is applied in order to induce biomass explosion. The steam explosion process defibrillates the cellulose bundles making the cellulose more accessible [16, 18]. The nature of the steam-exploded lignin is highly hydrophobic, with a low level of carbohydrate and biomass-extractive impurities. Its molecular mass is relatively low since some acid hydrolysis of lignin takes place during the steam explosion process [8, 16].

1.1.5 Organosolv Lignin

In the organosolv process, lignin is extracted from biomass in the presence of an organic solvent (e.g., ethanol) or a mixture of water and an organic solvent. The process is conducted at high temperatures and pressures [8, 16]. Organosolv pulping or pretreatment enables the production of high-quality cellulose and lignin [16, 22]. This type of lignin has a less modified structure than kraft lignin and is largely sulfur-free [8, 16].

1.1.6 Acid-Hydrolysis Lignin

Dilute and strong acid hydrolysis is among the most effective pretreatment methods for the fractionation of lignocellulosic biomass. Both inorganic and organic acids such as sulfuric, oxalic, or per-acetic acid can be used [16, 23]. Depending on the concentration of the acid used in the process, the acid fractionation methods are divided into dilute and strong acid hydrolysis. Strong hydrolysis of biomass is normally conducted in the presence of strong and concentrated mineral acids (e.g., H_2SO_4, H_3PO_4, or HCl) at temperatures lower than 160 °C. Batch reaction systems are preferred and high biomass loading can be processes [16, 24]. However, this process requires large amounts of acids causing corrosion problems to the equipment. Little is known about the structural change of the lignin upon the acid hydrolysis process. Evstigneyed et al. [25] studied the structure of a lignin produced by the industrial acid hydrolysis in a H_2O_2–H_2SO_4 system. This process leads to an opening of aromatic rings and probable formation of muconic acid derivatives. However, the chemical process has little effect on alkyl-aryl ether linkages (β-*O*-4) between lignin phenyl-propane subunits.

Dilute acid treatment is considered as a cheap and effective pretreatment method due to the low cost and easy availability of the acids [23]. The method is especially suitable for the pretreatment of biomass with low lignin content [16, 26]. It is also generally combined with steam explosion or enzymatic hydrolysis. In general, dilute sulfuric acid is sprayed on raw biomass, which is then heated up to 160–220 °C for a few minutes. Low acidities (e.g., concentration of $H_2SO_4 < 4$ wt%) and a biomass concentration around 5–10 wt% are used [27]. During acid hydrolysis of biomass, lignin can undergo chemical and structural changes, the most relevant being the cleavage of a pretreatment of the β-*O*-4 and the shift of the S/G units ratios [26].

1.1.7 Enzymatic-Hydrolysis Lignin

Enzymatic hydrolysis of biomass is normally performed after a first steam explosion or dilute hydrolysis step. In enzymatic hydrolysis, cellulolytic enzymes are used to hydrolyze the carbohydrate fraction, leaving behind a cellulose-enzyme-lignin residue. This process occurs at mild conditions producing some slight structural changes in lignin: decrease of the phenolic hydroxyl group content, increase of the β-*O*-4 linkages and increase the molecular weight of lignin [28]. In addition, enzymatic hydrolyzed lignin contains a considerable amount of the protein and carbohydrate impurities [29].

1.2 Lignin Thermochemical Conversion

In recent years, research on lignin thermochemical valorization into value-added chemicals and alternative fuels has grown rapidly. Generally, thermochemical lignin conversion processes fall in one of the following three categories: (1) those occurring in the presence of an oxidant, preferably O_2 or air although H_2O_2 is also used, (2) those occurring in the presence of a reducing agent, either H_2 or a hydrogen donor molecule, or (3) those occurring in a neutral environment [9, 30]. However, these categories do not consider the specific nature of all the process applied for lignin depolymerization. A more rational classification often employed by researchers working in the field divides lignin thermochemical conversion processes into oxidative, reductive, pyrolytic, hydrolytic, and gasification [9]. In some cases, hydrolytic processes are further divided into three different categories: acid catalyzed, base catalyzed, and liquid phase reforming [9].

Reductive lignin conversion involves those processes that are carried out in the presence of a reducing agent. Reaction temperatures typically range from 100 to 350 °C and the targeted products are hydrodeoxygenated monoaromatics such as phenols, benzene, toluene, and xylene. Oxidative lignin conversion processes are characterized by the used of oxidants and are generally carried out at lower temperatures of 0–250 °C [9]. The target compounds are alcohols, acids, and aldehydes although the latter, aldehydes, particularly vanillin, are the compounds of greatest interest [9]. Pyrolysis, also known as thermal cracking, is the breakdown of organic molecules by the effect of temperature in a non-oxidizing medium. Pyrolysis of lignin is typically carried out at 450–700 °C and is characterized, unlike other processes, for following a radical reaction pathway [9]. Hydrolytic processes refer to those methods applied for lignin conversion in which water is used to break down the lignin matrix. In this processes, acid or basic species are typically used as catalysts, although in some cases no catalyst is used. Finally, gasification refers to those lignin conversion processes that intent to produce synthesis gas (CO and H_2) from a range of real lignin feedstocks [9].

1.3 Lignin Model Compounds

As described above, lignin is a very complex molecule. Moreover, the chemical system generated in the conversion of this biopolymer results in an infinite number of chemical compounds, many times distributed in the solid, liquid, and gas phases, that interact in a highly intricate reaction network involving de-polymerization and re-polymerization reactions among others [9, 30]. For this reason, research often uses the so-called lignin model compounds, simpler molecules that are still representative of the lignin biopolymer, in order to simplify the reaction network. This strategy is typically applied for the elucidation of reaction mechanisms and for the exploration of novel lignin conversion approaches [7, 31]. The knowledge derived from the conversion of lignin model compounds is very helpful to understand the direct conversion of real lignin.

Lignin model compounds contain those functional groups and linkages found in the native and technical lignins; thus, their reactivity mimics the one of the lignin macromolecule. Model compounds are generally classified into two categories, i.e., monomers and dimers. Monomers refer to hydroxyl, methoxyl, and alkyl substituted monoaromatic compounds. Monomeric model compounds are generally used to study the chemical behavior of de-polymerized lignin fragments in lignin conversion processes. They are used to evaluate what the chemical behavior of the methoxy, hydroxy, and alkyl groups of the lignin macromolecule is in the conversion process under study and how these functional groups affect the reaction mechanism. The type of model compound used during reaction varies depending on the nature of the process, i.e., oxidative, reductive, etc., and the targeted reaction product. In hydrolytic, pyrolysis, and reductive processes, monomers such as guaiacol, anisole, or phenol are commonly used (Fig. 3). While the chemical nature of model compounds such as veratryl, vanillil, or cinnamyl alcohol is preferred to study the chemical transformations that occur in oxidative processes (Fig. 3).

Dimers refer to chemical compounds that contain those most common lignin linkages. These include C–O–C linkages, i.e., β-*O*-4, α-*O*-4, and 5-*O*-4, and C–C linkages such as 5–5/5´, β–β, β-1 and β-5 (Fig. 2). Thus, in comparison to monomeric lignin model compounds, dimeric compounds are more representative of the lignin polymer, i.e., more representative of a lignin fragment/linkage. When using dimeric lignin model compounds, lignin de-polymerization mechanism is the focus of study: they are mainly employed to evaluate the reactivity and chemical behavior of a specific lignin linkage. Dimeric lignin model compounds can also be used to study the chemical behavior of the produced lignin monomers; however, the complexity of these reaction systems may hinder drawing any clear conclusion about

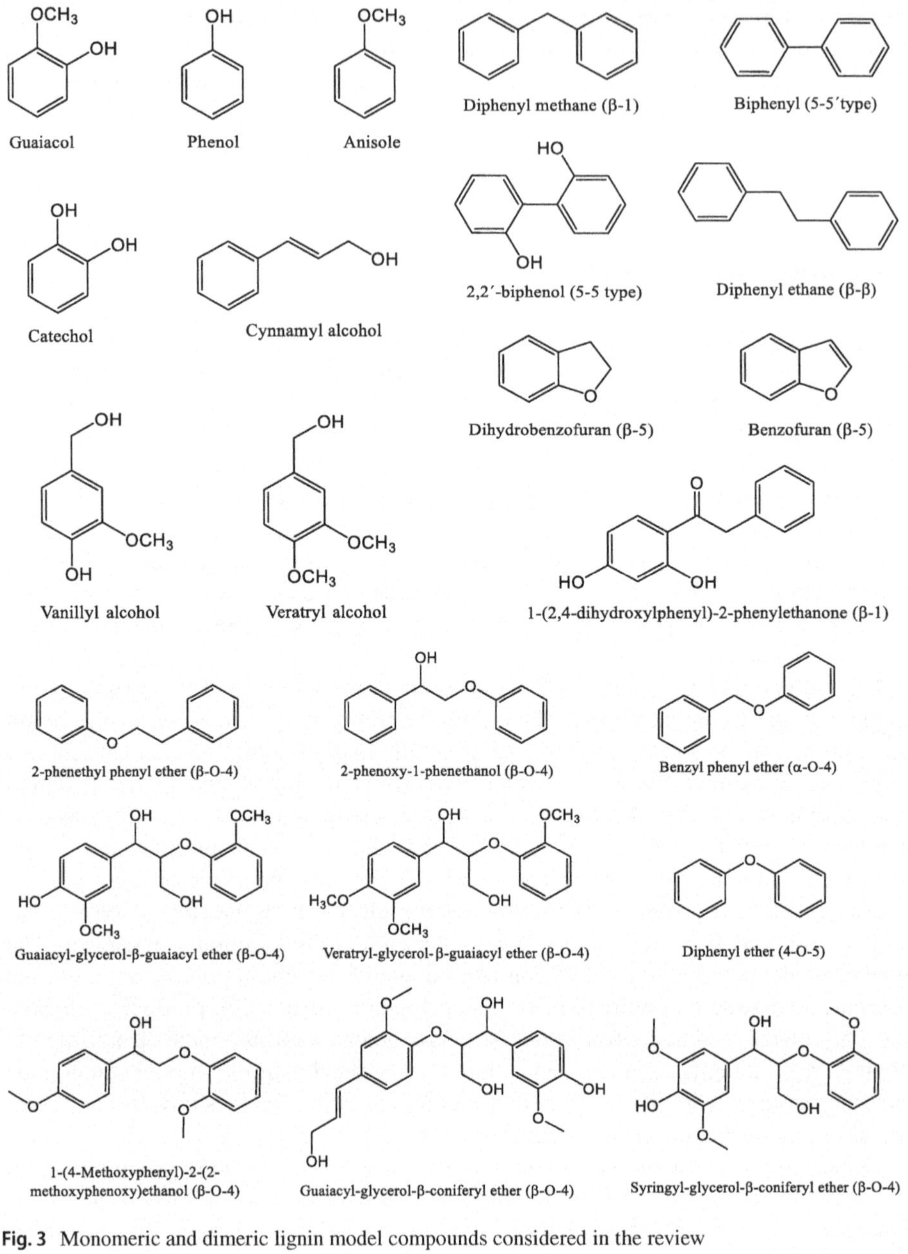

Fig. 3 Monomeric and dimeric lignin model compounds considered in the review

the reactivity of those functional groups attached to the benzene ring. For this reason, many researchers study the reactivity of both dimeric and monomeric model compounds in order to get a clearer picture of the possible chemical behavior of the lignin macromolecule. Dimeric model compounds include simple aromatic rings linked through C–O–C ether and C–C linkages, or more complex compounds containing additional hydroxyl, methoxyl, and alkyl functionalities (Fig. 3). 2-phenethyl phenyl ether (PPE) is the simplest β-*O*-4 model compound consisting of two aromatic rings linked through a β-*O*-4 linkage (Fig. 3). Other more complex β-*O*-4 model compounds include 2-phenoxy-1-phenethanol (PHPL), guaiacyl-glycerol-β-guaiacyl-ether (GGG) and veratryl-glyccerol-β-guaiacyl-ether (VGG) (Fig. 3). Benzyl phenyl ether (BPE) and diphenyl-ether (DPE) are the most studied α-*O*-4 and 4-*O*-5 lignin model compounds, respectively (Fig. 3). Examples of typical C–C dimeric compounds include 2,2′-biphenol (5–5 type), biphenyl (5–5′type), diphenylethane (β–β), diphenyl methane (β-1), 1-(4-hydroxylphenyl)-2-2phenylethanone (β-1). Dihydrobenzofuran (β-5) and benzofuran (β-5) (Fig. 3).

1.4 Heterogeneous Catalysis

Catalysis is an important aspect when designing chemical technology. Around 80% of the production of industrially important chemicals involve the direct use of catalysts [32]. A catalyst is a substance that usually allows alternative reaction pathways with lower activation barriers than in their absence. The use of a catalyst can facilitate better control on the selectivity of the process and can reduce both the process severity and the production of non-valuable sub-products. As a consequence, catalytic processes are generally more economically viable, environmentally friendly, and energy efficient than non-catalyzed or stoichiometric reactions [32].

Catalysts are generally classified into three categories: heterogeneous, homogenous, and biocatalysts or enzymes. Heterogeneous catalysts are generally the preferred choice in industrial applications. Compared to homogenous or biocatalysts, heterogeneous catalysts are easier to recover and recycle and, thus, are not released to the environment through waste streams [7, 32]. In addition, they normally require milder pH conditions; conventional mineral acids are often replaced by solid-acid catalysis, reducing handling and corrosion problems [7, 32]. Moreover, heterogeneous catalysts can typically have multiple types of active sites [7]. This represents a great advantage for the conversion of chemically complex reactants such as lignin. Finally, the tunable porosity of the catalyst supports can impose selectivity on the reaction products by restricting access of reactants to the active site, the formation of certain transition states, or the egress of certain products [33].

1.5 Overview and General Considerations for This Review

A large number of research articles have been published on thermochemical conversion of lignin and lignin model compounds with heterogeneous catalysts. In some cases, the results are conflicting even with similar experimental parameters such as same catalysts or solvents [34]. Zakzeski et al. [30] published in 2010 a thorough research on the conversion of lignin and lignin model compounds using both heterogeneous and homogeneous catalysts. In 2015, Li et al. [9] published a review on thermochemical conversion of real lignin with homogeneous and heterogeneous catalysts. They included most of the thermochemical processes applied for lignin deconstruction: reductive (referred as hydroprocessing), oxidative, pyrolytic, hydrolytic (base- and acid-catalyzed and liquid phase reforming), and gasification. Yang et al. [34] focused their review on thermal chemical reactions of lignin model compounds, however, they did not consider oxidative processes and included both homogeneous and heterogeneous catalysis.

In this work, we present a general overview of the latest advances in the main thermochemical processes applied for the conversion of lignin model compounds using heterogeneous catalysis. The review will focus on the most representative lignin conversion methods, i.e., reductive, oxidative, pyrolytic, and hydrolytic processes. An additional subchapter including lignin model compound reforming is also included. Special attention will be given to those research papers using "green" reactants (i.e., H_2 or renewable hydrogen donor molecules in reductive processes or air/O_2 in oxidative processes) and solvents, although less environmentally friendly chemicals will be also considered. The use of ionic liquids as solvents is also out of the scope of this review. Moreover, the scope of the review is limited to those most representative lignin model compounds and to those reaction products that are typically targeted in lignin valorization, e.g., vanillin in lignin oxidation.

2 Reductive Conversion Processes

2.1 Hydroprocessing of Lignin Model Compounds

Hydroprocessing involves thermal reduction of a compound using hydrogen. It is one of the most efficient strategies applied in deconstruction of lignin and for the conversion of lignin model compounds [9]. Lignin model compound hydroprocessing includes a series of chemical reactions such as hydrodeoxygenation (HDO), hydrogenation and hydrogenolysis. In the present work, the term (1) HDO will refer to the removal of the oxygen present in the methoxyl and hydroxyl groups, (2) hydrogenation will refer to the saturation of the aromatic C=C bonds, while (3) hydrogenolysis will be used to describe the scission of the interunit C–O–C ether and C–C bonds within the lignin dimers.

2.1.1 Monomers

Among the monomeric lignin model compounds used to study lignin hydroprocessing, guaiacol is the one that has attracted the most attention. Guaiacol contains the basic functional elements found in the lignin matrix, i.e., hydroxyl and methoxyl groups associated to an aromatic ring. Within the linkages associated to these two functional groups, the aliphatic C_{AL}–O bond in the methoxy group is the weakest (247 kJ/mol), while the aromatic C_{AR}–O bonds within the methoxy and hydroxyl groups are significantly stronger, of 356 kJ/mol and 414 kJ/mol, respectively [7]. Table 1 provides an overview of the latest studies on guaiacol hydroprocessing, summing up the solvents, catalysts, reaction conditions, main products and highest conversion levels.

Monometallic and bimetallic Mo-containing catalysts have been widely studied for guaiacol hydroprocessing. Monometallic Mo catalysts, in their oxide, sulfide,

Table 1 Summary of the analyzed literature for monomeric lignin model compounds hydroprocessing

Entry	Catalyst[a]	Solvent	R.A.[b]	Experimental conditions[c]	Model compound	Main Products (selectivity/yield in %)	Conv. (%)[d]	Refs.
1	MoO_2/C, Mo_2C/C	Tetralin	H_2	573 K, 3 MPa, and 3 h	Guaiacol	Phenol (72.0)[g], cresol (19.4)[g], anisole (6.2)[g]	98	[37]
2	MoO_3/SiO_2	Octane	H_2	613 K, 3.5 MPa, and 6 h	Guaiacol	Benzene (30.2)[h], toluene (20.1)[h], phenol (17.4)[h], methylphenol (14.8)[h], anisole (7.2)[h]	97.2	[38]
3	$NiMoS/Al_2O_3$, $CoMoS/Al_2O_3$	Tetradecane	H_2	573-673 K, 5 MPa, and 1 h	Guaiacol	Phenol (80)[g], cyclohexane (20)[g]	–	[35]
4	NiMoS, CoMoS, NiMoWS	–	H_2	673 K and 2.8 MPa, and 2.7 h^{-1} (WHSV)	Guaiacol	Phenol, cresol, catechol, benzene, toluene	99.9	[40]
5	$CoMoS/Al_2O_3$, $CoMoS/SiO_2$, $CoMoS/TiO_2$, $CoMoS/ZrO_2$	Toluene	H_2	613 K, 3 MPa, and 80 h^{-1} (FSV)	Guaiacol	Phenol (20)[g], catechol, *o*-cresol, cyclohexanol and methylcatechol	91	[41]
6	$NiMoS/K\text{-}Al_2O_3$, $CoMoS/K\text{-}Al_2O_3$	Xylenes containing ethylbenze	H_2	523 K, 5.5 MPa, and 4 h	Guaiacol	Catechol (~ 75)[h], cyclohexane (~ 35)[h], phrecienol (~ 30)[h]	~90	[39]
7	Pt/C, Ru/C, Pd/C,	Water	H_2	473 K, 2 MPa, and 2 h	Guaiacol	Cyclohexanol (~ 70)[g], methylcatechol (~ 25)[g]	~80	[42]
8	Pt/TiO_2, Pt/Al_2O_3, $NiMo/Al_2O_3$	Dodecane	H_2	573 K and 7.1 MPa[e]	Guaiacol	Cyclohexane (~ 60)[g], methane (~ 20)[g], methylcyclohexane (~ 8)[g], methylcyclopentane (~ 8)[g]	96	[36]
9	PtPd/Al-HMS, $PtPd/ZrO_2\text{-}SiO_2$, $PtPd/Al_2O_3\text{-}SiO_2$, Ru/Al-HMS, $Ru/ZrO_2\text{-}SiO_2$, $Ru/Al_2O_3\text{-}SiO_2$	Methanol	H_2	523 K, 5 MPa, and 3 h	Guaiacol	Cyclohexane (69)[g], methylcyclohexane (12)[g], cyclohexanol (5)[g]	97	[43]

Table 1 (continued)

Entry	Catalyst[a]	Solvent	R.A.[b]	Experimental conditions[c]	Model compound	Main Products (selectivity/yield in %)	Conv. (%)[d]	Refs.
10	Ni/SiO_2-Al_2O_3	–	H_2	573 K, 5 MPa, and 1 h	Guaiacol	Phenol, methyl guaiacol, benzene, anisole, cresol	87	[44]
11	Ni/MgO	Decahydronaphalene	H_2	433 K, 4 MPa, and 4 h	Guaiacol	Cyclohexanol (~98.3)[g]	~99.6	[45]
12	NiFe/CNT	–	H_2	573 K, 3 MPa, and 3 h^{-1} (WLHSV)	Guaiacol	Cyclohexane (99.9)[g]	99.8	[46]
13	CuNi/CNT	*n*-dodecane	H_2	573 K, 5 MPa, and 4 h	Guaiacol	Methylcyclohexanol (~60.0)[h], cyclohexanol (~28)[h], cyclohexanone (~5)[h], cyclohexane (~5)[h]	~100	[47]
14	Ag/TiO_2-A	Heptane	H_2	573 K, 3 MPa, and 0.5 h	Guaiacol	Phenol (~23)[h], cresols (~5)[h]	~40	[48]
15	NiMo/SiO_2	–	H_2	683 K, 0.1 MPa, and 1.5 h^{-1} (WHSV)	Anisole	Benzene (66.2)[g], toluene (24.5)[g], xylene (6.3)[g]	99.3	[49]
16	NiMo/SBA-15, NiMo/SAPO-11	Heptane	H_2	573 K and 1.5 MPa	Anisole	Methoxycyclohexane (57)[h], benzene (14)[h], cyclohexane (12)[h]	72	[50]
17	Ni/SBA-15, Ni/PTN	Hexadecane	H_2	553 K and 4.8 MPa	Anisole	Cyclohexyl methyl ether (35.4)[h], cyclohexanol (10.2)[h], cyclohexane	50	[51]
18	Pt/SiO_2, Pt/Al_2O_3, Pt/Na-B, Pt/NaHB	–		473–773 K and 0.1 MPa[i]	Anisole	Benzene, methylcyclohexane, toluene, xylenes, phenol	–	[52]
19	Pt/SiO_2, Pd/SiO_2, Rh/SiO_2, Ru/SiO_2, Ni/SiO_2, Co/SiO_2	–	H_2	573 K, 0.1 MPa, and 0.151 h^{-1}(W/F)	Phenol	Cyclohexanone (88.0)[g], cyclohexanol (11.0)[g]	17.1	[53]

Table 1 (continued)

Entry	Catalyst[a]	Solvent	R.A.[b]	Experimental conditions[c]	Model compound	Main Products (selectivity/yield in %)	Conv. (%)[d]	Refs.
20	Nb/SBA-16, Ru/SBA-16, Pd/SBA-16, Pt/SBA-16	Dodecane	H_2	363 K, 2.5 MPa, and 4 h	Phenol	2,4-dimetylhexane (51.3)[g], 1-methoxy-2-hexane (33.7)[g]	~100	[54]
22	$NiRu/Al_2O_3$, $Ni/Ru/TiO_2$, $NiRu/ZrO_2$	Dodecane	H_2	593 K and 5.4 MPa	Phenol	Cyclohexanol, cyclohexane, benzene, cyclohexene	–	[55]

[a]Catalysts considered in the review. The catalyst giving the highest conversion is in bold

[b]Reducing agent

[c]Temperature (K), total reaction pressure (MPa) and reaction time unless specified

[d]Maximun conversion achieved with the optimal catalyst (bold) and at the reaction conditions specified in the table

[e]5.4 ml/h feed rate of dodecane solution (3 wt% guaiacol)

[f]14.8 ml/h feed rate of heptane solution (2 wt% anisole)

[g]Selectivity (%) of the catalyst with the highest conversion (bold)

[h]Yield (%) for the catalyst with the highest conversion (bold)

[i]Space time (W/F) between 0.025 and 0.8 g_{cat} h/$g_{anisole}$

nitride, and carbide form, have been studied for the HDO of lignin and lignin model compounds as early the 1980s [9]. Zhe et al. [37] studied the HDO activity of activated carbon supported Mo oxide, MoO_2/C, and Mo carbide catalysts, Mo_2C/C, for the conversion of guaiacol at 573 K and 3 MPa of initial H_2 pressure in different solvents (tetradecane, dodecane, decalin, tetralin, and ethanol). The experiments were carried out in liquid phase using a stainless-steel batch reactor. The major difference between the oxide and carbide phases was on the final product distribution. While the MoO_2/C yielded mainly phenolic compounds such as phenol and cresols, the carbide Mo_2C/C catalysts yielded a higher amount of benzene. Thus, the results showed that the Mo_2C/C catalyst exhibited remarkable activity towards the scission of the stronger C_{AR}–O bond of the phenolic hydroxyl group. Concerning the type of solvent, tetralin and decalin gave the highest guaiacol conversion and phenol selectivities; this was attributed to the hydrogen donor nature. On the other hand, the worst conversion and phenol selectivity was obtained when ethanol was used as solvent. Zhang et al. [38] studied the activity of MoO_3/SiO_2 catalyst for guaiacol conversion in liquid phase at 613 K and 3.5 MPa total pressure, 0.5 MPa H_2, in octane using a 100-ml autoclave reactor. They found out that the presence of oxygen vacancies within the MoO_{3-x} favor the direct cleavage of the C_{AR}–O bonds of the hydroxyl and methoxyl groups, leading to high selectivity for arenes. Polycyclic aromatic products were detected when the experiments were carried out at high temperatures (i.e., 633 K), suggesting that polymerization occurred. The stability of the catalyst was not evaluated.

Sulfided NiMo (W) and CoMo (W) supported on Al_2O_3 have been the most widely studied catalytic systems for the HDO of guaiacol. Generally, traditional sulfided catalyst leads to fully deoxygenated products such as toluene, benzene, and cyclohexane, and to phenol as partially deoxygenated compound. However, highly undesirable reactions such as methyl substitution of the aromatic ring and the formation of high molecular weight compounds are also present; both of which are well-known coke precursors that rapidly deactivate the catalysts [39]. Lin et al. [35] studied the hydroprocessing activity of sulfided $NiMoS/Al_2O_3$ and $CoMoS/Al_2O_3$ catalysts in guaiacol HDO at 573–673 K and 5 MPa H_2 pressure in tetradecane solvent using a batch-type high-throughput system. After 40 min of reaction time at 673 K, phenol was the only product obtained with the $CoMoS/Al_2O_3$ catalyst. Conversely, the NiMoS catalyst exhibited a selectivity to cyclohexane of 80% and only 20% to phenol. Thus, they found out that NiMoS catalysts have a higher hydrogenation activity than CoMoS, thus favoring the aromatic ring saturation. A general guaiacol conversion mechanism was proposed, valid for both sulfided and noble metal catalysts (Fig. 4a). The stability of the catalyst was not within the scope of the study; the maximum reaction time studied was only 60 min. Tran et al. [40] compared the activity of different unsupported sulfided catalysts, i.e., NiMoS, CoMoS, and NiMoWS, for the liquid phase HDO of guaiacol in a down-flow fixed-bed reactor at 673 K, 2.8 MPa, and a WHSV = 2.7 h^{-1}. They also found that Ni-based catalysts exhibited higher activity than Co-based ones due to the higher hydrogenation activity of Ni. The stability of the catalyst was not considered.

Another key aspect that determines the activity of the traditional sulfided catalyst is the nature of the support. Sal´nikov et al. [41] studied the effect of different

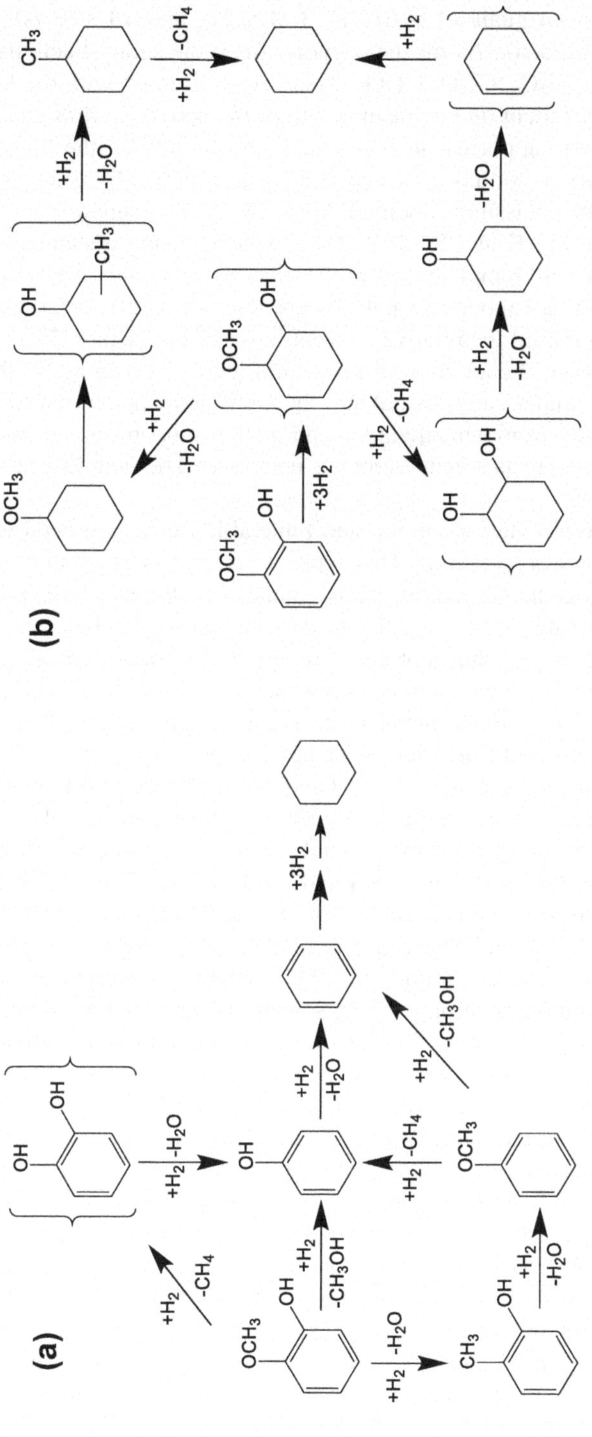

Fig. 4 **a** Adapted mechanism for guaiacol conversion proposed by Lin et al. [35] for Rh-based and sulfided CoMo and NiMo catalysts. **b** Adapted reaction network for guaiacol conversion over Pt catalyst proposed by He et al. [36]

supports, i.e., Al_2O_3, SiO_2, TiO_2, and ZrO_2, on the activity of CoMoS catalysts for the hydroprocessing of guaiacol at 613 K, 3 MPa H_2, and a FSV = 80 h^{-1} in toluene using a tubular reactor. As for the support effect, the catalyst activity decreased in the order $SiO_2 > Al_2O_3 > ZrO_2 \sim TiO_2$. This was associated with the higher HDO rate, which was dependent on the mean length of the active CoMoS phase particles. Besides, the catalyst supported on Al_2O_3 and ZrO_2 exhibited the highest stability with a degree of deactivation of 37% and 46%, respectively. Mora-Vergara et al. [39] studied the effect of potassium-modified Al_2O_3 (K–Al_2O_3) supports on NiMoS and CoMoS catalyst at 523 K and 5.5 MPa H_2 in xylenes using a stainless-steel batch reactor. Potassium was found to be an effective activity and selectivity modifier. Non-modified alumina supported catalyst were less active that potassium-modified ones since they increased the relative concentration of less active Co and Ni aluminate spinels. However, the addition of potassium led to an increase in the yields to products that may hinder catalysts coking: the addition of potassium to the support shifted the selectivity from demethylation and methyl substitution reactions to direct C_{AR}–OH bond scission and hydrogenation reactions. The stability of the catalyst was not considered.

Noble metal-based catalysts, mono and bimetallic, have also been widely studied for guaiacol hydroprocessing. This type of catalyst is generally characterized by their high hydrogenation activity. Thus, guaiacol hydroprocessing products with noble metal-based catalysts often fall into two categories: (1) benzene ring saturation products such as cyclohexanol and 2-methylcyclohexanol and (2) HDO products such as benzene, toluene, and cyclohexane [7]. Güvenatam et al. [42] studied the effect of the type of noble metal in different activated-carbon supported catalysts, i.e., Pt/C, Pd/C, and Ru/C, for guaiacol hydroprocessing at 473 K and 2 MPa in water using stainless-steel autoclaves. Guaiacol hydroprocessing mainly leads to benzene ring saturation products such as cyclohexane, cyclohexanol, and cyclohexanone, regardless of the type of noble metal. However, significant differences were observed regarding guaiacol conversion and product selectivity. After 4-h reaction time, the Pt/C catalyst exhibited the highest guaiacol conversion with a selectivity to cyclohexanol of 70% and selectivity to methyl-1,2-cyclohexanediol of ca. 25%. Under the same reaction conditions, the Pd/C catalyst converted guaiacol mainly into methyl-1,2-cyclohexanediol, ca. 60% selectivity, and cyclohexanol, 25% selectivity. With the Ru/C catalyst, cyclohexanol was the main product, although methyl-1,2-cyclohexanediol and phenol were also obtained. The recyclability and stability of the catalysts was not considered. He et al. [36] compared the activity of different Pt/TiO_2, Pt/Al_2O_3, and NiMo/Al_2O_3 catalysts for the conversion of guaiacol at 543 K, 7.1 MPa total pressure, and 5.4 ml/h continuous feed rate in dodecane (3 wt% guaiacol) under flow conditions. For all cases, the reactions proceed via aromatic ring hydrogenation, followed by demethylation and dihydroxylation to produce cyclohexane (Fig. 4b). The NiMo/Al_2O_3 and Pt/Al_2O_3 catalysts exhibit higher activity and selectivity to cyclohexane, which suggests that both metal and acid sites are required for a complete HDO reaction of guaiacol. Conversely, the Pt/TiO_2 yielded a higher amount of cyclohexanol. Besides, the Pt/TiO_2 catalysts exhibit higher stability than the Al_2O_3-based catalysts due to less coke formation. Rodulgina et al. [43] studied the activity of bimetallic PtPd and monometallic Ru catalysts supported on mesoporous aluminosilicates (Al-HMS) and mesoporous zirconia modified with silica (m-ZrO_2–SiO_2). Guaiacol liquid-phase hydrprocessing was carried out at 673 K and 5 MPa H_2 in methanol using a batch reactor. Conversion of guaiacol on PtPd/m–ZrO_2–SiO_2 catalyst was higher than that on PtPd/Al-HMS, in accordance with the larger number of acid sites on the catalyst surface. However, the undesirable heavy fraction of methylated by-products, which leads to coke formation, was also higher. Monometallic Ru catalysts exhibited the highest activity, even higher than bimetallic catalysts, and showed high selectivity towards fully hydrodeoxygenated products such as cyclohexane and methylcyclohexane.

Monometallic Ni catalysts, often supported on a solid acid, are frequently used as hydroprocessing catalysts for guaiacol conversion. Ni-based catalysts exhibit hydrogenation activities comparable to noble metal catalysts, although they often deactivate very fast [7]. Broglia et al. [44] studied the role of support acidity in Ni catalysts supported on silica-alumina (Ni/SiO_2–Al_2O_3) at 573 K and 5 MPa using a stainless-steel autoclave. They found out that those catalysts supported on high silica-content supports had a more acidic isoelectric point and a better overall catalytic performance. The higher silica content also leads to lower methane production, supporting the predominance of transalkylation reactions over acid sites. However, acidic sites can strongly adsorb poisoning compounds, such as organic deposits, leading to a faster deactivation of catalytic activity. The catalyst was regenerated and recycled in order to evaluate its stability. After regeneration, the catalyst had a carbon amount comparable to the fresh catalysts (0.5%); however, it exhibited slightly lower guaiacol conversion and HDO degree than the parent fresh oxide. The deactivation of the catalysts was attributed to Ni particle sintering, the predominance of larger aggregates, due to the drastic reaction conditions. In an attempt to reduce the deactivation of Ni-based catalysts, Long et al. [45] studied the effect of a basic support, MgO, in the liquid phase hydroprocessing of guaiacol at 433 K and 3 MPa of H_2 in decahydronaphalene using a stainless-steel autoclave. The selectivity of the Ni/MgO catalyst to cyclohexanol was compared with another Ni catalyst supported on SiO_2, SiO_2–ZrO_2- and HZSM-5. The results showed that the acid–base effect between the phenolic OH and MgO leads to a higher conversion and selectivity to cyclohexanol in comparison to the SiO_2, SiO_2–ZrO_2, and HZSM-5 supported catalysts. Moreover, the Ni/MgO catalysts showed excellent recyclability with no significant activity loss after four cycles. In an attempt to increase the activity and selectivity of Ni-based catalysts, many research groups have studied the promoting effect of different transition metals, e.g., NiFe and CuNi. Fang et al. [46] studied the gas-phase conversion of guaiacol over bimetallic NiFe catalysts supported on carbon nanotubes (CNTs) at 573 K, 3 MPa H_2, and a WLHSV = 6 h^{-1} using a conventional fixed-bed flow reactor. The results showed that the Ni/Fe ratio had a strong effect on the product selectivity. Higher Ni/Fe ratios lead to highly hydrogenated products such as cyclohexane and cyclohexanol, while at low Ni/Fe ratios, phenol was the main product. The selectivity-switchable performance of the catalysts was assigned to the synergism between Ni domains, where H_2 could be easily activated, and Fe domains, which exhibited strong oxophilicity. The catalytic performance in terms of conversion and product selectivity of the bimetallic NiFe catalysts differed considerably from the monometallic Ni/CNT and Fe/CNT catalysts. Furthermore, unlike the monometallic Ni/CNT catalyst, bimetallic NiFe/CNT catalysts exhibited no noticeable deactivation under the studied reaction conditions. Dongil et al. [47] studied the effect of Cu loading on Ni/CNT catalysts for the hydroprocessing of guaiacol at 573 K and MPa H_2 in *n*-dodecane using a batch reactor. The addition of Cu favored the stabilization of smaller Ni nanoparticles and decreased the reduction temperature of Ni. However, it also led to the formation of less reducible NiO species and the coverage of Ni sties with Cu resulting in a decrease of the overall activity of the catalysts. The recyclability and stability of the catalysts was not considered.

Precious metal catalysts based on Ag have also been studied for guaiacol hydroprocessing. Liu et al. [48] studied the HDO of Ag supported on anatase-TiO_2 at 573 K and 3 MPa H_2 in heptane using 50-ml batch reactors. The Ag/TiO_2-A catalyst was able to selectively cleave the C_{AR}–O bond of the methoxyl group in guaiacol but unable to cleave the C_{AR}–O bond of the hydroxyl-group. The weak hydrogenation activity of Ag favors the production of phenolics as benzene hydrogenation does not occur. Ag was found to provide dissociated hydrogen pool for hydrogen spillover on TiO_2-A surface, which creates oxygen vacancies that act as HDO sites. The Ag/TiO_2-A was recycled for repeated tests and prolonged reaction time. The results showed that the catalyst was stable under the reaction conditions.

Anisole is a lignin model compound commonly used to study the competitive reaction between de-alkylation and de-alkoxylation of the methoxyl groups within the lignin matrix. It is often studied together with guaiacol, since its simpler chemical structure, with only one methoxy functionality, facilitates the elucidation of the guaiacol hydroprocessing reaction mechanism. Table 1 provides an overview of the latest studies on anisole hydroprocessing, including used solvent, catalysts, reaction conditions, as well as obtained main products and highest conversion levels. He et al. [49] studied the selective gas-phase conversion of anisole into aromatics over NiMo/SiO_2 catalyst at 683 K, low H_2 pressure (0.1 MPa) and WHSV = 1.5 h^{-1} using a fixed-bed stainless-steel reactor. Anisole was converted in a 99.35% to mainly benzene and toluene, suggesting that the NiMo catalyst could selectively catalyze anisole de-alkoxylation, although it also exhibited significant catalytic activity towards de-alkylation and methyl transfer reaction. They also observed that anisole conversion and benzene, toluene, and xylene selectivities were maintained at a steady level only at the first 70–90 min of reaction. After that, the catalyst deactivated due to carbon deposition and the oxidation of Ni particles into NiO. However, the spent catalyst could be regenerated by combustion in air flow. Fourteen regeneration cycles were performed: it was observed that the anisole conversion and hydrocarbon selectivity increased slightly after the first regeneration cycle, and then reached a plateau. Thus, the regenerated catalysts exhibited practically stable performance for short time-on-streams. Hewer et al. [50] studied the gas-phase conversion of anisole over NiMo catalysts on acidic (SAPO-11 and Al_2O_3) and non-acidic (SBA-15) supports at 573 K, 1.5 MPa H_2, and a feed rate of 14.8 ml/h of a heptane solution (2 wt% anisole) using a fixed-bed flow reactor system. The results showed that the activities of the NiMo/SAPO-11 catalyst are higher than of the NiMo/SBA-15 and the NiMo/Al_2O_3 catalysts due to the presence of silanol groups and Al_3^+ Brønsted acid sites. The product distribution also varied considerably depending on the type of support. Over the Lewis acid support, i.e., Al_2O_3, the NiMo catalysts yielded aromatic phenols such as *o*-cresol and phenol as major products. In contrast, over the Brønsted acid, i.e., SAPO-11, and the non-acidic support, i.e., SBA-15, similar product selectivities were obtained: 1-methoxycyclohexane was the major product although significant amounts of benzene and cyclohexane were also detected. The evaluation of stability was not within the scope of the study.

Anisole conversion over monometallic Ni or noble metal catalysts has been the focus of many research groups. Vargas-Villagran et al. [51] studied the conversion of anisole over Ni catalysts supported on SBA-15 and protonated titanate nanotubes (PTN) at 553 K and 3.5 MPa H_2 pressure (4.8 MPa total pressure) in hexadecane using a 300-ml stainless-steel batch reactor. Without the promoting effect of Mo, Ni monometallic Ni/SBA-15 and Ni/PTN catalysts yield mainly hydrogenated products such as cyclohexyl methyl ether and cyclohexanol for a 50% conversion. Zanuttini et al. [52] studied the effect of the type of support, i.e., SiO_2, γ-Al_2O_3 and Na-Beta and NaH-Beta zeolites, using Pt as the active phase. The study was carried out at a 473-773 K, low H_2 pressure (0.1 MPa), and a space time (W/F) between 0.025 and 0.8 g_{cat} h/$g_{anisole}$ using a continuous-flow fixed-bed reactor. The Pt/γ–Al_2O_3 catalyst had the highest selectivity to deoxygenated compounds such as benzene, while with the Pt/Na-Beta and NaH-Beta more oxygenated products were obtained. All the catalysts studied presented deactivation due to coke deposition. After regeneration, the catalysts exhibited similar conversion levels to the fresh ones; however, the selectivity did change.

Phenol is another well-known lignin model compound that is often used to investigate the influence of the phenolic hydroxyl group in hydroprocessing of lignin and its derivatives. Table 1 provides and overview of the latest studies on phenol hydroprocessing under reported reaction conditions. Teles et al. [53] studied the influence of different metals such as Pt, Pd, Rh, Ru, Ni, and Co, supported on SiO_2 for the vapor-phase hydroprocessing of phenol at 573 K, low hydrogen pressure (0.1 MPa), and different space times (W/F) using a fixed-bed quartz reactor. They found out that the Pt, Pd, and Rh catalysts favor the formation of hydrogenated products such as cyclohexanone and cyclohexanol, whereas the Co, Ni, and mainly Ru catalysts exhibited significant formation of hydrodeoxygenated products such as benzene or cyclohexane. Feliczak-Guzik et al. [54] studied the conversion of phenol of different metals, Ru, Pd, Pt, and Nb, supported on SBA-16, which were tested at 363–403 K and 2.5–6 MPa H_2 in dodecane. In the reaction conditions applied, 100% of phenol conversion was achieved. The main reaction product with noble metal catalysts, i.e., Ru, Pd, and Pt, was highly hydrogenated compounds, e.g., 2,4-dimethylhexane, while when using Nb/SBA-16 the main product was 1-methoxy-2-hexene. Many groups have focused their research on the effect of different bimetallic catalysts for phenol hydroprocessing. Valdés-Martínez et al. [55] studied the conversion of phenol over NiRu catalysts supported on Al_2O_3, TiO_2, and ZrO_2 at 593 K and 5.4 MPa H_2 pressure in dodecane using a high-pressure batch reactor. The activity of the bimetallic RuNi catalysts was considerably higher than their corresponding monometallic Ni and Ru catalysts regardless the nature of the support. The presence of Ru enhanced nickel reducibility due to hydrogen dissociation on Ru^0 and spillover effect. The type of support had a considerable effect on the conversion level and product distribution. The highest degree of phenol conversion was achieved over the $RuNi/Al_2O_3$ catalysts, resulting in high cyclohexane selectivity. Similar product selectivities but lower conversion level was achieved using the $RuNi/TiO_2$ catalyst. Finally, the $RuNi/ZrO_2$ catalyst yielded a higher amount of hydrogenated products such as cyclohexanol. The results indicate that the catalytic activity was correlated with the strength of the surface acid sites. The recyclability and stability of the catalysts was not considered.

2.1.2 Dimers

Table 2 provides an overview of the latest studies on the hydroprocessing of dimeric lignin model compounds, summing up the solvents, catalysts, reaction conditions, main products, and highest conversion levels. β-*O*-4 is the most abundant linkage in the lignin structure and the scission of this bond has often been used as a model reaction for lignin depolymerization. As mentioned above, phenethyl-phenyl-ether (PPE) is the most simple β-*O*-4 lignin model compound since it does not contain any hydroxyl or methoxyl functionalities. Zhao et al. [56] studied the selective HDO of PPE over a combination of Pd/C and HZSM-5 catalysts at 473 K and 5 MPa H_2 in water using a batch autoclave reactor. After 2 h of reaction time, 100% of the PPE was converted to ethylcyclohexane and cyclohexane with selectivities of 54% and 46%, respectively. The study was further extended to different α-*O*-4, 4-*O*-5, 5–5, β-1, and β–β lignin model compounds. The results showed that the combination of Pd/C and HZSM-5 was extremely active for the hydrogenolysis of β-*O*-4, α-*O*-4, and 4-*O*-5 ether linkages, however, all the C–C linkages in 5–5, β-1, and β–β were preserved. The Pd/C and HZSM-5 combination also showed an extremely high selectivity in removing oxygen-containing groups, i.e., hydrogyl, methoyx, and ketone, and for the hydrogenation of the aromatic ring. The evaluation of the catalyst stability was not within the scope of the study. Luo et al. [57] studied the conversion of PPE over a bifunctional Ru catalysts supported in sulfated zirconia (Ru/SZ) at 513 K, low H_2 pressure (0.8 MPa), and 1.5 h in water using a batch autoclave reactor. The major reaction products were non-oxygenated aromatic compounds such as ethylbenzene, benzene, and toluene. The results indicated that at low H_2 pressures, the Ru/SZ catalyzes the selective cleavage of the β-*O*-4 ether bond without any aromatic ring hydrogenation. The recyclability and stability of the catalysts was not considered.

2-phenoxy-1-phenethanol (PHPL) is a slightly more complex β-*O*-4 lignin model compound that contains a hydroxyl group in the α-position. Barton et al. [58] studied the catalytic conversion of PHPL with bifunctional Ni/HZSM-5 catalysts at 523 K and 5 MPa of H_2 in a dilute aqueous solution of NaOH in water using a 300-ml batch reactor. After 30 min of reaction time, ~ 100% conversion of PHPL was achieved producing ethylbenzene and phenol as its two main products. The high selectivity to aromatic compounds (~ 60%), and the minimal selectivity for ring saturation and recombination reactions (both less than 10%), made this catalyst a good alternative for the production of aromatic monomers from lignin. In a novel approach, Li et al. [59] studied the conversion of PHPL over nitrogen-doped carbon supported iron catalysts (NFe/C) at 513 K, 1 MPa H_2, and 12 h in a water–THF mixture using a 50-ml Zr-alloy autoclave. The study was extended to different β-*O*-4 lignin model

Table 2 Summary of the analyzed literature for dimeric lignin model compounds hydroprocessing

Entry	Catalyst[a]	Solvent	R.A.[b]	Experimental conditions[c]	Model compound[d]	Linkage type	Main products (selectivity/yield in %)	Conv. (%)[e]	Refs.
1	Pd/C and HZSM-5	Water	H_2	473 K, 5 MPa, and 2 h	2-phenethyl phenyl ether	β-*O*-4	Ethylcyclohexane (54)[f] and cyclohexane (46)[f]	100	[56]
2	Ru/SZ	Water	H_2	513 K, 0.8 MPa, and 1.5 h	2-phenetyl phenyl ether	β-*O*-4	Ethylbenzene (~45)[f], benzene (~45)[f]	~100	[57]
3	Ni/HZSM-5	Water (0.1 M NaOH)	H_2	523 K, 5 MPa, and 0.5 h	2-phenoxy-1-phenylethanol	β-*O*-4	Ethylbenzene (~35)[g], phenol (~25)[g]	100	[58]
4	NFe/C	Water-THF	H_2	513 K, 1 MPa, and 12 h	2-phenoxy-1-phenylethanol	β-*O*-4	Phenol (43)[g], ethylbenzene (40)[g], and acetophenone (3)[g]	–	[59]
5	Pd/Zn/C	Methanol	H_2	423 K, 2 MPa, and 2 h	Guaiacylglycerol-β-guaiacyl ether	β-*O*-4	Guaiacol (85)[g], 2-methoxy-4-propylphenol (85)[g]	100	[60]
6	$CoMoS/Al_2O_3$	Dodecane	H_2	573 K, 5 MPa, and 4 h	Veratryl-glycerol-β-guaiacyl-ether	β-*O*-4	Phenol (9)[f], guaiacol (6)[f], dimethylphenol (5)[f]	100	[61]
7	Ru, Pd, Ni over Al_3O_3, ZrO_2, TiO_2, C, MWCNT	Methanol	H_2	423 K, 2.5 MPa, and 2 h	Benzyl phenyl ether	α-*O*-4	Toluene, phenol, cyclohexanes, cyclohexanols, saturated dimers	100	[62]
8	$Ru/SiO_2–Al_2O_3$	n-decane	H_2	523 K, 4 MPa, and 1 h	Benzyl phenyl ether	α-*O*-4	Cyclohexane (63.4)[f], methycyclohexane (27.4)[f], cyloheptane (4.9)[f]	100	[63]
9	Pt/C	Water	H_2	473 K, 2 MPa, and 2 h	Benzyl phenyl ether	α-*O*-4	Methylcyclohexane (33)[f], phenol (27)[f], cyclohexanol (11)[f], and benzyl alcohol (9)[f]	99	[42]

Table 2 (continued)

Entry	Catalyst[a]	Solvent	R.A.[b]	Experimental conditions[c]	Model compound[d]	Linkage type	Main products (selectivity/yield in %)	Conv. (%)[e]	Refs.
10	Pt/C	Water	H_2	473 K, 2 MPa, and 0.5 h	Diphenyl ether	4-*O*-5	Cyclohexane (~43)[f], cyclohexanol (~30)[f], cyclohexyl ether (~18)[f]	100	[42]
11	Ru/TiO_2	Octane	H_2	533 K, 1 MPa, and 4 h	Diphenyl ether	4-*O*-5	Cyclohexane (86.1)[f], bicyclohexane (13.9)[f]	99	[64]
12	Ru/H-Beta	Water	H_2	393 K, 4 MPa, and 3 h	Diphenyl ether	4-*O*-5	Cyclohexane (66.0)[f], cyclohexanol (30.5)[f]	70.4	[65]
13	$Ni/ZrNbPO_4$	n-decane	H_2	493 K, 0.5 MPa, and 5.5 h	Diphenyl ether	4-*O*-5	Benzene (62.1)[f], cyclohexane (29.6)[f], phenol (7.4)[f]	83.7	[66]
14	Pt/C	Water	H_2	473 K, 2 MPa, and 0.5 h	Diphenyl methane, biphenyl	β-1, 5–5′-type	Dicyclohexyl methane (~100)[f]	100	[42]
15	Pd/C and HZSM-5	Water	H_2	473 K and 5 MPa	2,2′-biphenol, 1-(4-hydroxylphenyl)-2-phenylethanone, 1,2-diphenylethane	5–5, β-1, β–β	Saturated C_{12}, C_{14} and C_{16} bicycloalkanes	100	[56]
16	Ru/HZSM-5	Water	H_2	473 K, 5 MPa, and 4 h	Dihydrobenzofuran, benzofuran	β-5	Ethylcyclohexanol, methylcyclohexanol, cylohexanol	~100	[67]

[a] Catalysts considered in the review

[b] Reducing agent

[c] Temperature (K), total reaction pressure (MPa), and reaction time unless specified

[d] Analyzed model compound

[e] Maximum conversion achieved with the optimal catalyst and at the reaction conditions specified in the table

[f] Selectivity (%) of the catalyst with the highest conversion (bold)

[g] Yield (%) for the catalyst with the highest conversion (bold)

compounds containing ketone, methoxy, and additional hydroxyl functionalities. They found out that the vicinal –OH groups were essential for the iron-catalyzed hydrogenolysis of the ether linkages. Those β-*O*-4 lignin model compounds with vicinal –OH groups were converted mainly into aromatic hydrocarbons, indicating the ability of the NFe/C catalyst for the selective cleavage of the ether bond without inducing any aromatic ring hydrogenation. The evaluation of the catalyst stability was not within the scope of the study.

Guaiacyl-glycerol-β-guaiacyl-ether (GGG) and veratryl-glycerol-β-guaiacyl-ether (VGG) are lignin model compounds that accurately resemble those β-*O*-4 linkages in native lignin. GGG resembles the β-*O*-4 linkage between two coniferyl alcohol units, while the VGG resembles the β-*O*-4 linkage between a coniferyl and a sinapyl alcohol unit. Parsell et al. [60] studied the hydroprocessing of GGG over Pd/Zn synergistic catalysts supported on carbon (Pd/Zn/C) at 423 K and 2 MPa H_2 in methanol using a stainless-steel batch reactor. After 2 h of reaction time, GGG was completely converted to two primary products, i.e., guaiacol and 2-methoxy-4-propylphenol in addition to a small amount of 3-(3-methoxy-4-hydroxyphenyl)-1-1propanol. The results show that the Pd/Zn/C catalyst selectively cleaves recalcitrant ether bonds, removes alcohol oxygen atoms on the alkyl chains, and maintains the valuable aromaticity of the starting material under the reaction conditions. The recyclability of the catalyst was tested by filtering and simply reusing it in a new reaction. Selectivity and activity were maintained for at least three consecutive cycles. Jongerius et al. [61] studied the conversion of VGG over traditional CoMo sulfided catalysts supported on Al_2O_3 (CoMoS/Al_2O_3) at 573 K and 5 MPa H_2 in dodecane using a 100-ml stainless-steel batch autoclave reactor. After 4 h of reaction time, complete conversion of VGG was observed, and only mono-aromatics such as phenol, guaiacol, and syringol-like products could be identified by GC analysis. The results indicate that the catalysts are able to cleave the ether C–O bond within the β-*O*-4 linkage and the C_{AR}–O bond within the methoxyl group, but could not cleave the stronger C_{AR}–O bond of the hydroxyl groups. The recyclability and stability of the catalysts was not considered.

Benzyl phenyl ether (BPE) is a widely studied lignin model compound representative for the α-*O*-4 linkage. The cleavage of this linkage is relatively easy due to its theoretical low bond dissociation enthalpy (167–184 kJ/mol) [7]. Gómez-Monedero et al. [62] studied the selective hydrogenolysis of BPE over several Ru, Pd, and Ni catalysts supported over different metal oxides, i.e., Al_2O_3, ZrO_2, TiO_2, C, and multiwall carbon nanotubes (MWCNT) at 423 K and 2.5 MPa H_2 in methanol using a stainless-steel batch reactor equipped with a Teflon liner. Ru-based catalysts were the most active ones followed by Pd and Ni; however, the selectivity to aromatic monomers increased in the reverse order (Ni > Pd > Ru). Concerning the type of support, those catalysts supported on Al_2O_3, C and MWCNT exhibited generally the highest activity and selectivity. Yoon et al. [63] studied the conversion of BPE over a silica-alumina-supported Ru catalyst (Ru/SiO_2–Al_2O_3) at 523 K, 4 MPa H_2 for 1 h in *n*-decane using a 160-ml Hastelloy batch reactor. BPE was converted to 2-benzylphenol and 4-benzylphenol by intramolecular rearrangement and phenol by the cleavage of ether bonds. The catalysts were allowed to be deactivated by long-term reaction; there was no H_2 supply to ensure a poor H_2 environment during the reaction. The BPE conversion increased from 18.9% (after 1 h) to 43.5% (after 100 h), indicating that deactivation of the solid acid catalyst may not be significant for the conversion of BPE. Güvenatam et al. [42] studied the conversion of BPE with a Pt/C catalyst at 473 K and 2 MPa H_2 in water using autoclave reactors. The hydrogenolysis of BPE was very fast: after 0.5 h, 81% of BPE conversion was achieved. Cyclohexanol, phenol, cresol, and methyl cyclohexanol were the main reaction products, indicating that besides its hydrogenolysis activity, the Pd/C catalysts exhibited considerable aromatic ring hydrogenation activity. The evaluation of the catalyst stability was not within the scope of the study.

As mentioned above, diphenyl-ether (DPE) is the most studied 4-*O*-5 lignin model compound. The 4-*O*-5 linkage is the ether bond with the highest bond dissociation energy within lignin (330 kJ/mol) [66]. Thus, its conversion generally results in a significant amount of dimeric saturated compounds. Güvenatam et al. [42] carried out the conversion of DPE over Pt/C catalysts at 473 K and 2 MPa H_2 in water using autoclave reactors. The results revealed that the direct hydrogenolysis of DPE to yield cyclohexanone and cyclohexane is in competition with the direct hydrogenation of the aromatic moieties to cyclohexyl ether, which can then undergo Pt-catalyzed hydrogenolysis reaction resulting in cyclohexanol and cyclohexane (Fig. 5). The recyclability and stability of the catalysts was not considered. Shu et al. [64] studied the conversion of DPE over highly dispersed Ru/TiO_2 catalysts at 533 K, 1 MPa H_2, and 4 h in octane using a stainless-steel autoclave reactor. The results were similar to the ones obtained by Güvenatam et al. [42]: the main reaction product was cyclohexane, 86.1% selectivity, with small amounts of cylohexyl ether, 13.9% selectivity. The results also suggested a competition between direct hydrogenation of aromatic moieties and noble metal-catalyzed hydrogenolysis of the 4-*O*-5 ether bond. The stability of the catalysts was not evaluated using DPE, but real lignin. They found out that the fresh and spent catalyst exhibited no significant difference on catalyst morphology and metal particle size, and that a small amount of carbon deposition was formed on the spent catalysts. Yao et al. [65] carried out the conversion of DPE over bifunctional Ru/H-beta catalysts at 393 K, 4 MPa H_2 for 3 h in water using high-pressure stainless-steel autoclaves. The activity of the Ru/H-Beta catalyst was compared with other Ru catalysts supported on acidic and non-acidic supports. The results showed that the combination of the Ru metal and the acidic H-Beta support decreased the amount of saturated dimeric products such as cyclohexyl ether. Cyclohexane and cyclohexanol were the main reaction products with 66.0% and 30.5% selectivity, respectively. The evaluation of the catalyst stability was not within the scope of the study. On a different approach, Jin et al. [66] studied the hydroprocessing of DPE over Ni catalysts supported on Zr-doped Nb-phosphate (Ni/$ZrNbPO_4$) at 493 K, low H_2 pressure (0.5 MPa), and 5.5 h in dodecane using a 50-ml stainless-steel autoclave. This novel catalytic approach led to considerable amounts of aromatic monomers in comparison to noble metal-based catalysts: benzene was the most abundant product with a selectivity of 62.1%, with cyclohexane and phenol as minor products with a 29.6% and 7.4% selectivity, respectively. The stability of the catalyst was evaluated 493 K, 0.5 MPa H_2, and 5.5 h. The conversion decreased slightly after four runs, due to the partial oxidation of nickel nanoparticles caused by unavoidable exposure to air. However, at the fifth run, the catalyst was reactivated in hydrogen and the activity of the catalyst, conversion level, was restored. Besides, the selectivity of oxygen-free aromatics and alkanes was kept constant for all runs.

Cleavage of C–C lignin linkages are much more difficult to cleave in comparison to C–O–C ether linkages. Hence, the conversion of dimeric lignin model compounds containing C–C interunit linkages often results in high yields of dimeric saturated compounds. Güvenatam et al. [42] carried out the conversion of diphenyl methane, a β-1 model compound, biphenyl, a 5–5′ type model compound, and other β-*O*-4 and 4-*O*-5 lignin model compounds with a Pt/C catalyst at 473 K and 2 MPa H_2 in water using a batch reactor. They found out that although β-*O*-4 and 4-*O*-5 lignin model compounds could be readily converted to monomeric species, cleavage of C–C bonds in diphenyl methane and biphenyl did not occur. The conversion of both diphenyl methane and biphenyl led only to complete and rapid hydrogenation of the aromatic rings. Zhao et al. [56] studied the conversion of 5–5, β-1, and β–β lignin model compounds over Pd/C catalysts at 473 K and 5 MPa H_2 in water using a batch autoclave reactor. After 2-h reaction time, the 5–5, β-1, and β–β lignin model compounds were converted to their corresponding non-oxygenated saturated dimers in high yields. The evaluation of the catalyst stability was not within the scope of the study. Zhang et al. [67] studied the hydroprocessing of two different β-5 lignin model compounds, dihydrobenzofuran and benzofuran, with a Ru/HZSM-5 catalyst at 473 K and 5 MPa in H_2 water using a stainless-steel batch reactor. After 4-h reaction time, both dihydrobenzofuran and benzofuran were completely converted yielding alkanes in high selectivities. Moreover, ethylcyclohexane, obtained by hydrogenation of aromatic ring followed by dehydration and hydrogenation reactions, was the main deoxygenated product. The evaluation of the catalyst stability was not within the scope of the study.

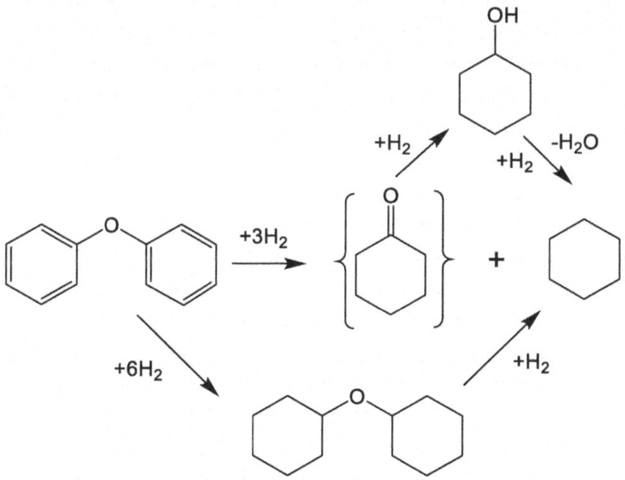

Fig. 5 Reaction network for DPE conversion on Pt/C catalyst proposed by Güvenatam et al. [42]

2.2 Catalytic Transfer Hydrogenation of Lignin Model Compounds

As mentioned above, in the reductive processes, the incorporated H_2 can be supplied by molecular hydrogen or by hydrogen-donating molecules, which generate or transfer hydrogen in situ to an acceptor. The latter are generally referred to as catalytic transfer hydrogenation reactions (CTH). CTH processes usually employ formic acid, isopropyl alcohol, or a hydrocarbon, such as decalin or tetralin, as the hydrogen source [68–71]. CTH reactions are considered to be more promising than hydroprocessing reactions for different reasons. Hydrogenating processes usually operate at high hydrogen pressures, which can result in safety issues. Moreover, molecular hydrogen is nowadays produced from fossil sources, which is in contradiction with the emerging environmental concern and the biorefinery concept. Thus, hydrogen donor molecules from renewable origin can be an alternative to overcome the issues related to traditional hydroprocessing [68]. Particularly, hydrogen donor molecules such as methanol, ethanol, and isopropanol are interesting candidates for CTH of lignin since they are widely used as (co-)solvent in the extraction of organosolv lignin [72]. However, their performance as hydrogen donors in the CTH of lignin is highly dependent on the reaction temperature. Methanol and ethanol are only dehydrogenated at high temperatures (supercritical conditions); thus, they are adequate hydrogen donor molecules for high-temperature processes [72]. On the other hand, isopropanol is usually dehydrogenated into acetone at much milder reaction conditions and its performance as hydrogen donor molecule is restricted to low-temperature processes.

2.2.1 Monomers

Shafaghat et al. [73] carried out the CTH conversion of guaiacol using Pt/C and Pd/C as catalysts and decalin and tetralin as hydrogen donor molecules. The best result was obtained using decalin and Pt/C at 548 K and under autogenous pressure: 95.1% guaiacol conversion and 63.7% selectivity to cyclohexanol and 19.6% selectivity to cyclohexanone (Table 3). Minor amounts of phenol, 1,2 cyclohexadiol, cyclopentanol, cyclopentanone, and anisole were also achieved. With the Pd/C catalyst, however, the conversion decreased to 49.8% for the same reaction conditions (Table 3). The major product in this case was not cyclohexanol but phenol. According to the authors, those hydrogen atoms released from the hydrogen donor molecules are less accessible in the Pd/C catalyst, resulting in lower conversion. Besides, when using tetralin as the hydrogen source, guaiacol conversion decreased to 62.0% with the Pt/C catalyst and 15.6% for the Pd/C catalyst (Table 3). The authors observed that tetralin and phenol compete for the catalytic active sites, which results in a reduced phenol conversion in comparison to decalin experiments. The authors also observed that replacing hydrogen donor molecules (i.e., tetralin and decalin) for molecular H_2 changed the product distribution. Thus, the results indicated that the reaction mechanism for guaiacol conversion, depicted in Fig. 6, could be initiated following four different routes: guaiacol demethoxylation (path I), guaiacol demethylation (path III), guaiacol dihydroxylation (path VI), and guaiacol hydrogenation (path VIII). They found out that guaiacol demethoxylation (path I) was the main reaction route regardless the hydrogen source (H_2, decalin, or tetralin). However, the relative contribution of the other reaction routes varied depending on the type of hydrogen source. Altogether, the results indicate that the interaction between the hydrogen donor (i.e., decalin), hydrogen acceptor (i.e., phenol), and the catalysts (i.e., Pt/C and Pd/C) is a key factor to consider in CTH processes. The mechanism by which hydrogen is transferred from the hydrogen donor molecules (tetralin or decalin) to the acceptor (phenolic compounds), however, is not clear. Two different possible mechanisms were proposed: (1) abstraction of the hydrogen atoms from the hydrogen donor molecules by activated acceptors, (2) formation of dihydrogen, which can be activated over the metal catalysts, which is later involved in the hydrogenation of the acceptors. Nevertheless, the presence of hydrogen donor molecules competes with the phenolic compounds for the active centers of the catalyst. That is why the overall reaction activity decreased with respect to H_2 gas (lower guaiacol conversion) [73].

Wang et al. [74] studied the CTH conversion of guaiacol using a Raney Ni catalyst and isopropanol (2-PrOH) as hydrogen donor at 353 K under autogenous pressure in closed glass vials. After 3-h reaction time, total guaiacol conversion was achieved. Cyclohexanol was the major product, 95% selectivity, although traces of 2-methoxycyclohexanol, 4% selectivity, and benzene, 1% selectivity, were also obtained (Table 3). Jae et al. [75] also studied the CTH conversion of guaiacol using 2-PrOH as hydrogen donor. These authors studied the activity of different catalysts, Pd/C, Ru/C and Ni/C, at 473 K for 5 h [76]. The highest guaiacol conversion was reached using Ru/C, 99.3%, yielding cyclohexanol (70.2% selectivity), 2-methoxycyclohexanol (11.4% selectivity) and 1,2-cyclohexanediol and cyclohexanone (2.1% selectivity for both compounds) as reaction products (Table 3). All these indicated that hydrogenation of the aromatic ring and demethoxylation readily occur through CTH when using Ru/C catalyst [76]. Therefore, the results obtained by Jae et al. [76] suggest that the CTH guaiacol conversion over Ru/C catalyst mainly follows path I and II (Fig. 6) of the reaction network proposed by Shafaghat et al. [73].

In a subsequent study, Jae et al. [76] evaluated the effect of replacing the C support for a Lewis acid (Al_2O_3) and a Brønsted acid support (H-beta zeolite) in order to fully deoxygenize guaiacol into cyclohexane. The Ru/Al_2O_3 catalyst showed similar guaiacol conversion (99.3%) and product distribution as the Ru/C, indicating that Lewis acids did not promote the dehydroxylation of cyclohexanol. With the Ru/H-BEA catalyst, a lower conversion was obtained: only 10% of the guaiacol was converted with 1-isopropoxy-2-methoxybenzene (77% selectivity) as the major product. These results suggest that Brønsted acids inhibit the CTH of guaiacol over Ru by catalyzing the undesired etherification of guaiacol and 2-propanol. Besides, Jae et al. [76] added oxophilic promoters, such as Re, to Ru/C to enhance its dehydroxylation activity. The experiment were carried out at 473 and 573 K in 2-propanol. The results showed that the RuRe/C catalysts were fairly active for the CTH conversion of guaiacol (70–99% conversion) and that they could further convert cyclohexanol to cyclohexane.

Phenol is also a widely studied monomer in CTH conversion. Shafaghat et al. [73] studied the CTH conversion of phenol using Pt/C and Pd/C as catalysts and decalin and tetralin as hydrogen donor molecules. The experiments were carried out

Table 3 Summary of the analyzed literature for the reductive conversion of lignin model compounds using hydrogen donor molecules

Entry	Catalyst	Solvent	H. D. M.[b]	Experimental conditions (K)[c]	Model compound	Products (selectivity/yield in %)	Conv. (%)[d]	Refs.
1	Pt/C	–	Decalin	548	Phenol	Cyclohexanol (69.0)[e], cyclohexanone (30.9)[e]	100	[73]
2	Pt/C	–	Tetralin	548	Phenol	Cyclohexanol (61.2)[e], cyclohexanone (38.8)[e]	85.4	[73]
3	Raney Ni	–	2-propanol	353	Phenol	Cyclohexanol (99.3)[e], cyclohexanone (0.7)[e]	100	[74]
4	Raney Ni	–	2-propanol	353	Guaiacol	Cyclohexanol (95.0)[e], 2-methoxycyclohexanol (4)[e], benzene (1)[e]	100	[74]
5	Raney Ni	–	2-propanol	353	Anisole	Benzene (78.0)[e], cyclohexane (18.0)[e], methoxycyclohexane (4.0)[e]	58	[74]
6	Raney Ni–B-zeolite	n-hexadecane	2-propanol	433	Phenol	Benzene (82.0)[e], cyclohexane (9.0)[e]	100	[77]
7	Raney Ni	–	2-propanol	353	Diphenyl-ether	Benzene (48.0)[e], cyclohexanol (48.0)[e], cyclohexane (3.0)[e]	100	[74]
8	Pt/C		Decalin	548	Guaiacol	Cyclohexanol (63.7)[e], cyclohexane (19.6)[e], phenol (7.9)[e]	95.15	[73]
9	Pd/C		Decalin	548	Guaiacol	Phenol (60.2)[e], cyclohexanol (25.1)[e], cyclohexane (13.9)[e]	49.80	[73]
10	Ru/C		2-propanol	473	Guaiacol	Cyclohexanol (70.2)[e], 2-methoxycyclohexanol (11.4)[e]	99.3	[76]
11	Ru/Al_2O_3		2-propanol	473	Guaiacol	Cyclohexanol (60.4)[e], 2-methoxycyclohexanol (12.9)[e]	99.3	[76]
12	RuRe/C		2-propanol	513	Guaiacol	Cyclohexane (57.0)[e], cyclohexanol (24.1)[e],	99	[76]
13	Ru/C	–	Isopropanol	433	Diphenyl-ether	Cyclohexanol (49.5)[f], benzene (33.2)[f], cyclohexane (16.3)[f]	99	[80]

Table 3 (continued)

Entry	Catalyst	Solvent	H. D. M.[b]	Experimental conditions (K)[c]	Model compound	Products (selectivity/yield in %)	Conv. (%)[d]	Refs.
14	Pd/C	EtOH/H_2O	$NaBH_4$	353	2-aryloxy-1-arylethanol ester (β-O-4)	Acetophenone (99.0)[f]	100	[78]
15	Pd/C	EtOH/H_2O	$NaBH_4$	433	Ethanolaryl ester (β-O-4)	Acetophenone (98.0)[f], phenol (1.0)[f]	100	[78]

All experiments were carried out under autogenous pressure

[a]Catalysts considered in the review. The catalyst giving the highest conversion is in bold

[b]Hydrogen donor molecule

[c]Temperature (K), and reaction time unless specified

[d]Maximun conversion achieved with the optimal catalyst (bold) and at the reaction conditions specified in the table

[e]Selectivity (%) of the catalyst with the highest conversion (bold)

[f]Yield (%) for the catalyst with the highest conversion (bold)

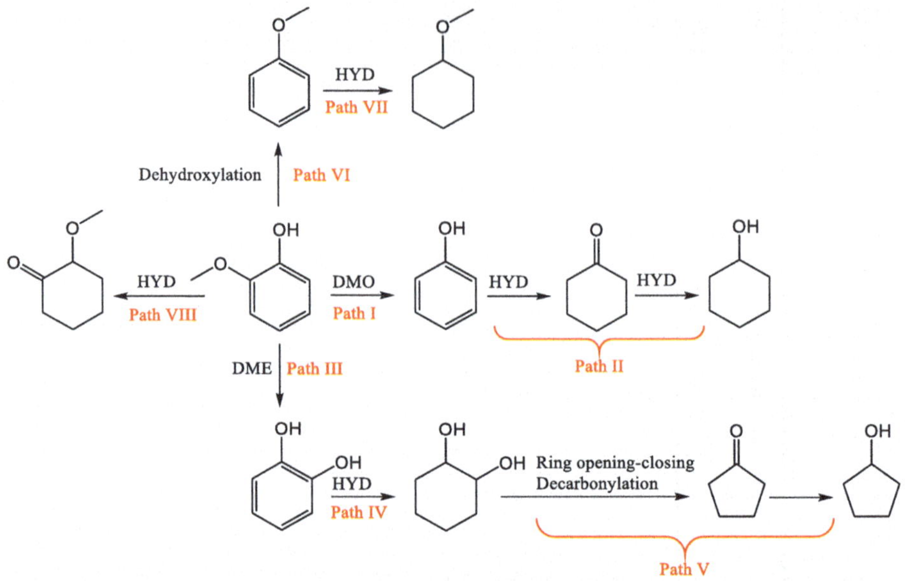

Fig. 6 Reaction mechanism of guaiacol hydrogenation over Pt/C and Pd/C using H_2 gas, decalin, and tetralin as hydrogen source, DMO demethoxylation, HYD hydrogenation, DME demethylation [73]

at 548 K under autogenous pressure for 4 h. As for guaiacol, the type of hydrogen donor and the type of catalyst significantly affects the process. When using decalin as hydrogen donor, total conversion of phenol was achieved for the Pt/C catalyst, with 30.95% selectivity to cyclohexanone and 60.0% to cyclohexanol (Fig. 7) [73]. However, with the Pd/C catalysts, the selectivity to cyclohexanol decreased significantly. When using tetralin as hydrogen donor, the conversion of phenol decreased significantly: down to 85.4% for the Pt/C catalysts and 36.2% for the Pd/C catalyst [73]. Still, higher cyclohexanol selectivities were again obtained with the Pt/C catalyst, suggesting that Pt/C is a more efficient catalyst for phenol hydrogenation.

Wang et al. [74] also studied phenol CTH over Raney Ni catalysts using a wide variety of hydrogen donor molecules: methanol (MeOH), ethanol (EtOH), 2-propanol (2-PrOH), butan-2-ol (2-BuOH), or 2-methyltetrahydrofuran (2-Me-THF). The tests were carried out in closed glass vials at 353 or 393 K for 3 h [74]. The best results were obtained using 2-PrOH and 2-BuOH as hydrogen donors at 353 K and under autogenous pressure: total conversion of phenol was achieved with a $<$99.3% cyclohexanol conversion. Benzene, cyclohexanone, and cyclohexane were also obtained as minor products (Table 3). The proposed reaction mechanism was similar to the one depicted in Fig. 7. However, benzene formation indicates that phenol also underwent catalytic hydrodeoxygenation, i.e., C_{AR}–O scission. Significantly, lower conversions were achieved when using MeOH, EtOH, and 2-Me-THF as hydrogen donors. The maximum conversion was achieved using those hydrogen donors and Raney Ni catalyst was ca. 18%. In a later work, Wang et al. [77] studied the CTH of phenol but this time using a Raney Ni supported over a solid-acid support, a β-zeolite (H-BEA). The activity tests were carried out at 433 K for 4 h under autogenous pressure. With the Raney Ni catalyst over H-BEA, benzene is the major reaction product, suggesting that phenol conversion occurs through a one-pot tandem reaction mechanism different from the one depicted in Fig. 7. The authors suggested [77] that phenol conversion results from summing up three different reactions (Fig. 8): (a) phenol hydrogenation to cyclohexanol, (b) dehydration of cyclohexanol to cyclohexene, and (c). Furthermore, they discovered that steps (a) and (c) become compatible in a one-pot procedure when the reactions are performed by CTH (Fig. 8). They also found out [77] that phenol conversion and benzene selectivity mainly depended on (1) the type of the solid acid (SA), (2) the molar 2-PrOH to substrate ratio (*n*), and (3) the weight SA/Raney Ni ratio (SA/Ni). At 433 K and with a *n* of 1.5 and a SA/Ni of 0.133, total phenol conversion and a 82% benzene yield was achieved (Table 3). Altogether, the results show that the use of a solid acid, as β-zeolite, favored the cyclohexanol deoxygenation that resulted in benzene production. In the absence of solid acids, as observed by the above-mentioned studies carried out by Shafaghat et al. [73] or Wang et al. [74], cyclohexanol deoxygenation and, thus, benzene production was not favored.

Wang et al. [74] also studied the CTH conversion of anisole using a Raney Ni catalyst and 2-PrOH as hydrogen donor at 353 K under autogenous pressure in closed glass vials. After 3-h reaction time, only 58% conversion was achieved. Benzene was the major product, 78% selectivity, with cyclohexane, 18% selectivity, and methoxycyclohexane, 4% selectivity, as the minor products (Table 3). The results show that the Raney Ni catalyst is more active for anisole hydrodeoxygenation, C_{AR}–O bond scission, than for anisole hydrogenation, i.e., saturation of the aromatic ring, when using hydrogen donor molecules.

2.2.2 Dimers

Galkin et al. [78] studied the CTH conversion of different 2-aryloxy-1-arylethanol ethers, β-*O*-4 lignin model compounds. Initially, they studied the effect of several hydrogen donor molecules and additives (i.e., acetic acid, 2-PrOH, NH_4HCO_2, and $NaBH_4$) in the conversion of 2-phenoxy-1-phenylethanol and compared these results with experiments carried out in the absence of hydrogen donor molecules and in the presence of molecular H_2. The experiments were carried out at 353 K, under autogenous pressure and for 1 h using a Pd/C catalyst and an ethanol/water solvent. The use of hydrogen donors increased considerably 2-aryloxy-1-arylethanol ether

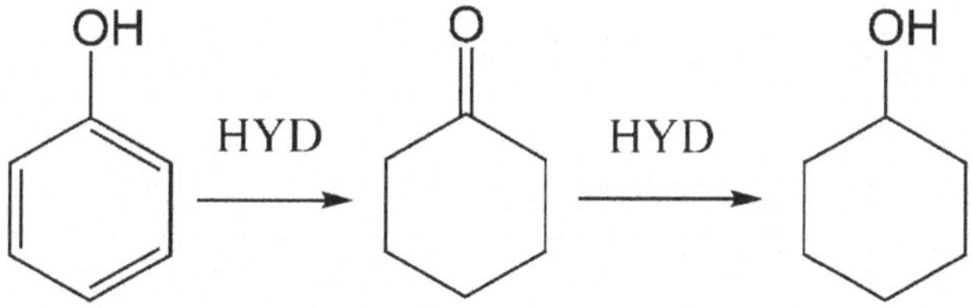

Fig. 7 Reaction mechanism of phenol hydrogenation over Pt/C and Pd/C using decalin and tetralin as hydrogen source; *HYD* hydrogenation [73]

conversion. The best results were obtained for $NaBH_4$: 100% conversion and 99% selectivity to acetophenone, the desired compound (Table 3). The results were comparable to the ones obtained when using molecular H_2. With the rest of the hydrogen donor molecules, the conversion was considerably lower, particularly with 2-PrOH, where no conversion was observed. Later, they evaluated the effect of methoxyl groups attached to the aryl rings on the conversion 2-aryloxy-1-arylethanol ethers. These experiments were carried out at the same reaction conditions as the former experiments, but using only $NaBH_4$ as hydrogen donor. The exact conversion of the 2-aryloxy-1-arylethanol ether depended on the position of the methoxy groups, but good selectivities to acetophenone and phenol were obtained in all cases. In addition, the authors proposed a new mechanism for the CTH conversion of 2-aryloxy-1-arylethanol ethers (Fig. 9). The reaction starts [78] with a reversible dehydrogenation of the 2-aryloxy-1-arylethanol ether (1) to generate a reactive intermediate (4). As a result, the intermediate 4 is adsorbed over palladium to generate the compound C. Later, the atomic hydrogen inserts into either the α or β position of the C compound following a Horiuti–Polanyi-type mechanism [79]. Two possible routes are suggested for the next step. In route I, atomic hydrogen inserts into the β position generating intermediate D. The ether bond in intermediate D is cleaved, the Pd is inserted into the C–O bond, giving intermediate E, which is later decomposed to acetophenone and phenol. In route II, atomic hydrogen inserts into the α position of the intermediate C to give intermediate F. Palladium is then inserted into the C–O

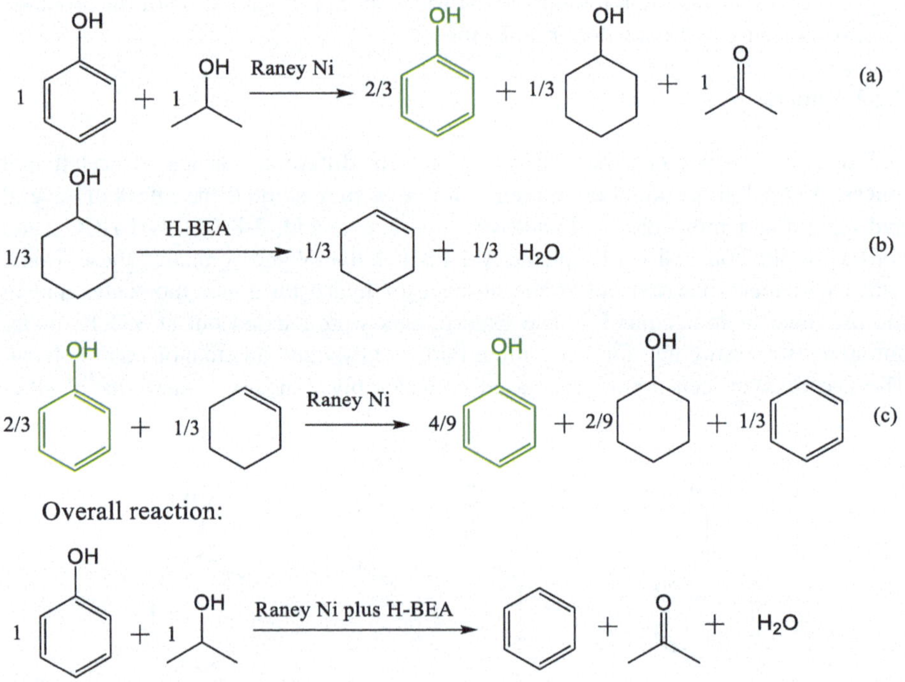

Fig. 8 Simplified pathway proposed for the conversion phenol to benzene by one-pot catalytic tandem reaction with Raney Ni and β-zeolite (H-BEA-35) [77]

bond to generate intermediate G, in which a carbene and an aryloxide are chemically adsorbed over the Pd. Reductive elimination of the aryloxide and carbene tautomerization leads to acetophenone and phenol.

Wu et al. [80] studied the CTH conversion of diphenyl ether (4-*O*-5) using 2-PrOH as hydrogen source and a wide variety of catalysts: Ru/C, Pt/C, Pd/C, Ru/MoS_2, Ru/TiO_2, Ru/SiO_2, and Raney Ni. The experiments were carried out at 303 K, 10 h of reaction time, 4 g of 2-PrOH, and 0.1 g of catalyst (0.05 in the case of the Pt/C catalyst). The reaction products varied depending on the type of catalyst used. All the catalysts, except Pd/C, yielded cyclohexane, benzene, and cyclohexanol as major products. Pt/C, Pd/C Ru/MoS_2, and Ru/SiO_2 also yielded small amounts of phenols. The best results were obtained for the Ru/C catalyst with 99% conversion (Table 3). Wu et al. [80] expanded the study to other dimeric compounds using the most active Ru/C catalyst: PPE (β-*O*-4), PHPL (β-*O*-4), BPE (α-*O*-4). Conversions over 99% were achieved for PPE and BPE, and only monomeric products were obtained. However, the PHPL conversion was only of 66.7%, suggesting that the hydroxyl group in the α position is detrimental for the conversion of β-*O*-4 bonds. Finally, Wu et al. [80] proposed a reaction mechanism for the CTH conversion of DPE using Ru catalysts and 2-PrOH as hydrogen donor (Fig. 10). They claimed that the ether bond of DPE was directly cleaved to obtain benzene and phenol through the formation of a weak coordinative bond between the phenoxide group and the Ru (Pathway I, Fig. 10). Benzene and phenol were subsequently hydrogenated with the aid of 2-PrOH. They also suggested an alternative reaction pathway that involved a first step where DPE was hydrogenated to cyclohexyl phenyl ether, which was subsequently converted to cyclohexane and phenol through the cleavage of the C(sp^3)–O bond on the surface of the Ru (pathway II, Fig. 10). The authors claimed that this reaction mechanism could be suitable for other diphenyl ether derivatives. Wang et al. [74] also studied the CTH conversion of different dimeric compounds using a Raney Ni catalyst and 2-PrOH as hydrogen donor at 353 K under autogenous pressure in closed glass vials. Dimers such as benzyl-phenyl-ether (α-*O*-4), diphenyl ether (4-*O*-5), benzofuran (β-5), and diphenyl-methane (β-1) were tested. For benzyl-phenyl-ether (α-*O*-4), total conversion was achieved, obtaining only cyclohexanol and toluene as products. In the case of diphenyl ether (4-*O*-5), total conversion was also achieved, and the products were again monomeric compounds such a cyclohexanol, benzene, and cyclohexane. For diphenyl-methane (β-1), total conversion was also achieved, however, no monomeric products were obtained. In this case, hydrogenation of the aromatic rings was only observed, suggesting that C–C bonds cannot be cleave using this catalytic system. Benzofuran (β-5) was also totally converted into a complex product mixture: dihydrobenzofuran, 2-methylcyclohexanol, toluene, methylcyclohexane, ethylbenzene, and ethylcyclohexane. These results again confirm the ability of the catalyst system to cleave ether bonds but not C–C bonds.

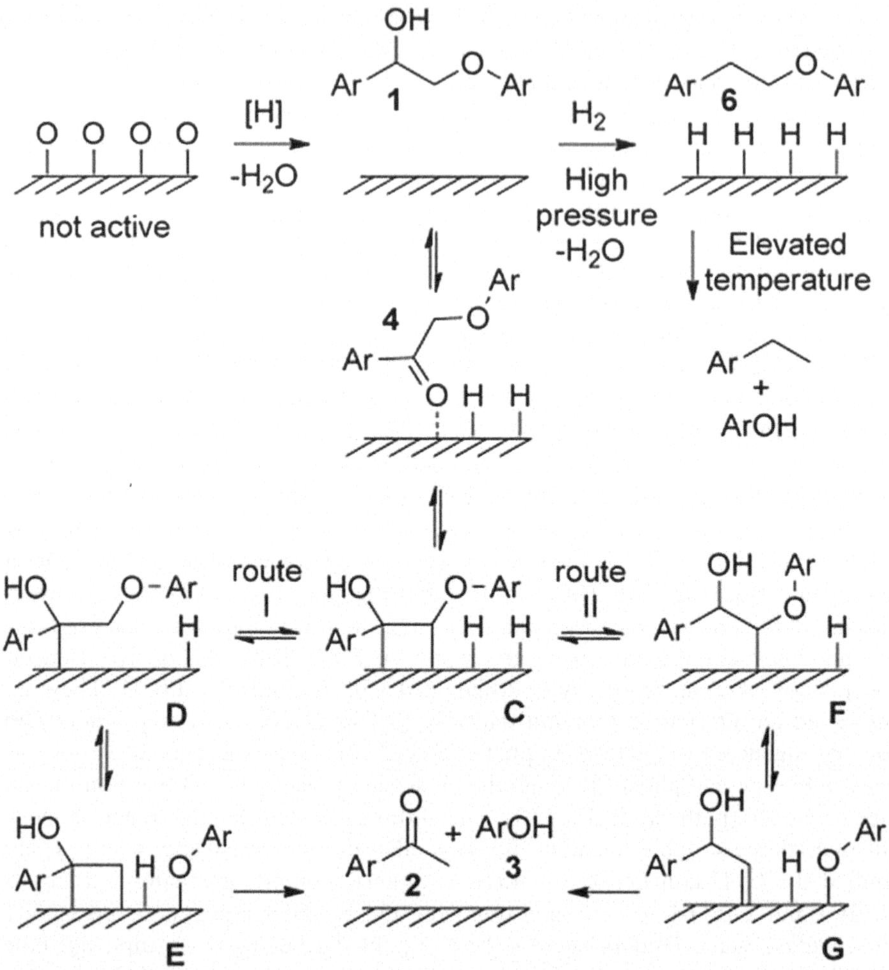

Fig. 9 Proposed reaction mechanisms for the redox neutral transfer hydrogenolysis of 1. Metal atoms are depicted by surface for clarity [78]

3 Oxidative Conversion Processes

Industrial conversion of lignin employs aerobic oxidation in alkaline media to produce aromatic aldehydes [81], particularly vanillin, which is the only commercial product derived from lignin, as mentioned above [82]. There are no industrial lignin-conversion processes that use other oxidants such as hydrogen peroxide or nitrobenzene [83]. For this reason, this section is focused on the oxygen-assisted oxidation reactions, i.e., oxygen or air-assisted, to produce phenolic aldehydes using heterogeneous catalysts. Oxygen or air-assisted oxidation presents some attractive advantages such as the employed oxidant is environmentally friendly, there is product accumulation, low energy consumption, mild conditions and, in general, high conversion [84].

Typically, catalytic oxidation of lignin model compounds is carried out in lab-scale autoclave-type reactors in liquid phase [85–87]. Veratryl (VA), vanillyl (VAL), and cinnamyl (COL) alcohols are the most common monomeric lignin model compounds used in catalytic oxidation to aromatic aldehydes. These model compounds allow simulating the propyl side-chain, phenolic and non-phenolic functional groups of lignin [86]. Dimers like PHPL [88], GGG [89], and VGG [90] are also employed as substrates.

3.1 Monomers

Veratryl alcohol (VA) oxidation to its corresponding aldehyde, veratraldehyde (VAld), represents the main product obtained from the cleavage of β-*O*-4 model compounds [91]. Thus, many studies employ this chemical as model compound for the study of lignin oxidation [86, 91]. Apart from the VAld, during VA oxidation veratrol and veratric acid (VAc) can be produced, as is reflected in Fig. 11. In this figure, the aerobic oxidation reaction pathway of vanillyl alcohol (VAL) is also shown. Veratrol can be formed through VAld decarbonylation, while VAc is generated by overoxidation of VAld [86, 91–93]. Table 4 provides an overview of the latest studies on VA oxidation, summing up the solvents, catalysts, reaction conditions, and conversion and selectivity levels.

Supported noble metal catalysts such as monometallic (Ru) and bimetallic (AuPd) and cobalt and copper oxide catalysts were found to be the most used catalytic systems. Moreover, the analyzed literature focused on the study and comparison of the influence of catalysts' properties and operating conditions on the activity and selectivity toward VAld. In this sense, the activity of Ru/Al_2O_3 and Ru/SiO_2 catalysts with their corresponding bare supports and with different Al_2O_3 supported metal catalysts (Co, Cu, Ag and Mn) for VA oxidation with air were compared [91]. The conversion to VAld increased considerably when using the supported Ru catalysts, indicating that Ru plays an important role in this process. The highest conversion was obtained for the Ru/Al_2O_3 catalyst, 93%, presenting a VAld yield of 67% at

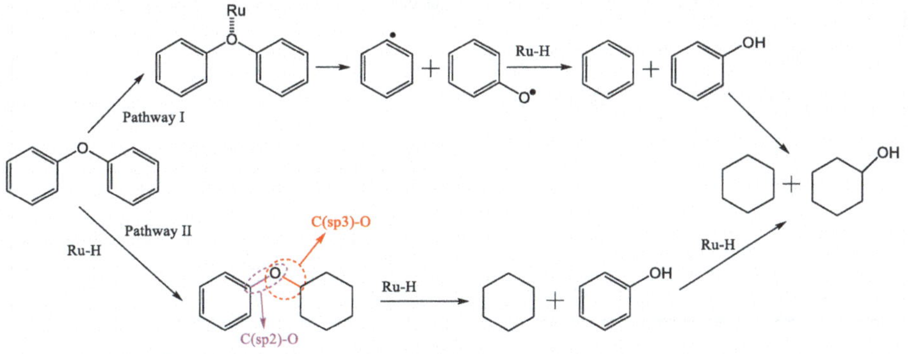

Fig. 10 Possible reaction pathway for the transfer hydrogenolytic cleavage of diphenyl ether [80]

433 K, 0.5 MPa, and 5 h of reaction time. This good behavior was attributed to lower RuO_x particle sizes detected for this catalyst. An increase in temperature improved the activity, achieving slightly higher conversions and yields [86, 91, 94]. However, a temperature excess could provoke a decrease of VAld yield due to disproportionation and overoxidation side reactions [94]. Regarding the solvent type, there is no consensus about the influence of using inorganic or organic polar solvents. Rodriguez et al. [91] obtained better catalytic performance with H_2O (inorganic polar solvent) when compared with methanol (organic polar solvent). However, it seems that organic polar solvents favor the dispersion of the catalysts in the reaction mixture and the solubility of the substrates in comparison to inorganic polar ones, facilitating substrate adsorption onto the catalyst [86].

The effect of metal loading (in the range of 0.4–4.1 wt%) on nanohydrotalcite (n-HT)-supported Ru catalyst and optimized the reaction conditions were also studied [86]. This study concluded that the 1.2 Ru/n-HT catalyst (1.2 wt% of metal) presented a complete conversion and the highest VAld selectivity (99%) at 353 K, 0.3 MPa, and 1 h. This good behavior was ascribed to the accessibility, number of Ru^0 active sites, basic properties and high surface area of the catalyst. This catalyst also showed good reusability, although it underwent considerable sintering after the third reuse. Besides the increase of reaction pressure and, thus, O_2 solubility reduced VAld selectivity due to its overoxidation [86, 94, 98, 99]. Moreover, they concluded that non-polar solvents provided better catalytic performance due to high catalyst dispersion [86] and O_2 solubility [98]. Another type of support and catalyst were studied in VA oxidation [94], being these the Ru/ZIF-8 (metal–organic framework (MOF)) and CuO catalysts. Both catalysts tested separately gave low VA conversions and VAld yields at 403 K, 0.5 MPa, and 0.92 h. However, a combination of both catalysts improved the conversion and the VAld yield, reaching 97% and 94.9%, respectively. This good behavior was associated to the synergistic effect between both catalysts. The Ru active phase reacted with the substrate to form a

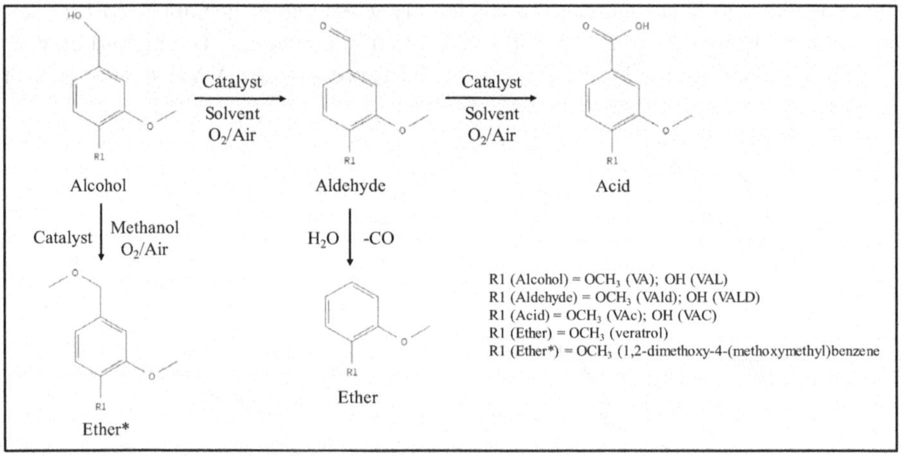

Fig. 11 Catalytic oxidation pathway of veratryl (VA) and vanillyl (VAL) alcohols (scheme adapted from literature [91, 93–97])

Ru-alcoholate intermediate that underwent β-hybrid elimination to yield the aldehyde and a Ru hybrid species. These Ru species reacted with O_2, completing the cycle. On the other hand, the CuO catalyzed the alcohol conversion into its corresponding aldehyde or ketone due to its dehydrogenation capacity. In conclusion, the presence of a second metal seems to improve the performance of the monometallic catalysts. Based on this, $Ce_{0.62}Zr_{0.38}O_2$ supported AuPd bimetallic catalyst with different metal loadings (with values in the range of 0.36–0.86 wt% for Au and 0.08–0.82 wt% for Pd) in VA oxidation at 353 K, 0.2 MPa of O_2, and 8-h reaction time were tested [93]. Under these operating conditions, the 1.4 AuPd/$Ce_{0.62}Zr_{0.38}O_2$ catalyst showed the best conversion, 72%, and VAld selectivity, 99%. The activity of this catalyst depended on the Au:Pd molar ratio, the size of Au and bimetallic AuPd particles, and co-existence of Pd^0 and oxidized $Pd^{\delta+}$ species.

As metal oxide, like CuO seems to participated in VA oxidation. Mate et al. [98, 99] employed non-noble metal oxides, such as Co_3O_4, for VA aerobic oxidation. They studied the influence of the preparation method and operating conditions on the activity and selectivity. Regarding the preparation method, the evaluated parameters were pH (8–14), calcination temperature (573–873 K), and time (1–6 h). They concluded that a pH of 8, a calcination time of 5 h, and a calcination temperature of 573 K were the optimum values to obtain a good VA conversion (38%) and VAld yield (100%, taking into account that yield was defined as moles of VAld produced per moles of VA reacted) at 413 K, 0.7 MPa, and 7 h of reaction time [98]. According to these authors, the selected preparation conditions provided the catalyst with a high surface area, oxidizing ability, and resistance to metal leaching. Conversion of 85% and VAld selectivity of 96% were obtained when the pressure was increased to 4.0 MPa. The catalyst presented good reusability, exhibiting no appreciable deactivation or leaching.

As mentioned above, VAL is one of the most used phenolic type lignin model compound to simulate oxidative conversion processes. Its selective oxidation of sp^3 C–H bond to produce vanillin (VN) has attracted considerable attention due to the numerous applications of this compound in food additives, perfumes, cosmetics, cleaning products, and so on [95, 100]. In VAL oxidation process, vanillic acid (VAC) (see Fig. 1) is the main byproduct. It must also be mentioned that the catalysts tested for VAL oxidation are similar to the ones used for VA oxidation. Table 5 shows an overview of the latest studies on VAL oxidation, summing up the solvents, catalysts, reaction conditions, and conversion and selectivity levels. Although it has been proven the high activity of noble metals for oxidation reactions [101], the use of transition metal catalysts such as Cu, Co, Mn, or Cr are the reported ones in the present review. Jha and Rode [101] investigated the activity of Co_3O_4 nanoparticles due to the strong capability of Co for dioxygen activation and cost effectiveness [85, 98, 99, 102]. The experiments were carried out at 353 K, 0.7 MPa, and 6-h reaction time using a mixture of VAL, isopropanol, and NaOH. They found out that the conversion and selectivity increased when NaOH was added to the reaction media. This increase was associated to the deprotonation of the phenolic –OH to form the phenoxy anion, which is the reactive species. Moreover, they carried out the same experiment with and without NaOH in the reaction media and the conversion and the selectivity to VN decreased from 80 to 25% and from 98 to 88%, respectively.

Table 4 Summary of the analyzed literature for VA oxidation

Entry	Catalyst	Solvent	Oxidant	Exp. conditions	Conv. (%)	Selec. to VAld (%)	Refs.
1	Ru/Al_2O_3	H_2O	Air	433 K, 0.5 MPa, and 5 h	93.0	67.0[a]	[91]
2	1.2Ru/n-HT	Toluene	O_2	353 K, 0.3 MPa, and 1 h	100.0	99.0	[86]
3	Ru/ZIF-8 + CuO	H_2O	O_2	403 K, 0.5 MPa, and 0.92 h	98.0	94.9[a]	[94]
4	$1.4AuPd/Ce_{0.62}Zr_{0.38}O_2$	Xylene	O_2	353 K, 0.2 MPa, and 8 h	75.0	99.0	[93]
5	Co_3O_4	H_2O	O_2	413 K, 0.7 MPa, and 5 h	38.0	100.0[a]	[98]
6	Co_3O_4	H_2O	O_2	413 K, 4.0 MPa, and 5 h	85.0	96.0	[98, 99]

[a]Yield

The activity of the Co_3O_4 nanoparticles was comparable to the one exhibited by Co homogeneous catalysts, and considerably higher than commercial Co_3O_4 catalysts. The results suggest that small size, high surface area, and Co^{3+} exposure are behind this high activity. They also checked that the catalyst could be reused during three runs without any significant conversion and selectivity drop. Yuan et al. [96] used nitrogen-doped reduced-graphene oxide-supported Mn_3O_4 nanoparticles (N-RGO/Mn_3O_4). They studied the influence of the reaction conditions, type of solvents (i.e., DMF, acetonitrile, isopropanol), and catalyst loadings. They found that *N,N*-dimethylformamide (DMF) was the best solvent, obtaining 92.5% of VAL conversion and 91.4% selectivity to VN after 12 h of reaction at 393 K. They observed that lower temperatures resulted in similar selectivities to VN, but also lower VAL conversions, while at higher temperatures the selectivity to VN decreased. They also found out that the higher the catalyst loading the higher the conversion (92.5% with 40 wt% of the N-RGO/Mn_3O_4), however, regardless the catalyst loading the selectivity was surprisingly kept at a constant value of 91%. Finally, they confirmed that the catalyst was stable even after six runs.

Due to the good ability to store, release, and transport oxygen by CeO_2, Mn and Zr-doped CeO_2 catalysts were prepared [95, 100]. Thus, it has been widely applied in processes such as automotive three-way catalysis, CO oxidation, and soot oxidation. They prepared nanosized mixed-oxide samples with different cerium and manganese contents ($Ce_{0.8}$ $Mn_{0.2}O_{2-\delta}$ – CM82– and $Ce_{0.7}$ $Mn_{0.3}O_{2-\delta}$ – CM73–). CM73 showed superior catalytic activity in terms of conversion (89%) and selectivity to VN (95%). In terms of the reaction conditions, the highest conversion and selectivity to VN were obtained at 413 K and 2.0 MPa after 4 h of reaction. They observed that decreasing the temperature resulted in similar selectivities to VN, but lower VAL conversions, while the selectivity decreased at higher temperatures. This was corroborated by Yuan et al. [96]. In terms of O_2 pressure, they observed that low O_2 pressures resulted in low conversion and high selectivity. On the other hand, high O_2 pressures resulted in high conversion but lower selectivity. Among the tested solvents (isopropanol, acetonitrile, methanol, toluene, and water), acetonitrile showed the best results. With the optimum solvent and reaction conditions, 89% conversion and 95% selectivity to the desired VN product were achieved. In the case of Zr-doped CeO_2 catalysts, the effect of Ce/Zr molar ratio was also studied [95]. A series of Ce catalysts were synthesized with different Ce and Zr contents. The best results were achieved with the $Ce_{0.8}Zr_{0.2}O_2$ catalyst. In terms of the reaction conditions, 413 K, 2.0 MPa, and 5 h were found to be the optimum ones, as corroborated also by Ramana et al. [100] and Yuan et al. [96]. Regarding the solvent, they observed that its nature plays a significant role; i.e., non-polar solvents, such as toluene and hexane, resulted in poor VAL solubility and therefore the conversion decreased to ca. 23–30%. Meanwhile, the use of polar solvents increased VAL conversions; the higher the polarity, the higher the conversion. Thus, ethyl acetate yielded a 52% conversion, isopropanol 65%, 1,4-dioxane 80%, and finally acetonitrile 98% conversion [86].

As the presence of ZrO_2 in the catalysts seems to improve catalyst activity and selectivity, and the different oxidation states of copper confer a broad range of activities to catalyze oxidation, activity of perovskite type $CuZrO_3$ catalysts were studied [103].

Table 5 Summary of the analyzed literature for VAL oxidation

Entry	Catalyst	Solvent	Oxidant	Exp. conditions	Conv. (%)	Selec. to VN (%)	Refs.
1	Co_3O_4 nanoparticles	Isopropanol (in alkali media)	Air	353 K, 0.7 MPa, and 6 h	80.0	98.0	[101]
2	N-RGO/Mn_3O_4	DMF	O_2	393 K, 0.1 MPa, and 12 h	92.5	91.4	[96]
3	$Ce_{0.7}Mn_{0.3}O_{2-\delta}$	Acetonitrile	O_2	413 K, 2.0 MPa, and 4 h	89.0	95.0	[100]
4	$Ce_{0.8}Zr_{0.2}O_2$	Acetonitrile	O_2	413 K, 2.0 MPa, and 5 h	98.0	99.0	[95]
5	$CuZrO_3$ (Cu:2Zr)	Acetonitrile	Air	393 K, 2.1 MPa, and 2.5 h	91.0	76.0	[103]
6	$Fe_3[Fe(CN)_6]_2$	Isopropanol	O_2	393 K, 2.1 MPa, and 2.5 h	99.9	99.0	[97]

They prepared three types of catalysts: Cu:Zr, Cu:2Zr, and 2Cu:Zr. The highest conversion and selectivity (91% and 76%, respectively) were achieved with Cu:2Zr catalyst at 393 K, 2.1 MPa air pressure, and 2.5-h reaction time using acetonitrile as a solvent. Zheng et al. [97] suggested the use of Prussian Blue Analogues ($Fe_3[Fe(CN)_6]_2$) (PBAs) as oxidating catalysts, which are used in dyeing, paints, and as an antidote. These analogues are very easy to prepare and are inexpensive and nontoxic. These authors prepared various PBA by replacing metal species by other transition metals yielding different $M_3^{II}[M^{III}(CN)_6]_2$ analogues (M^{II} = Mn, Fe, Co, Ni, Cu, and Zn; M^{III} = Fe, Co). They evaluated their activity in VAL oxidation to VN at 2.1 MPa and 393 K using isopropanol as solvent. They observed that the highest production yields were obtained for FeFe analogues (99.9% of conversion and 99.0% of selectivity to VN after 2.5 h of reaction time).

Cinnamyl alcohol (COL) is another important lignin model compound studied in oxidative processes. Its corresponding aldehyde, cinnamaldehyde (CAL), is a high-value intermediate/product for the pharmaceutical, agrochemical, and petrochemical industries [104, 105]. Table 6 shows an overview of the latest studies on COL oxidation, summing up the solvents, catalysts, reaction conditions, and conversion and selectivity levels. However, it can be observed how the majority of the studies employ toluene as solvent, although this solvent is not fully "green" [106, 107]. Figure 12 reflects the most reported reaction pathways of COL oxidation.

Chen et al. [87] conducted wide research in alcohol oxidation using Pd-supported on an acidic MOF. Two different heterogeneous catalysts were tested (1) Pd nanoparticles deposited on an acidic MOF (MIL-101) and, (2) ethylenediamine grafted MIL-101. With the first catalyst, they varied Pd loading from 0.15 to 0.94 wt% achieving the best results (99% of conversion and 99% of selectivity towards cinnamaldehyde with the 0.35 wt% Pd/MIL-101 catalyst, using toluene as solvent, at 353 K and 0.1 MPa. Afterwards, they screened several solvents (isooctane, DMSO, *o*-xylene) under the same reaction conditions using the 0.35Pd/MIL-101 catalyst. They determined that toluene was the best solvent for this transformation. Chen et al. [87] conclude that the conversion increased with the decrease in polarity of the solvent. The very low activity registered when using isooctane was ascribed to poor COL solubility. The amine-grafted catalyst 0.35% Pd/En-MIL-101 showed a lower activity for COL oxidation (45% yield of CAL), even with prolonged reaction times. They concluded that the amine was grafted on the uncoordinated Cr sites of the MIL-101 support, which suggests that open chromium sites may play an important role in the oxidation of COL.

The catalysts with the same active phase, Pd, and different support, ultrathin Al_2O_3 monolayers grafted onto an ordered mesoporous SBA-15 silica framework (Al-SBA-15) were also studied [104]. Its activity was compared to the activity Pd catalysts supported on mesoporous Al_2O_3 and SiO_2. The experiments were carried out at 363 K and atmospheric pressure (bubbling 3 ml/min of molecular O_2), using 1 wt% Pd catalyst and toluene as a solvent. They concluded that the Al-SBA-15 support facilitates the dispersion of Pd nanoparticles, resulting in superior activity for the aerobic selective oxidation of COL to CAL. After 30 min of reaction time, 95% of COL was converted (100% after 3 h), obtaining 64% CAL yield (66% after 3 h), while the conversion decreased to 56% and 21% when using mesoporous Al_2O_3 and SBA-15 as supports, respectively. Pt noble metal was also tested as a catalyst in COL oxidation to obtain CAL. Durndell et al. [105] controlled the dispersion of Pt nanoparticles and their oxidation state via the use of mesoporous SiO_2 supports. They concluded that the catalyst activity is a strong function of Pt dispersion, which in turn controls the concentration of surface PtO_2, the active site responsible for the desired selox (selective oxidation) chemistry. However, selectivity reflects a delicate balance of oxidation and competing hydrogenation/hydrogenolysis pathways driven by in situ hydrogen production and associated with nanoparticle reduction.

Apart from Pd, Ru-supported catalysts were employed in COL oxidation [86, 109] due to the good performance of these catalysts in VA oxidation. Concretely, Mori et al. [109] studied the oxidation process of COL using toluene as solvent, at atmospheric pressure and 343 K. In this case, the chosen catalyst was an environmentally friendly heterogeneous 10Ru/C catalyst. At these reaction conditions, they achieved 79% CAL yield. Baguc et al. [86] also studied the activity of COL to CAL under the same operation conditions that they used for the oxidation of VAR to VAld (353 K, 0.3 MPa, toluene as solvent, 1.2Ru/n-HT and 1 h of reaction time). In such conditions, 100% conversion was achieved with > 99% of selectivity to CAL.

Regarding AuPd catalysts, Wu et al. [108] investigated the activity of these catalysts supported in TiO_2 for VAL oxidation. This catalytic system was already investigated for VA oxidation by Olmos et al. [93], as mentioned above. In this case, the addition of Au enhances the activity in the selective oxidation of COL and benzyl

Table 6 Summary of the analyzed literature for COL oxidation

Entry	Catalyst	Solvent	Oxidant	Exp. conditions	Conv. (%)	Selec. to CAL (%)	Refs.
1	0.35% Pd/MIL-101	Toluene	O_2	298 K, 0.1 MPa, and 0.5 h	99	99	[87]
2	Pd/Al-SBA-15	Toluene	O_2	363 K, 0.1 MPa, and 0.5 h	95	64[a]	[104]
3	0.05 Pt/SiO_2	Toluene	O_2	363 K, 0.1 MPa, and 0.5 h	100	65–25[a]	[105]
4	10Ru/C	Toluene	O_2	343 K, 0.1 MPa, and 24 h	–	79[a]	[109]
5	1.2Ru/n-HT	Toluene	O_2	353 K, 0.3 MPa, and 1 h	100	> 99	[86]
6	AuPd (5:95)/TiO_2	Toluene	O_2	373 K, 0.4 MPa, and Cont.[b]	40	65	[108]
7	AuPd (5:95)/TiO_2	Toluene	Air	373 K, 0.4 MPa, and Cont.[b]	65	60	
8	Mn_3O_4 nanoparticles	DMF	O_2	353 K, n.a. MPa, and 24 h	80	75[a]	[110]
9	Co-ZIF-9	Toluene	O_2	423 K, 0.5 MPa, and 4 h	34	30[a]	[85]

[a] → yield

[b] → 10 mg of catalyst; gas flow, 2.0 ml min^{-1} at STP; 0.5 M cinnamyl alcohol in toluene, 0.020 ml min^{-1}

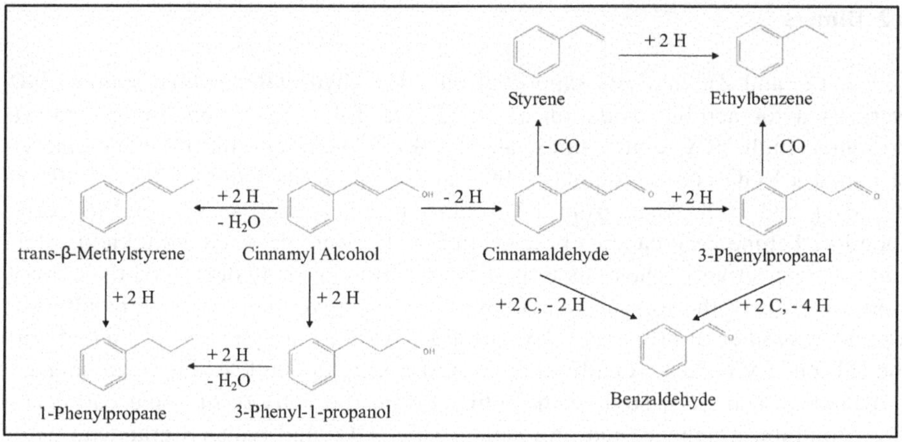

Fig. 12 Catalytic oxidation pathway of cinnamyl alcohol (COL) [105, 108]

alcohols, when compared to monometallic catalysts. They studied the stability of a 1 wt% bimetallic Au–Pd (5:95)/TiO_2 catalyst in a packed-bed capillary continuous microreactor under various reaction conditions (0.4 MPa, 373 K, 10 mg of catalyst; 2.0 ml min^{-1} at STP of gas flow; 0.020 ml min^{-1} of 0.5 M cinnamyl alcohol in toluene, were kept constant). The catalyst pre-reduction with COL or H_2 before reaction was one of the studied parameters, since some discussion exists about the suitability of using metallic Pd or palladium oxide as catalyst [111–113]. They observed that the pre-reduction step improved the initial activity but did not affect catalyst deactivation. The use of higher O_2 concentration (100 vs. 21%) did not affect the rate of CAL formation but led to lower initial alcohol conversion (40 vs. 65% in 1 h on stream) and higher CAL selectivity (65 vs. 60%). However, catalyst deactivation was observed in all the studied operation conditions, having detected Pd leaching phenomena ascribed to the presence of O_2, showing that the role of O_2 is very complex. Other authors, like Miyamura et al. [114], evaluated the utilization of a heterogeneous bifunctional chiral catalyst consisting of metal nanoparticles and a chiral Jørgensen–Hayashi-type organocatalyst (OC) supported on different polymers. In this case, these authors determined that Au–Pd alloy nanoparticles with 1:1 ratio were the one offering the best results.

Other authors such as Sun et al. [110] and Zakzeski et al. [85] investigated the use of non-platinum-group metals for COL oxidation. In this sense, Sun et al. [110] employed Mn_3O_4 nanoparticles as catalyst, achieving 80% of COL conversion and 75% CAL yield after 24 h, using DMF as solvent and at 353 K. Zakzeski et al. [85] employed Co-ZIF-9 as catalyst due to the high catalytic activity demonstrated by the homogeneous cobalt complexes. The experiments were carried out at 0.5 MPa and 423 K using toluene. However, the achieved results showed only ca. 34% conversion and 30% CAL yield.

3.2 Dimers

V, Cu, Fe, and Zn catalysts supported on a HT (hydrotalcite)-type support [90] were used for aerobic oxidation of VGG (see Table 7). V-containing catalysts presented high VGG conversions, above 70%. Concretely, the HT-CuV catalyst provided a VGG conversion above 99% and a VAld yield around 38% at 408 K, 0.6 MPa, and 16 h. Apart from VAld, other products such as VAc and its corresponding ketone, as it can be observed in Fig. 13, were detected, with yields of 31 and 5%, respectively. The reduction of the reaction pressure decreased the conversion. In terms of the solvent, toluene was less selective than pyridine, resulting in a greater number of products. The reusability of the catalysts was evaluated with the HT-ZnCuV catalyst, exhibiting poor reusability due to leaching phenomena.

Blandez et al. [89] studied the influence on the activity of some carbocatalyst (graphite, graphene, and, graphene oxide (GO) and reduced graphene oxide (rGO)) and of the operating conditions in GGG dimer oxidation. In this case, the main products were guaiacol and 2-methoxiketone, followed by VN and VAC. The GO catalyst provided the highest guaiacol and 2-methoxiquinone selectivity (87 and 15%, respectively) for a complete conversion at 413 K, 0.5 MPa, and 24 h of reaction time. In this case, the VN and VAC selectivities were 7% and 5%, respectively. In general, the solvent type, the use of O_2 or air as oxidants, and the selection of a suitable temperature were the main parameters that influenced the guaiacol and 2-methoxiquinone selectivities, while the VN and VAC selectivities remained almost constant.

On the other hand, Deng et al. [88] studied the Pd catalysts supported on different metal oxides (Al_2O_3, SiO_2, MgO, and CeO_2) for the aerobic oxidation of PHPL at 458 K, 0.1 MPa, and 24 h of reaction time using methanol as solvent. The Pd/CeO_2 catalyst exhibited the best behavior, reaching a PHPL conversion of 64% and phenol, acetophenone, methyl benzoate, and 2-phenoxy-1-phenylethanone yields of 48, 38, 14, and 12%, respectively. They concluded that Pd nanoparticles over CeO_2 play an important role in the production of 2-phenoxy-1-phenylethanone intermediate product, which undergo subsequent cleavage of β-*O*-4 and C_α–C_β bond to yield aromatic monomers. The products resulting from C_α–C_β bond cleavage are the major ones when CeO_2 bare support was used.

4 Pyrolytic Processes

As mentioned above, pyrolysis, also known as thermal cracking, is the breakdown of organic molecules by the effect of temperature in a non-oxidizing medium. It is a process that has been used on an industrial scale for many years, mainly in the processing of fuels. Some classic examples are the production of charcoal from wood or the manufacturing of metallurgical coke from coal. Pyrolysis can also be performed in the presence of catalysts, normally over acid solid catalysts, in which case it is called catalytic cracking. For several years, pyrolysis has been studied as a technique for the valorization of different types of waste, including municipal solid

Table 7 Summary of the analyzed literature for oxidation of dimers (β-O-4 lignin model compounds)

Entry	Catalyst	Solvent	Oxidant	Exp. conditions	Model compound	Conv. (%)	Products	Yield (%)	Refs.
1	HT-CuV	Pyridine	O_2	408 K, 0.6 MPa, and 16 h	VGG	99.0	VAld	38.0	[90]
							VAc	31.0	
							KN	5.0	
							EEE	3.0	
2	GO	Acetonitrile	O_2	413 K, 0.5 MPa, and 24 h	GGG	100	VN	7^a	[89]
							VAC	5^a	
							GAL	87^a	
							MK	15^a	
3	Pd/CeO_2	MeOH	O_2	458 K, 0.1 MPa, and 24 h	PHPL	64	PP-one	12.0	[88]
							BA	0.0	
							MB	14.0	
							AP	38.0	
							Phol	48.0	

KN ketone, *EEE* enol ether enone, *ArAcid* aromatic acid, *DE* diphenyl eter, *ATVL* acetovanillone, *MBQ* 2-methoxybenzoquinone, *GAL* guaiacol, *MQ* 2-methoxiquinone, *PP-one* 2-phenoxy-1-phenylethanone, *BA* benzoic acid, *MB* methyl benzoate, *AP* acetophenone, *Phol* phenol

[a]Selectivity

waste (MSW), plastic waste, waste of electric and electronic equipment (WEEE), and waste tires [115–118]. More recently, it has also been proposed as another alternative within the concept of biorefinery for the thermal processing of by-products derived from biomass, including lignin [119, 120].

One of the main problems in defining the chemical systems in pyrolysis processes of complex polymeric molecules like all those mentioned above is the difficulty of elucidating the reaction mechanisms involved. This is because pyrolysis is a sequential chemical process that generates a system of multiple reactions, which at the same time depend on the reaction temperature. At low temperatures, the initial radical cracking reactions of the polymeric substances occur. At higher temperatures, the products generated in such first cracking reactions are further cracked, and combination reactions between all these products happen. At very high temperatures, gasification reactions primarily forming CO/CO_2 and H_2 are predominant. For this reason, the utilization of model compounds that represent molecules belonging to one of the stages mentioned above is of particular relevance in the field of pyrolysis. In the specific case of lignin, these stages could be divided into the following temperature ranges: 423–673 K (first stage), 673-873 K (second stage), and > 873 K (third stage) [120]. In this way, research carried out with model compounds could be classified according to the size of the molecule used: polymeric molecules (normally dimers) represent those compounds typical of the first pyrolysis stages, while monomers are more representative of molecules generated in the pyrolysis of such polymeric species (dimers), that is, secondary products of the first cracking.

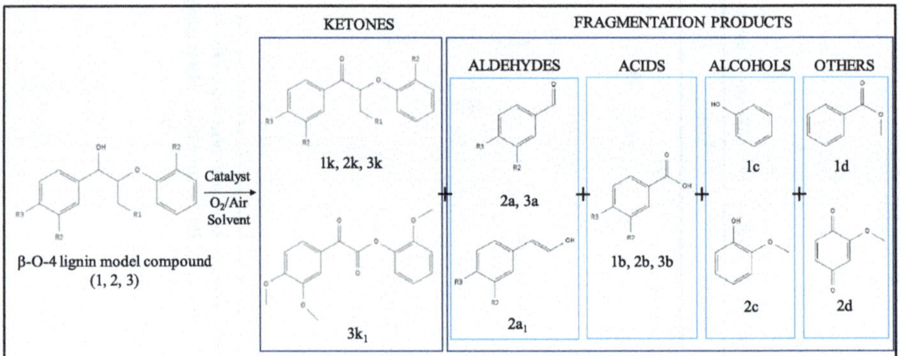

2-Phenoxy-1-phenylethanol (without R1, R2 and R3) **(1)**→ 2-phenoxy-1-phenylethanone **(1k)** + benzoic acid **(1b)** + phenol **(1c)** + methyl benzoate **(1d)**

Guaiacyl-glycerol-β-guaiacyl ether (R1, R3 = OH and R2 = OCH_3) **(2)** → 2-hydroxy-1-(4-hydroxy-3-methoxyphenyl)-2-(2-methoxyphenoxy)propan-1-one **(2k)** + VN **(2a)** + coniferyl aldehyde **($2a_1$)** + VAC **(2b)** + guaiacol **(2c)** + 2-methoxyquinone **(2d)**

Veratryl-glycerol-β-guaiacyl ether (R1 = OH and R2, R3 = OCH_3) **(3)** → 1-(3,4-Dimethoxyphenyl)-3-hydroxy-2-(2-methoxyphenoxy)propan-1-one **(3k)** + enol ether enone **($3k_1$)** + Vald **(3a)** + VAc **(3b)** +

Fig. 13 Catalytic oxidation pathway of β-*O*-4 dimer lignin model compounds (adapted from literature [88–90])

4.1 Monomers

Guaiacol is the most representative product of the pyrolysis of lignin at low temperatures [18]. The most recent investigations agree that the main mechanism of guaiacol thermal decomposition is the homolytic cleavage of the O–CH_3 bond (homolytic demethylation) to form methyl radical and hydroxyphenoxyl radical, which is at the same time an important pathway to form catechol [70, 121–123]. The decomposition of guaiacol has also been studied in the presence of HZSM-5 zeolite, in batch operation and vapor phase contact, giving rise to significant quantities of CO, light hydrocarbons, and aromatic hydrocarbons [124]. The authors of this work reported that zeolite promotes ring-opening reactions, which would be the reason of the formation of light hydrocarbons. In fact, they proposed a reaction pathway based on the cracking and deoxygenation of phenolic compounds forming light hydrocarbons that are later recombined to form aromatic products, being the latter and the compounds arising in deoxygenation (CO, CO_2, H_2O) the main products of the process. Catechol decomposition has also been studied in thermal and catalytic conditions. It is a highly stable compound that needs temperatures around 1023 K for its conversion to be greater than 50%. Its thermal decomposition mechanism has been defined as the combination of H-migration and ring-opening reactions to form CO and C4 hydrocarbons, mainly 1,3-butadiene [124, 125]. When decomposed in the presence of zeolite HZSM-5 (in batch operation and vapor phase contact) two fundamental differences were observed. On the one hand, zeolite accelerates decomposition reactions, achieving a conversion greater than 50% at 923 K. On the other hand, a greater yield to aromatic compounds is produced from the recombination of hydrocarbons [124].

As it has been shown in the previous paragraph and will be shown in 4.2 subsection, phenol is a characteristic compound evolving from pyrolysis of phenolic compounds and polymeric model compounds [70, 121–123, 125, 126]. For this reason, the study of the pyrolysis mechanism of phenol is of great importance. According to the work published by Scheer et al. [127], the primary product of the phenol thermal decomposition is cyclohexadienone, formed by enol/keto tautomerization (Fig. 14). Subsequently, CO and cyclopentadiene are generated from the decarbonylation of cyclohexadienone. In the next step, the cyclopentadiene loses an atom of hydrogen to become cyclopentadienyl radical, which in turn decomposes into acetylene and propargyl radical at high temperatures.

Cyclopentadiene and CO were also identified as the main products of phenol pyrolysis in a recently published work [128], together with benzene and smaller quantities of acetylene, naphthalene, methane, and methylcyclopentadiene. In such work, it was suggested that the resulting cyclopentadiene may continue decomposing to generate polyaromatic hydrocarbons (PAH). Yang et al. [124] have recently published a paper comparing the thermal pyrolysis of several phenolic compounds with catalytic pyrolysis in the presence of HZSM-5 zeolite (in batch operation and vapor phase contact). Cyclopentadiene, benzene, and CO were reported to be the main products of phenol thermal pyrolysis, while the production of CO and aromatic hydrocarbons (especially benzene) was much higher in the presence of this catalyst in comparison with the thermal counterpart. This result suggests that the presence of the acid catalyst promotes decarbonylation of phenol. A summary of the analyzed literature for monomeric lignin model compounds pyrolysis is shown in Table 8.

4.2 Dimers

The works analyzing the cracking of polymeric lignin model compounds are focused on the cleavage mechanism of the typical bonds found in lignin. Most of them studied the behavior of the β-*O*-4 linkage in pyrolysis, as it is the most common linkage type in the lignin molecule [129–137]. There seems to be some agreement in the scientific community that the initiation reaction of pyrolysis of β-*O*-4-containing model compounds is the homolytic cleavage of the $C_β$–O bond, at least at low temperatures [134–137]. This is because the energy needed for the $C_β$–O homolytic cleavage is less than the one needed for the $C_α$–$C_β$ homolytic scission and for the $C_β$–O concerted decomposition, respectively [134]. However, at high temperatures, these last two reaction mechanisms are also competitive. Akazawa et al. [135] also proposed a pathway involving the abstraction of the hydroxyl radical from $C_α$ in the pyrolysis of GGG, guaiacylglycerol-β-coniferyl ether, and *p*-hydroxyphenylglycerol-β-coumaryl ether, obtaining the same final products as those obtained by the $C_β$–O cleavage pathway. In addition, Wang and Liu [138] reported that $C_β$–O bond energy may vary in the presence of the *o*–CH_3O group, as demonstrated by comparing the pyrolysis mechanisms of PPE and *o*-methoxy phenethyl phenyl ether (*o*–CH_3O–PPE). Specifically, they showed that the decrease in bond energy caused the decomposition of *o*–CH_3O–PPE to be four times faster than that of PPE. These results highlight the influence that the specific structure of the lignin molecule in question can have on its thermal decomposition mechanisms.

Park et al. [140] studied the catalytic decomposition of PPE in the presence of cesium-exchanged heteropolyacids, with and without the incorporation of Pd, in

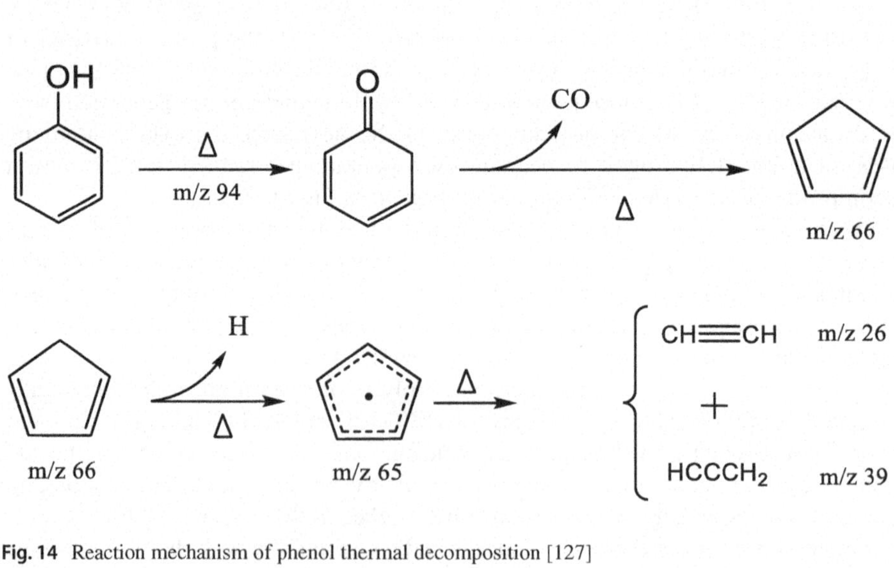

Fig. 14 Reaction mechanism of phenol thermal decomposition [127]

liquid phase contact (PPE in dissolution). The identified reaction mechanism was the same as in the case of thermal pyrolysis of PPE (the homolytic cleavage of the C_{β}–O bond), as well as the main products, i.e., styrene and phenol, although they reported that styrene was subsequently converted to toluene, benzene, and ethylbenzene due to secondary reactions on the surface of the catalyst (Fig. 15). However, the main difference lies in the fact that the pyrolysis reaction was carried out at 473 K, obtaining PPE conversions around 68% with the catalyst without Pd, and around 79% in the presence of Pd. These conversions are much higher than that reported in the thermal pyrolysis of PPE, that is, around 62% at 773 K [137]. Therefore, despite not influencing the reaction pathway, it is evident that these acid catalysts decrease the activation energies of the reactions involved in the mechanism.

The second most studied type of linkage in pyrolysis is the α-*O*-4 type, both noncyclic and cyclic ones [70, 140–142]. In this case, there is also quite a consensus that the lowest bond energy is that of the C_{α}–O bond, so the cracking mechanism of the molecules that present this linkage begins with the homolytic rupture of this bond, in both thermal and catalytic pyrolysis [70, 140, 141]. The competitive mechanisms with this main pathway are the homolytic cleavages of the C_{α}–C_{β} and C_{α}–$C_{aromatic}$ bonds. With regard to the cyclic α-*O*-4 link, typical of coumarone-contained linkages in lignin, Huang et al. [142] proposed four possible reaction pathways: the cleavage of C_{α}–C_{β} bond, the cleavage of C_{α}–O bond, the cleavage of C_{α}–$C_{aromatic}$ bond, and the cleavage of C_{β}–C_{γ} bond, concluding the C_{α}–O and C_{β}–C_{γ} homolytic cleavage were the main reaction mechanisms. Park et al. [140] also studied the pyrolysis of BPE in the presence of inorganic acids, $H_3PW_{12}O_{40}$-based heteropolyacids, identifying phenol and toluene as the main reaction products. Thermal pyrolysis studies of this same compound do show phenol as one of the main products, however toluene presents much lower concentrations than compounds such as 6*H*-benzo[c]chromene, 2-benzylphenol, or [1,1′-biphenyl]-2,2′-diol [70]. The structural complexity of these chemicals suggests the severe catalytic action of the acid catalysts employed in the work of Park et al., which are capable of cracking these compounds into simpler molecules.

In addition, this difference between thermal and catalytic pyrolysis products occurs at very different temperatures, in this case 473 K in the presence of the catalysts and 873 K without them. Therefore, as in the case of PPE, these inorganic acids demonstrate their high cracking capacity, probably based on their strong acid strength, which is higher than that of conventional solid acids [140]. A summary of the analyzed literature for dimeric lignin model compounds pyrolysis can be seen in Table 9.

4.3 Reforming of Lignin Pyrolysis Oil Model Compounds

Lignocellulosic biomass pyrolysis liquids are identified as a sustainable and CO_2-neutral feedstock [143] for producing hydrogen for current industrial processes, currently based on fossil hydrogen [144], and/or future energetic applications [143–147]. Nevertheless, pyrolysis liquids or bio-oils involve a huge amount of oxygenated and non-oxygenated molecules [143, 144, 146], with aromatic/phenolic chemicals

Table 8 Summary of the analyzed literature for monomeric lignin model compounds pyrolysis

Entry	Model compound	Experimental set-up	Carrier gas	Pyrolysis temperature	Main stable products (yield in %)	Refs.
1	Guaiacol	Reactor + SVUV-PIMS[a] Py-GC/MS[b]	Ar He	723–998 K 513 K	At 873 K: catechol, 2-methylcyclopenta-2,4-dienol, cyclopenta-2,4-dienone, 2-hydroxybenzaldehyde, cyclopenta-2,4-dienol, cyclohexa-3,5-diene-1,2-dione, phenol	[70]
2	Guaiacol	Simulation[c]	–	700–1500 K	At 1275 K: phenol, carbon monoxide, hydrogen, methane, ethane, formaldehyde, 2-hydroxybenzaldehyde	[121]
3	Guaiacol	Reactor + SVUV-PIMS[a]	Ar	650–800 K	At 800 K: carbon monoxide, catechol, methane, phenol, cyclopenta-2,4-dienone, cyclohexanone, cyclobutadiene	[122]
4	Guaiacol	Simulation[d]	–	–	Catechol, methane, phenol, *o*-cresol, 2-hydroxybenzaldehyde, coke	[123]
5	Guaiacol + HZSM-5	Reactor[e]	He	823–1223 K	At 923 K: benzene (16.5), carbon monoxide (13.6), methane (7.9), ethane (5.3), naphthalene (5.0), toluene (4.8), propene (2.1), styrene (1.5), ethane (1.3), anthracene (1.1)	[124]
6	Catechol	Reactor[f]	He	823–1223 K	At 1223 K: carbon monoxide (27.0), water (5), phenols, aromatic hydrocarbons, light-oxygenated compounds, methane, C2–C5	[125]
7	Catechol + HZSM-5	Reactor[e]	He	823–1223 K	At 1223 K: carbon monoxide (27.3), benzene (13.8), toluene (0.9), naphthalene (5.0), ethane (2.1), methane (6.4), water (3.8), propene (0.05), styrene (0.8)	[124]
8	Phenol	Reactor + SVUV-PIMS[a]	Ar	823–1223 K	Carbon monoxide, cyclopentadiene, cyclohexadienone, acetylene	[127]
9	Phenol	–	–	–	Carbon monoxide, cyclopentadiene, benzene, acetylene, naphthalene, methane and methylcyclopentadiene	[139]
10	Phenol + HZSM-5	Reactor[e]	He	823–1223 K	At 923 K: benzene (17.5), carbon monoxide (12.6), naphthalene (8.9), styrene (4.0), ethane (3.1), toluene (2.5), propene (1.7), methane (1.5)	[124]

[a]Reactor + Synchrotron vacuum ultraviolet-photoionization mass spectrometry

[b]Pyrolyzer connected to gas chromatography–mass spectrometry

[c]Gaussian 09

[d]Density functional theory methods

[e]U-shaped two-stage tubular reactor connected to gas chromatography

[f]Two-stage tubular reactor

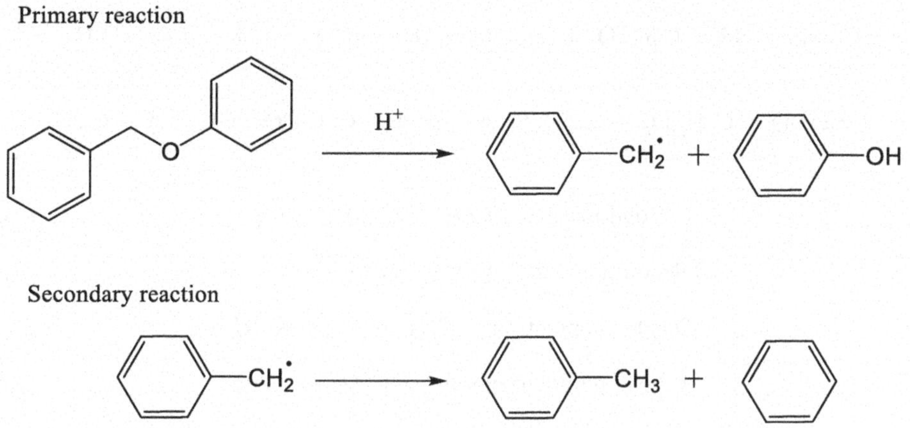

Fig. 15 Catalytic decomposition mechanism of PPE in the presence of cesium-exchanged heteropolyacids [140]

being the main lignin depolymerization products, as summarized by Sun et al. [148]. Therefore, model compounds are frequently used for getting a better understanding of the reactions involved in the reforming processes and for simplifying the experimental and modeling procedures [146]. Accordingly, reforming works of lignin depolymerization model compounds, such as benzene [149–155], toluene [152, 154, 156–164], naphthalene [154, 165], or xylene [152, 166, 167], are usually found in the literature.

Most of the hydrogen production processes share some common ground. Therefore, those processes and the catalysts used on them are discussed together for avoiding tedious repetitions. Research on hydrogen production from hydrocarbons is focused on studying the behavior of the developed catalysts for different feedstocks, instead of the development of catalysts for specific molecules. The rationale behind this is that the final application of the catalysts will be the hydrogen production from multi–component feedstock, such as lignocellulosic biomass pyrolysis liquids.

A reforming catalyst should be able to resist the deactivation by removing carbonaceous species, while promoting the breakage of C–C, C–H, and O–H bonds [146] and being highly active and stable towards hydrogen. Usually, reforming catalysts are nickel-based because this metal presents high bond breaking and water-gas shift reaction (WGSR) activity, promoting the hydrogen production [146]. Moreover, nickel-based catalysts are cheaper than noble metal-based ones [146]. For these reasons, nickel-based catalysts [150, 151, 154, 155, 159–162, 166, 167] are the most common catalysts for reforming or biomass gasification, as shown in Table 10. Similarly, iron is also known for its activity breaking aromatic hydrocarbons and performing the water-gas shift reaction, being potentially effective in different oxidation states [156]. Therefore, iron-containing catalysts [149, 151, 156] are also frequent in the literature.

Nevertheless, nickel particles present in the catalyst are commonly sintered during steam reforming (SR) (Eq. 1) [146] and tend to form carbon deposits by means of some undesired reactions (Eqs. 2–8), which can be substantial if the tar content fed is high [164]. In addition, nickel-based catalysts are also active for methanation [144, 145], which also decreases the hydrogen yield.

$$\text{Complete SR}: \ C_nH_mO_k + (2n - k)\,H_2O \leftrightarrow nCO_2 + (2n + m/2 - k)\,H_2. \tag{1}$$

$$\text{Cracking}: \ C_nH_mO_k \rightarrow C_XH_YO_Z + \text{gas}\left(H_2,\ CO,\ CO_2,\ CH_4 \cdots\right) + C\,(s). \tag{2}$$

$$\text{Boudouard}: \ 2\,CO \rightarrow C\,(s) + CO_2. \tag{3}$$

$$\text{CO decomposition}: \ CO + H_2 \leftrightarrow C\,(s) + H_2O. \tag{4}$$

$$CH_4\text{decomposition}: \ CH_4 \leftrightarrow C\,(s) + 2H_2. \tag{5}$$

$$\text{Ethene polymerization}: \ n\,C_2H_4 \rightarrow C\,(s). \tag{6}$$

$$\text{Methanation of CO}: \ CO + 3\,H_2 \leftrightarrow CH_4 + H_2O. \tag{7}$$

$$\text{Methanation of } CO_2: \ CO_2 + 4\,H_2 \leftrightarrow CH_4 + 2H_2O. \tag{8}$$

On the other hand, noble metals (Pt, Pd, Rh, or Ru)-based catalysts are known to provide higher activities during reforming processes, while carbon deposition on them is reduced. Nevertheless, their high cost makes this kind of catalyst prohibitive [145]. Accordingly, works based on precious metals-based catalysts are scarce in the literature. Thus, with the aim of addressing catalyst deactivation issues [145, 146], different alternatives such as modifications in the preparation method, support or the addition of promoters [160, 162] are studied. Thereby, e.g., Boldrin et al. [168] reported that rhodium promotion of nickel catalysts hinders the diffusion of oxygen and especially carbon in the metal. Therefore, the relative rate of formation of C–C bonds is decreased with respect to C–O bonds. Thus, the carbon oxidation is promoted instead of nickel deposit formation [168].

Therefore, different hydrogen production processes are described in the literature for avoiding carbon deposits. Thus, in processes such as partial oxidation (POX), autothermal reforming (ATR), or oxidative steam reforming (OSR), small amounts of oxygen are fed to the reactor for oxidizing carbon deposits.

$$\text{POX reaction}: \ C_nH_mO_k + (n/2 - \mathrm{k})\,O_2 \rightarrow n\,CO + m/2\,H_2. \tag{9}$$

$$\begin{aligned}\text{OSR reaction}: \ & C_nH_mO_k + \mathrm{p}\,O_2 + (2n - k - 2\mathrm{p})\,H_2O \\ & \rightarrow n\,CO_2 + (2n + m/2 - \mathrm{k} - 2\mathrm{p})\,H_2.\end{aligned} \tag{10}$$

Similarly, new reforming processes are being developed from producing high-purity hydrogen streams. Thereby, SR process can be combined with the selective adsorption of CO_2 using CaO-like adsorbents. Accordingly, in sorption-enhanced steam reforming (SESR) processes, the reforming reaction is displaced towards hydrogen production, while subsequent hydrogen purification equipment can be ideally avoided.

Currently, chemical looping reforming (CLR) processes are also being developed. The rationale behind this process is that carbon deposition cannot be completely avoided. Thus, this reaction system uses two reactors, as in FCC process. To this end, in one reactor, a reforming process is carried out, while in the second

Table 9 Summary of the analyzed literature for dimeric lignin-model compound pyrolysis

Entry	Model compound	Experimental set-up	Carrier gas	Pyrolysis temperature	Main stable products (yield in %)	Refs.
1	1-(4-methoxyphenyl)-2-(2-methoxyphenoxy) ethanol	Py-GC/MS[a]	He	573, 773 and 1073 K	At 1073 K: phenol, 2-hydroxybenzaldehyde, 4-vinylphenol, 4-methoxyacetophenone, 4-methoxystyrene, 2-hydroxytoluene	[134]
2	Guaiacyl-glycerol-β-guaiacyl ether	Py-GC/MS[a]	–	673–873 K	At 873 K: guaiacol, guaiacylethenyl-β-guaiaicyl ether, 2-methoxybenzaldehyde, salicylaldehide, 2-methylphenol, catechol, vanillin	[135]
3	Guaiacyl-glycerol-β-coniferyl ether	Py-GC/MS[a]	–	673–873 K	At 873 K: E-coniferylalcohol, E-coniferylaldehyde, 4-vinylguiaiacol, 4-propylguaiacol, guaiacol, vanillin	[135]
4	Syringyl-glycerol-β-guaiacyl ether	Py-GC/MS[a]	He	423–1123 K	Guaiacol, catechol, phenol, 3,5-dimethoxybenzaldehyde, 4,6-dimethoxy-1-indanone, 3,5-dimethoxyacetophenone	[136]
5	Phenethyl phenyl ether	Reactor[b]	N_2	773–973 K	At 973 K: phenol (19.4), styrene (13.2), benzene (5.9), toluene, ethylbenzene (1.3), phenanthrene, dibenzofuran, 1,2-diphenylethylene, naphthalene	[137]
6	*o*-methoxy phenethyl phenyl	Simulation[c]	–	–	Guaiacol, styrene, catechol, toluene, salicyla-ldehyde	[138]
7	Phenethyl phenyl ether + $CS_xH_{3.0-x}PW_{12}O_{40}$	Reactor[d]	N_2	473 K (10 atm)	Phenol, styrene, benzene, toluene, ethylbenzene. (Total selectivity: 80.6%, total yield: 54.9%)	[140]
8	Phenethyl phenyl ether + $Pd/Cs_{2.5}H_{0.5}PW_{12}O_{40}$	Reactor[d]	N_2	473 K (10 atm)	Phenol, styrene, benzene, toluene, ethylbenzene. (Total selectivity: 76.8%, total yield: 64–1%)	[140]
9	Benzyl phenyl ether + $CS_xH_{3.0-x}PW_{12}O_{40}$	Reactor[d]	N_2	473 K (10 atm)	Phenol, benzene, toluene. (Total selectivity: 55.5%, total yield: 40.9%)	[140]
10	Benzyl phenyl ether + $Pd/Cs_{2.5}H_{0.5}PW_{12}O_{40}$	Reactor[d]	N_2	473 K (10 atm)	Phenol, benzene, toluene. (Total selectivity: 54.8%, total yield: 42.7%)	[140]

Table 9 (continued)

Entry	Model compound	Experimental set-up	Carrier gas	Pyrolysis temperature	Main stable products (yield in %)	Refs.
11	α-*O*-4 linkage lignin dimer	Simulation[e]	–	–	Guaiacol, *p*-hydroxyphenyl-ethanol, *p*-hydroxyphenyl-acetaldehyde and 2-hydroxybenzaldehyde	[141]
12	Coumarone-contained lignin model compound	Simulation[e]	–	–	4-methyl-2-methoxy-phenol, 3-methoxy-2-hydroxylphenyl-ethanal, guaiacol, 8-methoxy-coumarone, formaldehyde	[142]

[a]Pyrolyzer connected to gas chromatography–mass spectrometry

[b]Fixed-bed reactor (quartz tube) + condenser

[c]Gaussian 09

[d]Autoclave reactor

[e]Gaussian 03

Table 10 Summary of the analyzed literature for reforming of pyrolysis oil lignin-model compounds

Entry	Catalyst	Solvent (type of reforming)	Experimental conditions	Model compound	Max. H_2 Y, S or C (%)	Refs.
1	Ni/M_xO_y–Al_2O_3 (M = Ce, La or Mg), Ni/Olivine, Ni/Zr sand	H_2O (SR)	873–1073 K, 0.1 MPa and S/C = 5	*m*-cresol	Y = 88	[170]
2	Biomass char	CO_2 (dry reforming)	973 K	Toluene	S = 92	[157]
3	Ni/Olivine	H_2O (SR)	833–1123 K, 0.1 MPa, and S/C = 1.1–3.4	Toluene	Y = 84	[164]
4	Commercial dolomite	H_2O (SR) with and without CO, CO_2, CH_4, H_2 and N_2	638–1123 K, 0.2 MPa, and S/C = 0.2–7.1	Toluene	N.I.[a]	[152]
5	Ni/Hydrotalcite	H_2O (SR)	923–1173 °C, 0.1 MPa, and S/C = 1.5	Toluene	C = 30	[154]
6	Fe/Olivine	H_2O (SR) with CO, CO_2, CH_4, H_2	1098 K and S/C = 2	Toluene	C = 50	[156]
7	Ni ion exchanged coal	H_2O (SR)	623–698 K and S/C = 1–30	Toluene	Y = 65	[158]
8	Ni/Activated carbon	H_2O (SR)	773–973 K and S/C = 2	Toluene	C = 70	[159]
9	$NiAl_2O_4$, $RuNiAl_2O_4$, Pt/γ-Al_2O_3	H_2O (SR)	723–1173 K and S/C = 1–2	Toluene	C = 55	[160]
10	Ni/Ru-Mn/Al_2O_3, Ni-based commercial	H_2O (SR)	673–1073 K and S/Toluene = 25	Toluene	C = 21	[162]
11	Ni/Olivine with Ce and Mg	H_2O (SR)	1003–1063 K, 0.1 MPa, and S/C = 3.5–6.5	Toluene	C = 66	[161]
12	Fe/CaO	H_2O (SERS)	973 K, 0.1 MPa, and S/C = 2	Toluene	N.I.[b]	[173]
13	Ni/Al_2O_3	H_2O (SR)	1293–1446 K, 0.1 MPa, and steam 10 vol%	Benzene	N.I.[b]	[150]
14	Commercial Dolomite	H_2O (SR) with and without CO, CO_2, CH_4, H_2, and N_2	638–1123 K, 0.2 MPa, and S/C = 0.2–7.1	Benzene	N.I.[b]	[152]
15	Ni/Hydrotalcite	H_2O (SR)	1196–1446 K, 0.1 MPa, and S/C = 1.5	Benzene	C = 35	[154]
16	Ni/Ceramic foam	H_2O (SR)	973–1173 K and S/C = 0–3	Benzene	Y = 182 g H_2/kg benzene	[155]

Table 10 (continued)

Entry	Catalyst	Solvent (type of reforming)	Experimental conditions	Model compound	Max. H_2 Y, S or C (%)	Refs.
17	Iron-containing minerals	CO, CO_2, CH_4, and H_2	973–1073 K and S/C = 1 (benzene + CH_4)	Benzene	N.I.[b]	[149]
18	NiFe/SiC, Fe/SiC	CO_2 (chemical looping dry reforming)	573–1113 K	Benzene	Y = 7	[151]
19	La/Sr/Fe/ZnO	H_2O (chemical looping steam reforming) + gasification gas mixture	1296–1446 K, 0.1 MPa, and steam = 25 mol%	Benzene	C = 11	[153]
20	Ni/M_xO_y–Al_2O_3 (M = Ce or La)	H_2O (SR)	873–1073 K, 0.1 MPa, and S/C = 5	*m*-xylene	Y = 80	[166]
21	Commercial dolomite	H_2O (SR) with and without CO, CO_2, CH_4, H_2, and N_2	638–1123 K, 0.2 MPa, and S/C = 0.2–7.1	Xylene	C = 7.5 diluted in N_2	[152]
22	Ni/Al_2O_3	H_2O (SR)	573–1073 K, 0.1 MPa, and S/C = 6	Xylene	S = 79	[167]
23	Ni/Hydrotalcite	H_2O (SR)	1196–1446 K, 0.1 MPa, and S/C = 1.5	Naphthalene	C = 22	[154]
24	None	H_2O (SR)	1273–1673 K and steam: 0–40 vol%	Naphthalene, acenaphthylene, phenanthrene, fluoranthene, and pyrene	Y = 68	[165]

Y yield, S selectivity, C mol% on gas stream

[a]Not indicated

reactor, the carbon-containing catalyst is regenerated before sending it to the first reactor. So, the hot catalyst provides, at least partially, the heat required for the reaction. Therefore, this process could increase the durability of the catalysts.

Regarding the support materials, alumina is one of the most common ones. Nevertheless, it is common to find supports with oxide modifications, such as cerium oxide. CeO_2 favors carbon gasification and WGS reaction due to its high oxygen mobility [144]. Thus, the stability of the catalyst is improved [144]. Similarly, the presence of MgO in the Al_2O_3 support could enhance water adsorption, surface –OH group mobility, and neutralize the acidic sites [144], where the carbon deposits occur, promoting the catalytic activity and stability [161].

On the other hand, catalytic formulations based on natural materials can also be found in the literature [149, 152, 156, 161]. Materials such as olivine ($(Mg,Fe)_2SiO_4$) or dolomites, among others, are characterized as being cheap and abundant [149, 156] and they are active catalysts for tar reduction [161]. Among them, olivine presents a higher attrition resistance, which is to be considered especially using fluidized bed reactors [161, 164]. Moreover, when nickel is impregnated in olivine, the presence of MgO at the nickel–olivine interface enables the steam adsorption and increases surface carbon gasification rates [164]. Therefore, olivine can be considered a suitable reforming support material.

Among the different hydrogen production processes, SR has been widely studied (see Table 10) during the last years [144], with the fixed-bed reactor [149, 152, 154–156, 158–160, 162] being the most widely used reactor configuration. Authors tested catalytic activities from room temperature up to 1173 K, at atmospheric pressure conditions. However, for achieving high conversion and hydrogen yield values, temperatures above 873 K are necessary for this catalytic process. This temperature is in good agreement with the high stability of aromatic molecules in lignin depolymerization products. Accordingly, several authors reported that the non-catalytic thermal decomposition of benzene occurs above 1073–1173 K [149, 152].

Regarding S/C ratios, the authors are keen on using low ratios for avoiding excessive energy consumption. Accordingly, Nabgan et al. [169] indicated that a high steam-to-carbon ratio (S/C), higher than 10, is needed for avoiding extensive coking during phenolic molecules SR. However, the use of a lower S/C ratio is preferable for reducing the energy balance, as reported by Spragg et al. [145]. Thereby, S/C ratios up to 7 [146, 152, 154–156, 159–162, 166] can be found in the literature for SR of tar model compounds, while higher values, as in the work of Kim et al. [158], are not frequent.

Thus, looking at individual model compounds, Bizkarra et al. [166] and García–García et al. [170] performed SR experiments of *m*-xylene and *m*-cresol, respectively, using the same operational conditions, using nickel-based and alumina-supported catalysts. In those works, even if the reaction temperature was reduced from 1073 to 873 K, the decrease in catalytic activity was not permanent, as observed when the reaction returned to 1073 K, because catalytic activities were recovered. Wu and Liu [171] also studied the SR of *m*-cresol at 1123 K at $S/C=5$, using a Ni/MgO catalyst. In their work, the catalyst showed a steady performance for 6 h, without deactivation evidences.

Similarly, Trane–Restrup and Jensen reformed guaiacol at 857 K for almost 4 h at $S/C = 5$ [172]. In those conditions, the nickel catalyst they used was not deactivated, even if carbon was present on the catalyst.

Nabgan et al. [174] studied the SR of phenol at 873 K at a steam-to-phenol molar ratio of 9 ($S/C < 2$) using nickel- and cobalt-based bimetallic catalysts. The highest H_2 yields were achieved using a 3:1 nickel-to-cobalt ratio. Those catalysts did not present deactivation evidence during 6 h on stream. Nabgan et al. [175] also studied the SR of phenol using bimetallic nickel and cobalt catalysts on different supports. Firstly, they studied the effect of the reaction temperature from 773 to 1073 K, at the previously indicated steam-to-phenol ratio. These results showed that for achieving a hydrogen yield higher than 50%, a temperature above 873 K was necessary. Afterwards, they performed stability tests at 773, 873, and 1073 K, for 6 h. Nevertheless, stable results were only obtained for the experiments performed at 1073 K. Meng et al. [176] studied the SR of phenol from 1023 to 1223 K at S/C ratios from 0.3 to 1.2, using Ni/Fe/olivine catalysts. Afterwards, they performed stability tests at 1123 K and $S/C = 1$ for 20 h. In those conditions, after an irregular first 5 h, the catalysts achieved steady values for reaction products. Artetxe et al. [177] studied the SR of phenol using a Ni/Al_2O_3 catalyst from 923 to 1073 K, using a S/C ratio of 13. According to their results, for achieving a potential hydrogen yield higher than 40%, a reaction temperature of 1073 K is needed. Thus, these results show that even when using similar reaction conditions, different authors found different activity results, highlighting the difficulty of operating with lignin-derived model compounds.

On the contrary, in other work of Bizkarra et al. [146], the presence of a more aromatic feedstock produced a permanent catalyst deactivation by coke deposition, even if the reaction temperature was only decreased to 973 K. Moreover, when a real lignocellulosic bio-oil was used for producing hydrogen, catalysts (even palladium and platinum promoted ones) were quickly deactivated at 1073 K due to the encapsulation of nickel by amorphous coke [146]. Similar findings were presented by Valle et al. [143], who showed that bio-oil conversion quickly dropped when steam reforming experiments were performed at 973 K and $S/C = 6$. Remiro et al. [178] studied the oxidative steam reforming of raw bio-oil at $S/C = 6$ and oxygen to carbon of 0.34. Nevertheless, the oxygen feeding did not avoid the catalyst deactivation during the first 2 h on stream.

Furthermore, those mixtures of molecules are thermodynamically unstable. Thus, during its production and downstream processes, decomposition and cocking reaction occur, giving rise to pipeline blockage, reactor fouling, and catalyst deactivation [179], decreasing the system efficiency [143, 146] and increasing the cost of the process [143, 164].

However, advanced processes have been developed for increasing the hydrogen yield, such as sorption-enhanced steam reforming (SESR) [180] and chemical looping reforming (CLR) [145, 151]. SESR uses a solid CO_2 sorbent (as CaO) in the process for removing CO_2 from the reaction environment. Accordingly, WGS reaction is displaced towards H_2 production, favoring the achieved hydrogen yield and its purity [145]. Additionally, the net process energy balance is decreased, as the highly endothermic steam reforming is thermally neutralized by the exothermic carbonation reaction [181]. Nevertheless, long-term stability still remains a challenge [145].

Chemical looping reforming (CLR) is a novel approach for producing hydrogen from conventional or renewable feedstocks and steam, making use of the redox properties of an oxygen carrier (OC) [182]. This process takes place in two different steps. Firstly, the oxygen carrier reacts with the fuel fed to the reactor, producing reforming products (H_2, CO, CO_2, etc.). Then, the steam reacts with the reduced oxygen carrier for replenishing the lattice oxygen of the OC, which leads to the production of more hydrogen [183]. Therefore, an autothermal process can be achieved [145]. Nevertheless, due to the partial oxidation of the feedstock, hydrogen yield is partially reduced [145]. Additionally, depending on the amount of carbon deposited on the OC, a regeneration stage may be necessary. Thus, the heat produced by the combustion of carbon could reduce the external energy requirements of the process.

Ni, Cu, Co, Mn, and/or Fe oxides are considered good OC materials [183]. Those metal oxides are usually incorporated on support material for improving the catalytic properties. Chemical looping processes are usually carried out in fluidized bed reactors [183]. Thus, the incorporation of metals on attrition resistant material (e.g., olivine) would be an important factor for the overall performance. The synthesis of the OC should be economically feasible, while providing enough oxygen transport capacity, good thermal properties, high reactivity, and mechanical and chemical stability under operational conditions [183].

Moreover, these two novel processes can be combined into a SE-CLSR. Thus, a high-purity hydrogen stream is produced (avoiding expensive gas purification units [184]), while limiting the energy consumption by an autothermal reforming process [145].

5 Hydrolytic Processes

As mentioned above, in this section, we will review those processes in which water is used to break down different lignin model compounds [185]. The hydrolysis reactions of lignin model compounds in the absence of catalysts in water occur at sub- (573–643 K, SubW) and super- (663–723 K, SupW) critical conditions. Supercritical water features many advantages, as it is miscible with light gases or hydrocarbons, and stable under organic decomposition and catalysis conditions. As for reductive and pyrolytic processes, guaiacol is the most common lignin monomeric model compound used in hydrolysis studies. Guaiacol thermally degrades in SupW via cleavage of ether and methylene bridges. Guaiacol hydrolysis forms catechol, phenol, and cresol as main products, and its fast decomposition rates can yield side condensation reactions to larger fragments [186]. On the other hand, SubW guaiacol conversion leads predominantly to the production of phenol below 663 K, but higher temperatures under SupW show a clear shift into catechol and cresol formation [187].

Such reaction temperatures can be considerably reduced by the proper design of solid Lewis-acid catalysts, which have been tested for guaiacol [188] and anisole [189] hydrolysis. The use of metal triflates tune the pH of the catalysis solution, and among them indium triflates produce catechol with yields up to 80% when using guaiacol as the model compound. This confirms that such hydrolysis reactions are mediated via C_{AL}-O cleavage. These yields are higher than homogeneous catalysts such as HCl given the lower reaction temperatures used for such solid catalysts. These studies show a direct correlation between hydrolysis activity and metal features using guaiacol [188], which are also observed for anisole [189], attributed to strong metal and substrate interactions. These findings represent a significant advancement into the search for green processes that operate at low reaction temperatures, reducing energy consumption, and thus avoiding the use of homogeneous catalysts.

Dimer model compounds, such as diphenylether (DPE), also show effects of temperature under hydrothermal conditions [190]. Low water densities yield polycondensation products that form via radical chemistry, but higher water densities convert DPE much faster to phenol via proton-catalyzed intermediates. Such temperature-mediated acid-catalyzed consequences can be tuned by adding basic carbonates under hydrothermal conditions [191], which considerably decrease DPE conversion when high temperatures are not used (< 573 K), and which show phenol and benzyl alcohol as major products. High molecular weight undesired products still form, however, when severe reaction temperatures are used. These studies suggest that DPE hydrolysis in SubW is facilitated under base conditions and in SupW catalyzed by catalysts containing acid–base functions. The hydrolysis of DPE produces phenol as the main compound under supercritical conditions in the presence of dissolved NaCl salt conditions [190], but forms a mixture of hydrocarbon products when subcritical conditions are used. These effects were attributed to the formation of acid H^+:Cl^- pairs at high salt concentrations, favoring an acid-catalyzed mechanism that selectively cleaves to phenol as the main hydrolysis product.

Overall, these studies show that DPE hydrolysis, even in the presence of catalysts, requires harsh hydrothermal conditions. This is evident from the low ether bond hydrolysis yields at subcritical conditions in water (523 K) even when homogeneous acid catalysts (formic or phosphoric acid) are used [192]. These yields can be improved when the supercritical aqueous conditions are mixed with CO_2 [193], leading to more acidic mixtures that can be combined with metal functions within solid catalysts to give phenol as the main product from DPE. Such drawbacks can also be overcome using Ni-based catalysts and in the presence of H_2 [194], which facilitate DPE cleavage. Moreover, water-tolerant Lewis-acid catalysts, among them indium triflate, yield phenol as hydrolysis product from DPE, with selectivities reaching around 40% [188]. Such low phenol formation rates were attributed to undesired reactions to oligomer side-products from DPE. However, the hydrolysis rate of model compounds such as BPE is much faster using such triflates, owing to a small activation energy of C_α–O cleavage in the presence of triflate catalysts. These catalysts show phenol as the predominant hydrolysis product together with benzyl alcohol at 423 K, and product distribution varies towards benzyl alcohol as the reaction temperature increases to 473 K.

Within other heterogeneous systems tested at relatively mild temperatures (<453 K), hydrolysis reactions of model monomers, which occur as side reactions of hydrogenation or hydrogenolysis reactions, are often catalyzed by metal and acid catalysts. In these cases, the hydrolysis of methoxy groups, derived from model phenolic monomers (phenol, guaiacol, and syringol) [195], leads to methanol as the main product.

In summary, these data indicate that the hydrolysis of lignin model monomer or dimer compounds strongly depends on the acidic nature of the catalysts and the reaction medium, and as reported for the un-catalyzed systems, especially on the temperature used during reaction.

6 Overview and Future Perspectives

Several thermochemical processes have been explored for the conversion of lignin-rich residual materials into fuels or phenols: numerous reductive, oxidative, pyrolytic, and hydrolytic processes have been developed for the production of lignin-derived value-added chemicals and fuels. However, lignin conversion processes are generally complex, involving several simultaneous chemical reactions and resulting in complex product mixtures. Hence, to facilitate the study of lignin conversion processes, research tends to replace lignin for simpler but still representative molecules, the so-called lignin model compounds.

This review has summarized the most important heterogeneously catalyzed processes to transform lignin model compounds into value-added chemicals. The review is divided into categories according to the type of conversion process. Only those processes utilizing heterogeneous catalysts are considered, with a special focus on those studies that use "green" reactants. Processes using ionic liquids as solvent are discharged to limit the scope of the review.

Conversion in reductive atmospheres is one of the most popular and efficient strategies to transform lignin model compounds into monoaromatic chemicals and cyclic alkanes. Both lignin model compound hydroprocessing, i.e., in the presence of H_2, and CTH of lignin model compounds, i.e., using hydrogen donor molecules, have been considered. Lignin model compound hydroprocessing is generally carried out at mild temperatures and high hydrogen pressures, although the conditions vary considerably depending on the type of model compound, targeted product, and type of catalyst used. For monomeric lignin model compounds, the reaction temperature is mostly held between 473 and 673 K [35–43, 46, 48, 50, 51, 53, 55, 196, 197]. In terms of reaction pressure, monomeric compound hydroprocessing can be classified into (1) high-pressure processes, typically 3-5 MPa of H_2 [35, 37, 38, 41, 43–46, 48, 51, 196], and (2) low-pressure processes, around 0.1 MPa H_2 [49, 52, 53, 197]. Low pressures generally lead to more hydrogenated products, i.e., ring-saturated products, while higher pressures lead to higher amounts of aromatics. In addition to hydrogen pressure, the type of catalyst plays a key role in achieving the desired product distribution. Noble metal and Ni-based catalysts can lead to hydrogenated products (i.e., aromatic ring hydrogenation) even at low hydrogen pressures [53, 197], while traditional NiMo and CoMo sulfided and non-sulfided catalysts induce hydrodeoxygenation mechanisms leading to functional group removal (i.e., methoxy and hydroxyl) even at high reaction pressures and, as a consequence, the production of monoaromatic compounds [35, 39]. Batch reactors are generally used for the conversion of monomeric compounds. Non-polar solvents such as dodecane and heptane are generally preferred [48, 54, 55, 196]. Lignin reductive de-polymerization mechanism, lignin bond cleavage, is usually studied using dimeric lignin model compounds. The conversion of dimeric compounds is only studied using batch reactors [42, 56–66] in the presence of polar solvents, preferably water [42, 56–58, 65] but also in methanol [60, 62] or nonpolar solvent [61, 63, 64, 66]. Typical temperatures for the conversion of dimeric compounds range from 473 to 573 K with hydrogen pressures that usually vary between 1 and 5 MPa hydrogen [42, 56–60, 62–66]. Dimeric compounds bearing ether C–O–C linkages, i.e., β-*O*-4, α-*O*-4, and 4-*O*-5 linkages, are relatively easy to cleave in comparison to C–C-bearing dimeric compounds. For any tested catalyst and under the same reaction conditions, the conversion of the ether dimeric compounds decreases, based on their bond dissociation energy, as follows: α-*O*-4 > β-*O*-4 > 4-*O*-5. Noble metal- and nickel-based catalysts are generally used due to their high hydrogenolysis activity, although traditional sulfided catalysts have also been applied. When using noble metal catalysts, hydrogenated products such as cyclohexane, cyclohexanol, and methylcyclohexanol are generally obtained, while with Ni-based and traditional sulfided catalyst, hydrodeoxygenated monoaromatics are the major products. On the other hand, hydroprocessing of dimeric model compounds bearing C–C linkages, 5–5/5′, β–β, β-1, and β-5 linkages, result in dimeric saturated compounds. No, or insignificant, C–C linkage cleavage is observed regardless of the type of catalyst and reaction conditions used. The evaluation of the stability of the catalyst is not considered in most of the analyzed literature, both in the case of monomeric and dimeric compounds. In the future, research efforts should focus on the evaluation of the activity and stability of the catalytic systems in continuous reactors and on the replacement of non-renewable solvents (i.e., dodecane and heptane) with "greener" alternatives such as ethyl acetate or γ-valerolactone.

CTH can be used in the conversion of any lignin model compound ranging from the most simple molecules as phenol, guaiacol, and others, to more complex molecules such as diphenyl ether or even more complex and realistic lignin model compounds containing β-O-4 bonds and α-*O*-4 or 4-*O*-5 dimers. Therefore, it seems that CTH technology can be a suitable alternative to be applied in lignin conversion. Different types of hydrogen donor molecules can be used and MeOH, 2-BuOH, 2-PrOH formic acid, isopropanol, decalin, and tetralin are the most studied hydrogen donor molecules. It is remarkable how many of these hydrogen donor compounds have a renewable character and the environmental advantages that this entails. All cases studied have been carried out in discontinuous reaction systems employing low-severity operating conditions in comparison to other reductive conversion processes. In CTH processes, catalysts with hydrogen affinity such as noble metals (e.g., Pt, Pd, and Ru) and transition metals such as Ni are needed. Ni Raney catalysts exhibit a great potential for CTH of lignin model compounds. It is generally more active than noble metals at milder reaction temperatures, although its activity differs depending on the lignin model compound and the hydrogen donor molecule used. Neutral carbon is the most studied support in comparison to other acid or basic supports. However, the use of basic or acid supports favors the cleavage of C–C bonds, while the neutral carbon does not favor this cleavage. Therefore, the support used should be chosen taking into account the desired reaction products. Despite the positive results reported for CTH of lignin model compounds, it still is necessary to evaluate the activity and stability of the catalysts in continuous reaction systems.

Oxidation of lignin model compounds can be a good alternative to produce their corresponding aldehydes. Oxidation of lignin model compounds to yield aromatic aldehydes requires milder operating conditions in comparison to the rest of the conversion processes, being the optimum temperature, pressure, and reaction time ranges 353–433 K, 0.5–4 MPa, and 0.92–24 h, respectively. In general, the increase of temperature and pressure increases the conversion but reduces the selectivity due to secondary reactions (disproportionation and overoxidation). However, it must be mentioned that all the studies except one were carried out in very small scale batch experiments. For this reason, future work should be focused on testing the most promising catalytic systems in continuous reaction systems in order to study the stability of the developed catalysts. The used solvents must favor the solubility of the substrates and the oxidant solubility, and catalyst dispersion in the reaction mixture in order to improve the catalytic performance. Although many of the used solvents, such as H_2O, isopropanol, and methanol, are considered green solvents, other used solvents, especially for VAL and dimers oxidation, such as acetonitrile, DMF, and pyridine are not recommendable from an environmental point of view. Other solvents such as toluene (especially employed in COL oxidation) and xylene are not considered as green solvents but depend of the followed guide. They can be considered to be in between green and hazardous solvents [106, 107]. In these cases, future work should be focused on the substitution of these solvents by a more sustainable one, keeping or even improving the functionality of the process. The catalysts employed were supported noble (Ru, Pd, Pt, Au) and non-noble (Co, Cu, V, Mn, Fe) catalysts. The good behavior of these catalysts was ascribed to the high dispersion, accessibility, and number of the metal active phases, high surface area, and basic nature, in some cases, of the used supports, and the reusability in more than one cycle. However, in the particular case of VA and COL, the most used catalysts belongs to noble catalysts. In this sense, it seems reasonable to make an effort to synthesize and employ non-noble catalysts, with similar properties to non-noble ones with the consequent economic viability.

The work dedicated to the pyrolysis of lignin model compounds can be divided into two types: those based on simulation and those that carry out real pyrolysis experiments. The simulation works are dedicated to the study of reaction mechanisms through the calculation of energetic pathways, offering probabilities of appearance of products, but in most cases without determining the yields or selectivities of these reaction products. On the other hand, in the experimentation work, the inverse path is usually followed, i.e., from the yields of certain products, an attempt is made to deduce the mechanism that has led to their appearance. In these last works, the usual configurations are pyrojector + GC/MS or reaction in fixed-bed reactor + analysis of products, in which different techniques are used. In addition, the experimental approach on the pyrolysis of lignin model compounds can be classified according to the size of the molecule used: dimers generally represent those compounds typical of the first pyrolysis stage, while monomers are more representative of molecules generated in the pyrolysis of polymeric species, that is, secondary products of the first cracking. Pyrolytic conversion of monomeric compounds results in a complex reaction mixture comprising monoaromatic compounds, cyclic hydrocarbons, light hydrocarbons, and gases such as CO, CO_2, or H_2O. The nature and relative abundance of the pyrolysis products mainly depends on the reaction temperature, type of carrier gas, residence time, and the catalysts used. Solid acid catalysts are generally used to increase the process efficiency, especially zeolites (HZSM-5). Zeolites tend to accelerate decomposition reactions, achieving a greater conversion and yield compared to aromatic compounds from the recombination of hydrocarbons and CO. Pyrolysis of dimeric lignin model compounds focuses on the conversion of ether-containing dimers, i.e., β-*O*-4 and α-*O*-4 lignin model compounds. At low temperatures, pyrolysis of dimeric β-*O*-4 lignin model compounds proceeds via the homolytic cleavage of the C_β–O bond. However, at high temperatures, both C_β–O homolytic cleavage and C_α–C_β cleavage occur. Catalytic pyrolysis with heteropolyacids of β-*O*-4 model compounds seems to proceed through the same reaction mechanism and yielding a similar product distribution as in the case of thermal pyrolysis. However, catalytic pyrolysis occurs at significantly lower reaction temperatures. In the case of α-*O*-4, the reaction mechanism proceeds through the homolytic C_α–O cleavage in both thermal and catalytic pyrolysis. Regarding aspects that should be studied in the future, the main one is the in-depth study of the catalytic pyrolysis of lignin model compounds. This includes the study of different types of catalysts (not only acid catalysts) and the study of their stability and capacity for regeneration and reuse. Apart from this, it is also necessary to correlate studies about pyrolysis of lignin model compounds with pyrolysis studies of lignin containing real biomass, since the final objective must be the understanding and prediction of compounds of interest from real biomass pyrolysis.

In the last decades, there has been growing interest in pyrolysis oil reforming to produce renewable hydrogen. Given the complexity of the pyrolysis oils, model compounds are frequently used for getting a better understanding of the reactions involved. Catalysts are generally applied in pyrolysis oil model compounds reforming in order to increase their conversion and selectivity to hydrogen. Ni and Fe catalysts are generally preferred due to their low cost, high bond breaking ability, and WGS reaction activity. Noble metals, despite their higher activity, are only considered as promoters in Ni-based catalysts due to their high cost. Alumina modified by CeO_2 or MgO is the most common support used for the reforming of pyrolysis oil model compounds. CeO_2 favors carbon gasification and WGS reaction due to its high oxygen mobility, while MgO enhances water adsorption, surface –OH group mobility, and neutralizes the acidic sites, reducing coke formation. SR using fixed-bed reactors is the preferred choice among the different hydrogen production processes. High temperatures of 857 K or even higher are necessary in SR reforming of pyrolysis oil model compounds in order to limit carbon deposits that lead to catalyst deactivation. Other advanced processes such as SESR and CLR have also been tested with the aim of increasing the hydrogen yield. SERS processes are based on the displacement of the WGS reaction with the aid of a CO_2 adsorbent. In the case of CLR, an oxygen carrier, Ni, Cu, Co, Mn, and/or Fe oxides, are used to produce reforming products (H_2, CO, CO_2, etc.) and in part of the steam into more hydrogen. Moreover, these two advanced processes can be combined into a SE-CLSR. However, the stability and durability of these novel hydrogen production processes and reactor designs remain unclear, especially when real molecule mixtures are used as feedstock. Therefore, future efforts should be focused on developing and testing highly stable and active catalysts for these novel processes, paying special attention to experiments without model compounds.

The literature on the hydrolytic conversion of lignin model compounds is scarce, particularly when it comes to catalyze hydrolytic processes. Most of the literature focuses on the hydrolytic conversion of real lignin. According to the hydrolysis studies using lignin model monomer or dimer compounds, un-catalyzed systems require harsh hydrothermal or supercritical conditions to achieve significant hydrolysis yields. When such hydrolysis occurs under catalyzed conditions, rates show a strong dependence on the acidic nature of homogeneous or heterogeneous catalysts, as well as the reaction medium used for catalysis. Overall, reported studies indicate that the reaction system needs to be accurately designed to achieve the desired selectivity to hydrolysis products and avoid side-reactions. A critical measure that will determine the commercial feasibility of the hydrolysis reaction systems is the mechanical or chemical stability of the catalysts used in the process. Harsh reaction conditions usually limit their potential application, so the specific search for low-temperature catalysts is of utmost urgency. In addition to this, the concept of future biorefinery should not only focus on replacing fossil-based feedstock by biomass-derived compounds, such as lignin, but also focus on using green technology. A key aspect that should be investigated in the future is related to the use of sustainable solvents that can facilitate the chemistry and the post-catalysis purification stages. Such solvents can indeed reduce deactivation phenomena within hydrolysis catalysts. Research efforts into this direction will certainly improve the current limitations during lignin hydrolysis.

References

1. Popa VI (2018) Biomass for fuels and biomaterials. Biomass as renewable raw material to obtain bioproducts of high-tech value. Elsevier, Amsterdam
2. International Energy Agency (2012) World energy outlook 2012. Executive summary. Int Energy Agency Books Online. https://doi.org/10.1049/ep.1977.0180
3. Wahyudiono Sasaki M, Goto M (2008) Recovery of phenolic compounds through the decomposition of lignin in near and supercritical water. Chem Eng Process Process Intensif. https://doi.org/10.1016/j.cep.2007.09.001
4. Zhou CH, Xia X, Lin CX et al (2011) Catalytic conversion of lignocellulosic biomass to fine chemicals and fuels. Soc Rev, Chem. https://doi.org/10.1039/C1CS15124J
5. Brandt A, Gräsvik J, Hallett JP, Welton T (2013) Deconstruction of lignocellulosic biomass with ionic liquids. Green Chem. https://doi.org/10.1039/c2gc36364j
6. Gosselink RJA, De Jong E, Guran B, Abächerli A (2004) Co-ordination network for lignin—standardisation, production and applications adapted to market requirements (EUROLIGNIN). In: Industrial crops and products. https://doi.org/10.1016/j.indcrop.2004.04.015

7. Yang L, Seshan K, Li Y (2017) A review on thermal chemical reactions of lignin model compounds. Catal Today 298:276–297. https://doi.org/10.1016/j.cattod.2016.11.030
8. Calvo-Flores FG, Dobado JA (2010) Lignin as renewable raw material. Chemsuschem. https://doi.org/10.1002/cssc.201000157
9. Li C, Zhao X, Wang A et al (2015) Catalytic transformation of lignin for the production of chemicals and fuels. Chem Rev 115:11559–11624. https://doi.org/10.1021/acs.chemrev.5b00155
10. Baskar C, Baskar S, Dhillon RS (2012) Biomass conversion: the interface of biotechnology, chemistry and materials science. https://doi.org/10.1007/978-3-642-28418-2
11. Yue F, Lu F, Sun R, Ralph J (2012) Synthesis and characterization of new 5-linked pinoresinol lignin models. Chem A Eur J. https://doi.org/10.1002/chem.201201506
12. Lange H, Decina S, Crestini C (2013) Oxidative upgrade of lignin—recent routes reviewed. Eur Polym J. https://doi.org/10.1016/j.eurpolymj.2013.03.002
13. Shahbazali E (2013) Biorefinery: from biomass to chemicals and fuels. Green Process Synth. https://doi.org/10.1515/gps-2012-0094
14. Oregui-Bengoechea M, Miletić N, Hao W et al (2017) High-performance magnetic activated carbon from solid waste from lignin conversion processes. 2. Their use as NiMo catalyst supports for lignin conversion. ACS Sustain Chem Eng. https://doi.org/10.1021/acssuschemeng.6b02796
15. Lora JH (2008) Industrial commercial lignins: sources, properties and applications. Monom Polym Compos Renew Resour. https://doi.org/10.1016/j.eurpolymj.2013.03.002
16. Oregui-Bengoechea M (2017) Formic acid aided catalytic lignin conversion in ethanol and water media. The University of Bergen, Bergen. http://hdl.handle.net/1956/15655
17. Albright L (2013) Albright's chemical engineering handbook. Choice Rev Online. https://doi.org/10.5860/choice.46-4467
18. Li C, Zhao X, Wang A et al (2015) Catalytic transformation of lignin for the production of chemicals and fuels. Chem Rev 115:11559–11624. https://doi.org/10.1021/acs.chemrev.5b00155
19. Valment Forward (2019). https://www.valmet.com/pulp/other-value-adding-processes/lignin-separation/lignoboost-process/. Accessed 15 Jul 2019
20. Oregui Bengoechea M, Hertzberg A, Miletić N et al (2015) Simultaneous catalytic de-polymerization and hydrodeoxygenation of lignin in water/formic acid media with Rh/Al_2O_3, Ru/Al_2O_3 and Pd/Al_2O_3 as bifunctional catalysts. J Anal Appl Pyrolysis. https://doi.org/10.1016/j.jaap.2015.04.020
21. Gupta VG, Tuohy MG, Kubicek CP et al (2014) Bioenergy research: advances and applications. Elsevier, Amsterdam
22. Suhas Carrott PJM, Ribeiro Carrott MML (2007) Lignin—from natural adsorbent to activated carbon: a review. Bioresour Technol 98:2301–2312. https://doi.org/10.1016/j.biortech.2006.08.008
23. Chaturvedi V, Verma P (2013) An overview of key pretreatment processes employed for bioconversion of lignocellulosic biomass into biofuels and value added products. 3 Biotech. https://doi.org/10.1007/s13205-013-0167-8
24. Sun Y, Cheng J (2002) Hydrolysis of lignocellulosic materials for ethanol production: a review. Bioresour Technol. https://doi.org/10.1016/S0960-8524(01)00212-7
25. Evstigneyev EI, Yuzikhin OS, Gurinov AA et al (2016) Study of structure of industrial acid hydrolysis lignin, oxidized in the H_2O_2–H_2SO_4 system. J Wood Chem Technol. https://doi.org/10.1080/02773813.2015.1137945
26. Pu Y, Hu F, Huang F et al (2013) Assessing the molecular structure basis for biomass recalcitrance during dilute acid and hydrothermal pretreatments. Biotechnol Biofuels. https://doi.org/10.1186/1754-6834-6-15
27. González G, López-Santín J, Caminal G, Solà C (1986) Dilute acid hydrolysis of wheat straw hemicellulose at moderate temperature: a simplified kinetic model. Biotechnol Bioeng. https://doi.org/10.1002/bit.260280219
28. Abdulkhani A, Karimi A, Mirshokraie A et al (2010) Isolation and chemical structure characterization of enzymatic lignin from *Populus deltoides* wood. J Appl Polym Sci. https://doi.org/10.1002/app.32428
29. Holtman KM, Chang HM, Kadla JF (2004) Solution-state nuclear magnetic resonance study of the similarities between milled wood lignin and cellulolytic enzyme lignin. J Agric Food Chem. https://doi.org/10.1021/jf035084k
30. Zakzeski J, Bruijnincx PCA, Jongerius AL, Weckhuysen BM (2010) The catalytic valorization of lignin for the production of renewable chemicals. Chem Rev. https://doi.org/10.1021/cr900354u

31. Gao R, Li Y, Kim H et al (2018) Selective oxidation of lignin model compounds. Chemsuschem. https://doi.org/10.1002/cssc.201800598
32. Joshi SS, Ranade VV (2016) Industrial catalytic processes for fine and specialty chemicals. Focus Catal. https://doi.org/10.1016/j.focat.2016.01.106
33. McLean M (2014) Encyclopedia of materials science and technology. Int Met Rev. https://doi.org/10.1179/imtr.1986.31.1.289
34. Yang L, Seshan K (2017) A review on thermal chemical reactions of lignin model compounds. Catal Today 298:276–297. https://doi.org/10.1016/J.CATTOD.2016.11.030
35. Lin YC, Li CL, Wan HP et al (2011) Catalytic hydrodeoxygenation of guaiacol on Rh-based and sulfided CoMo and NiMo catalysts. Energy Fuels. https://doi.org/10.1021/ef101521z
36. He Z, Wang X (2014) Highly selective catalytic hydrodeoxygenation of guaiacol to cyclohexane over Pt/TiO_2 and $NiMo/Al_2O_3$ catalysts. Front Chem Sci Eng. https://doi.org/10.1007/s11705-014-1435-9
37. Cai Z, Wang F, Zhang X et al (2017) Selective hydrodeoxygenation of guaiacol to phenolics over activated carbon supported molybdenum catalysts. Mol Catal. https://doi.org/10.1016/j.mcat.2017.07.024
38. Zhang X, Liu Q, Zhang Q et al (2019) Aromatic fuel production from phenolics by catalytic hydrodeoxygenation over novel Mo-based catalyst. Energy Procedia. https://doi.org/10.1016/j.egypro.2019.01.240
39. Mora-Vergara ID, Hernández Moscoso L, Gaigneaux EM et al (2018) Hydrodeoxygenation of guaiacol using NiMo and CoMo catalysts supported on alumina modified with potassium. Catal Today. https://doi.org/10.1016/j.cattod.2017.07.015
40. Tran CC, Stankovikj F, Garcia-Perez M, Kaliaguine S (2017) Unsupported transition metal-catalyzed hydrodeoxygenation of guaiacol. Catal Commun. https://doi.org/10.1016/j.catcom.2017.07.029
41. Sal'nikov VA, Minaev PP, Mozhaev AV et al (2018) Effect of support of CoMoS catalysts on hydrodeoxygenation of guaiacol as a model compound of biopetroleum. Russ J Appl Chem. https://doi.org/10.1134/s1070427218020155
42. Güvenatam B, Kurşun O, Heeres EHJ et al (2014) Hydrodeoxygenation of mono- and dimeric lignin model compounds on noble metal catalysts. Catal Today. https://doi.org/10.1016/j.cattod.2013.12.011
43. Roldugina EA, Naranov ER, Maximov AL, Karakhanov EA (2018) Hydrodeoxygenation of guaiacol as a model compound of bio-oil in methanol over mesoporous noble metal catalysts. Appl Catal A Gen. https://doi.org/10.1016/j.apcata.2018.01.008
44. Broglia F, Rimoldi L, Meroni D et al (2019) Guaiacol hydrodeoxygenation as a model for lignin upgrading. Role of the support surface features on Ni-based alumina-silica catalysts. Fuel. https://doi.org/10.1016/j.fuel.2019.01.157
45. Long J, Shu S, Wu Q et al (2015) Selective cyclohexanol production from the renewable lignin derived phenolic chemicals catalyzed by Ni/MgO. Energy Convers Manag. https://doi.org/10.1016/j.enconman.2015.08.016
46. Fang H, Zheng J, Luo X et al (2017) Product tunable behavior of carbon nanotubes-supported Ni–Fe catalysts for guaiacol hydrodeoxygenation. Appl Catal A Gen. https://doi.org/10.1016/j.apcata.2016.10.011
47. Dongil AB, Bachiller-Baeza B, Rodríguez-Ramos I et al (2016) The effect of Cu loading on Ni/carbon nanotubes catalysts for hydrodeoxygenation of guaiacol. RSC Adv. https://doi.org/10.1039/c6ra00041j
48. Liu K, Yan P, Jiang H et al (2019) Silver initiated hydrogen spillover on anatase TiO_2 creates active sites for selective hydrodeoxygenation of guaiacol. J Catal. https://doi.org/10.1016/j.jcat.2018.11.033
49. He T, Liu X, Ge Y et al (2017) Gas phase hydrodeoxygenation of anisole and guaiacol to aromatics with a high selectivity over $Ni–Mo/SiO_2$. Catal Commun. https://doi.org/10.1016/j.catcom.2017.09.011
50. Hewer TLR, Souza AGF, Roseno KTC et al (2018) Influence of acid sites on the hydrodeoxygenation of anisole with metal supported on SBA-15 and SAPO-11. Renew Energy. https://doi.org/10.1016/j.renene.2017.12.044
51. Vargas-Villagrán H, Flores-Villeda MA, Puente-Lee I et al (2018) Supported nickel catalysts for anisole hydrodeoxygenation: increase in the selectivity to cyclohexane. Today, Catal. https://doi.org/10.1016/j.cattod.2018.07.057

52. Zanuttini MS, Lago CD, Gross MS et al (2017) Hydrodeoxygenation of anisole with Pt catalysts. Ind Eng Chem Res. https://doi.org/10.1021/acs.iecr.7b00521
53. Teles CA, Rabelo-Neto RC, de Lima JR et al (2016) The effect of metal type on hydrodeoxygenation of phenol over silica supported catalysts. Catal Letters. https://doi.org/10.1007/s10562-016-1815-5
54. Feliczak-Guzik A, Szczyglewska P, Nowak I (2019) The effect of metal (Nb, Ru, Pd, Pt) supported on SBA-16 on the hydrodeoxygenation reaction of phenol. Catal Today. https://doi.org/10.1016/j.cattod.2018.06.046
55. Valdés-Martínez OU, Suárez-Toriello VA, Reyes JA et al (2017) Support effect and metals interactions for $NiRu/Al_2O_3$, TiO_2 and ZrO_2 catalysts in the hydrodeoxygenation of phenol. Catal Today. https://doi.org/10.1016/j.cattod.2017.04.007
56. Zhao C, Lercher JA (2012) Selective hydrodeoxygenation of lignin-derived phenolic monomers and dimers to cycloalkanes on Pd/C and HZSM-5 catalysts. ChemCatChem. https://doi.org/10.1002/cctc.201100273
57. Luo Z, Wang Y, He M, Zhao C (2016) Precise oxygen scission of lignin derived aryl ethers to quantitatively produce aromatic hydrocarbons in water. Green Chem. https://doi.org/10.1039/c5gc01790d
58. Barton RR, Carrier M, Segura C et al (2018) Ni/HZSM-5 catalyst preparation by deposition-precipitation. Part 2. Catalytic hydrodeoxygenation reactions of lignin model compounds in organic and aqueous systems. Appl Catal A Gen. https://doi.org/10.1016/j.apcata.2018.06.012
59. Li J, Sun H, Liu J et al (2018) Selective reductive cleavage of C–O bond in lignin model compounds over nitrogen-doped carbon-supported iron catalysts. Mol Catal. https://doi.org/10.1016/j.mcat.2018.03.014
60. Parsell TH, Owen BC, Klein I et al (2013) Cleavage and hydrodeoxygenation (HDO) of C–O bonds relevant to lignin conversion using Pd/Zn synergistic catalysis. Chem Sci. https://doi.org/10.1039/c2sc21657d
61. Jongerius AL, Jastrzebski R, Bruijnincx PCA, Weckhuysen BM (2012) CoMo sulfide-catalyzed hydrodeoxygenation of lignin model compounds: an extended reaction network for the conversion of monomeric and dimeric substrates. J Catal. https://doi.org/10.1016/j.jcat.2011.10.006
62. Gómez-Monedero B, Ruiz MP, Bimbela F, Faria J (2017) Selective hydrogenolysis of A–O–4, B–O–4, 4–O–5 C–O bonds of lignin-model compounds and lignin-containing stillage derived from cellulosic bioethanol processing. Appl Catal A Gen. https://doi.org/10.1016/j.apcata.2017.04.022
63. Yoon JS, Lee Y, Ryu J et al (2013) Production of high carbon number hydrocarbon fuels from a lignin-derived α-O-4 phenolic dimer, benzyl phenyl ether, via isomerization of ether to alcohols on high-surface-area silica-alumina aerogel catalysts. Appl Catal B Environ. https://doi.org/10.1016/j.apcatb.2013.05.039
64. Shu R, Lin B, Zhang J et al (2019) Efficient catalytic hydrodeoxygenation of phenolic compounds and bio-oil over highly dispersed Ru/TiO_2. Fuel Process Technol. https://doi.org/10.1016/j.fuproc.2018.11.004
65. Yao G, Wu G, Dai W et al (2015) Hydrodeoxygenation of lignin-derived phenolic compounds over bi-functional Ru/H-Beta under mild conditions. Fuel. https://doi.org/10.1016/j.fuel.2015.02.035
66. Jin S, Chen X, Li C et al (2016) Hydrodeoxygenation of lignin–derived diaryl ethers to aromatics and alkanes using nickel on Zr–doped niobium phosphate. ChemistrySelect. https://doi.org/10.1002/slct.201601107
67. Zhang W, Chen J, Liu R et al (2014) Hydrodeoxygenation of lignin-derived phenolic monomers and dimers to alkane fuels over bifunctional zeolite-supported metal catalysts. ACS Sustain Chem Eng. https://doi.org/10.1021/sc400401n
68. Gandarias I, Arias PL, Fernández SG et al (2012) Hydrogenolysis through catalytic transfer hydrogenation: glycerol conversion to 1,2-propanediol. Catal Today. https://doi.org/10.1016/j.cattod.2012.03.067
69. Vriamont CEJJ, Chen T, Romain C et al (2019) From lignin to chemicals: hydrogenation of lignin models and mechanistic insights into hydrodeoxygenation via low-temperature C–O bond cleavage. ACS Catal 9:2345–2354. https://doi.org/10.1021/acscatal.8b04714
70. Dai G, Zhu Y, Yang J et al (2019) Mechanism study on the pyrolysis of the typical ether linkages in biomass. Fuel 249:146–153. https://doi.org/10.1016/J.FUEL.2019.03.099
71. Chauvier C, Cantat T (2017) A viewpoint on chemical reductions of carbon-oxygen bonds in renewable feedstocks including CO_2 and biomass. ACS Catal 7:2107–2115. https://doi.org/10.1021/acscatal.6b03581

72. Zhang J (2018) Catalytic transfer hydrogenolysis as an efficient route in cleavage of lignin and model compounds. Green Energy Environ. https://doi.org/10.1016/j.gee.2018.08.001
73. Shafaghat H, Rezaei PS, Daud WMAW (2016) Using decalin and tetralin as hydrogen source for transfer hydrogenation of renewable lignin-derived phenolics over activated carbon supported Pd and Pt catalysts. J Taiwan Inst Chem Eng 65:91–100. https://doi.org/10.1016/j.jtice.2016.05.032
74. Wang X, Rinaldi R (2012) Exploiting H-transfer reactions with RANEY® Ni for upgrade of phenolic and aromatic biorefinery feeds under unusual, low-severity conditions. Energy Environ Sci 5:8244–8260. https://doi.org/10.1039/c2ee21855k
75. Kim M, Ha JM, Lee KY, Jae J (2016) Catalytic transfer hydrogenation/hydrogenolysis of guaiacol to cyclohexane over bimetallic RuRe/C catalysts. Catal Commun 86:113–118. https://doi.org/10.1016/j.catcom.2016.08.022
76. Kim M, Ha JM, Lee KY, Jae J (2016) Catalytic transfer hydrogenation/hydrogenolysis of guaiacol to cyclohexane over bimetallic RuRe/C catalysts. Catal Commun 86:113–118. https://doi.org/10.1016/j.catcom.2016.08.022
77. Wang X, Rinaldi R (2013) A route for lignin and bio-oil conversion: dehydroxylation of phenols into arenes by catalytic tandem reactions. Angew Chemie Int Ed 52:11499–11503. https://doi.org/10.1002/anie.201304776
78. Galkin MV, Dahlstrand C, Samec JSM (2015) Mild and robust redox-neutral Pd/C-catalyzed lignol β-O-4′ bond cleavage through a low-energy-barrier pathway. Chemsuschem 8:2187–2192. https://doi.org/10.1002/cssc.201500117
79. Zaera F (2013) Key unanswered questions about the mechanism of olefin hydrogenation catalysis by transition-metal surfaces: a surface-science perspective. Phys Chem Chem Phys 15:11988–12003. https://doi.org/10.1039/c3cp50402f
80. Wu H, Song J, Xie C et al (2018) Efficient and mild transfer hydrogenolytic cleavage of aromatic ether bonds in lignin-derived compounds over Ru/C. ACS Sustain Chem Eng 6:2872–2877. https://doi.org/10.1021/acssuschemeng.7b02993
81. Gu X, Kanghua C, Ming H et al (2012) La-modified SBA-15/H_2O_2 systems for the microwave assisted oxidation of organosolv beech wood lignin. Maderas Cienc y Tecnol 14:31–41. https://doi.org/10.4067/S0718-221X2012000100003
82. Tarabanko VE, Tarabanko N (2017) Catalytic oxidation of lignins into the aromatic aldehydes: general process trends and development prospects. Int J Mol Sci. https://doi.org/10.3390/ijms18112421
83. Behling R, Valange S, Chatel G (2016) Heterogeneous catalytic oxidation for lignin valorization into valuable chemicals: what results? What limitations? What trends? Green Chem 18:1839–1854. https://doi.org/10.1039/c5gc03061g
84. Lin F, Liu C, Wang X et al (2019) Catalytic oxidation of biorefinery corncob lignin via zirconium(IV) chloride and sodium hydroxide in acetonitrile/water: a functionality study. Sci Total Environ. https://doi.org/10.1016/j.scitotenv.2019.04.224
85. Zakzeski J, Dbczak A, Bruijnincx PCA, Weckhuysen BM (2011) Catalytic oxidation of aromatic oxygenates by the heterogeneous catalyst Co-ZIF-9. Appl Catal A Gen 394:79–85. https://doi.org/10.1016/j.apcata.2010.12.026
86. Baguc IB, Celebi M, Karakas K et al (2017) Nanohydrotalcite supported ruthenium nanoparticles: highly efficient heterogeneous catalyst for the oxidative valorization of lignin model compounds. ChemistrySelect 2:10191–10198. https://doi.org/10.1002/slct.201700824
87. Chen G, Wu S, Liu H et al (2013) Palladium supported on an acidic metal-organic framework as an efficient catalyst in selective aerobic oxidation of alcohols. Green Chem 15:230–235. https://doi.org/10.1039/c2gc36618e
88. Deng W, Zhang H, Wu X et al (2015) Oxidative conversion of lignin and lignin model compounds catalyzed by CeO_2-supported Pd nanoparticles. Green Chem 17:5009–5018. https://doi.org/10.1039/c5gc01473e
89. Blandez JF, Navalõn S, Alvaro M, Garcia H (2015) Graphenes as metal-free catalysts for the oxidative depolymerization of lignin models. ChemCatChem 7:3020–3026. https://doi.org/10.1002/cctc.201500477
90. Mottweiler J, Puche M, Räuber C et al (2015) Copper- and vanadium-catalyzed oxidative cleavage of lignin using dioxygen. Chemsuschem 8:2106–2113. https://doi.org/10.1002/cssc.201500131

91. Melián-Rodríguez M, Saravanamurugan S, Kegnæs S, Riisager A (2015) Aerobic oxidation of veratryl alcohol to veratraldehyde with heterogeneous ruthenium catalysts. Top Catal 58:1036–1042. https://doi.org/10.1007/s11244-015-0472-z
92. Mate VR, Shirai M, Rode CV (2013) Heterogeneous CO_3O_4 catalyst for selective oxidation of aqueous veratryl alcohol using molecular oxygen. Catal Commun 33:66–69. https://doi.org/10.1016/j.catcom.2012.12.015
93. Olmos C, Chinchilla L, Cappella A et al (2018) Selective oxidation of veratryl alcohol over Au–Pd/Ce0.62Zr0.38O_2 catalysts synthesized by sol-immobilization: effect of Au: Pd molar ratio. Nanomaterials 8:669. https://doi.org/10.3390/nano8090669
94. Fan H, Yang Y, Song J et al (2014) One-pot sequential oxidation and aldol-condensation reactions of veratryl alcohol catalyzed by the Ru@ZIF-8 + CuO/basic ionic liquid system. Green Chem 16:600–604. https://doi.org/10.1039/c3gc41363b
95. Reddy PRGN, Rao BG, Rao TV, Reddy BM (2019) Selective aerobic oxidation of vanillyl alcohol to vanillin catalysed by nanostructured Ce–Zr–O solid solutions. Catal Lett 149:533–543. https://doi.org/10.1007/s10562-019-02658-1
96. Yuan Z, Chen S, Liu B (2017) Nitrogen-doped reduced graphene oxide-supported Mn_3O_4: an efficient heterogeneous catalyst for the oxidation of vanillyl alcohol to vanillin. J Mater Sci 52:164–172. https://doi.org/10.1007/s10853-016-0318-5
97. Zheng MW, Lai HK, Lin KYA (2018) Valorization of vanillyl alcohol by pigments: Prussian blue analogue as a highly-effective heterogeneous catalyst for aerobic oxidation of vanillyl alcohol to vanillin. Waste Biomass Valorization. https://doi.org/10.1007/s12649-018-0280-3
98. Mate VR, Jha A, Joshi UD et al (2014) Effect of preparation parameters on characterization and activity of CO_3O_4 catalyst in liquid phase oxidation of lignin model substrates. Appl Catal A Gen 487:130–138. https://doi.org/10.1016/j.apcata.2014.08.023
99. Mate VR, Shirai M, Rode CV (2013) Heterogeneous CO_3O_4 catalyst for selective oxidation of aqueous veratryl alcohol using molecular oxygen. Catal Commun 33:66–69. https://doi.org/10.1016/j.catcom.2012.12.015
100. Ramana S, Rao BG, Venkataswamy P et al (2016) Nanostructured Mn-doped ceria solid solutions for efficient oxidation of vanillyl alcohol. J Mol Catal A Chem 415:113–121. https://doi.org/10.1016/j.molcata.2016.01.028
101. Jha A, Rode CV (2013) Highly selective liquid-phase aerobic oxidation of vanillyl alcohol to vanillin on cobalt oxide (CO_3O_4) nanoparticles. New J Chem 37:2669–2674. https://doi.org/10.1039/c3nj00508a
102. Rinesch T, Bolm C (2018) Cobalt-catalyzed oxidation of the β-O-4 bond in lignin and lignin model compounds. ACS Omega 3:8386–8392. https://doi.org/10.1021/acsomega.8b00994
103. Saha S, Abd Hamid SB (2017) CuZrO3 nanoparticles catalyst in aerobic oxidation of vanillyl alcohol. RSC Adv 7:9914–9925. https://doi.org/10.1039/c6ra26370d
104. Parlett CMA, Durndell LJ, Machado A et al (2014) Alumina-grafted SBA-15 as a high performance support for Pd-catalysed cinnamyl alcohol selective oxidation. Catal Today 229:46–55. https://doi.org/10.1016/j.cattod.2013.11.056
105. Durndell LJ, Parlett CMA, Hondow NS et al (2013) Tunable Pt nanocatalysts for the aerobic selox of cinnamyl alcohol. Nanoscale 123:5412–5419. https://doi.org/10.1039/c3nr00184a
106. Capello C, Fischer U, Hungerbühler K (2007) What is a green solvent? A comprehensive framework for the environmental assessment of solvents. Green Chem 9:927–934. https://doi.org/10.1039/b617536h
107. Prat D, Hayler J, Wells A (2014) A survey of solvent selection guides. Green Chem 16:4546–4551. https://doi.org/10.1039/c4gc01149j
108. Wu G, Brett GL, Cao E et al (2016) Oxidation of cinnamyl alcohol using bimetallic continuous flow packed bed microreactor. Catal Sci Technol 6:4749–4758. https://doi.org/10.1039/c6cy00232c
109. Mori S, Takubo M, Makida K et al (2009) A simple and efficient oxidation of alcohols with ruthenium on carbon w. ChemComm. https://doi.org/10.1039/b908451g
110. Sun HY, Hua Q, Guo FF et al (2012) Selective aerobic oxidation of alcohols by using manganese oxide nanoparticles as an efficient heterogeneous catalyst. Adv Synth Catal 354:569–573. https://doi.org/10.1002/adsc.201100666
111. Lee AF, Wilson K (2004) Structure–reactivity correlations in the selective aerobic oxidation of cinnamyl alcohol: in situ XAFS. Green Chem 6:37–42. https://doi.org/10.1039/b310505a
112. Lee AF, Hackett SFJ, Hargreaves JSJ, Wilson K (2006) On the active site in heterogeneous palladium selox catalysts. Green Chem 8:549–555. https://doi.org/10.1039/b601984f

113. Keresszegi C, Bürgi T, Mallat T, Baiker A (2002) On the role of oxygen in the liquid-phase aerobic oxidation of alcohols on palladium. J Catal 211:244–251. https://doi.org/10.1016/S0021-9517(02)93723-1
114. Miyamura H, Choo GCY, Yasukawa T et al (2013) A heterogeneous layered bifunctional catalyst for the integration of aerobic oxidation and asymmetric C–C bond formation. Chem Commun 49:9917–9919. https://doi.org/10.1039/c3cc46204h
115. Panda AK, Singh RK, Mishra DK (2010) Thermolysis of waste plastics to liquid fuel. A suitable method for plastic waste management and manufacture of value added products—a world prospective. Renew Sustain Energy Rev 14:233–248. https://doi.org/10.1016/j.rser.2009.07.005
116. Martínez JD, Puy N, Murillo R et al (2013) Waste tyre pyrolysis—a review. Renew Sustain Energy Rev 23:179–213. https://doi.org/10.1016/j.rser.2013.02.038
117. Yang X, Sun L, Xiang J et al (2013) Pyrolysis and dehalogenation of plastics from waste electrical and electronic equipment (WEEE): a review. Waste Manag 33:462–473. https://doi.org/10.1016/j.wasman.2012.07.025
118. Chen D, Yin L, Wang H, He P (2014) Pyrolysis technologies for municipal solid waste: a review. Waste Manag 34:2466–2486. https://doi.org/10.1016/j.wasman.2015.01.022
119. Roy P, Dias G (2017) Prospects for pyrolysis technologies in the bioenergy sector: a review. Renew Sustain Energy Rev 77:59–69. https://doi.org/10.1016/j.rser.2017.03.136
120. Chio C, Sain M, Qin W (2019) Lignin utilization: a review of lignin depolymerization from various aspects. Renew Sustain Energy Rev 107:232–249. https://doi.org/10.1016/j.rser.2019.03.008
121. Nguyen TTP, Mai TV-T, Huynh LK (2018) Detailed kinetic modeling of thermal decomposition of guaiacol—a model compound for biomass lignin. Biomass Bioenerg 112:45–60. https://doi.org/10.1016/J.BIOMBIOE.2018.02.006
122. Liu C, Ye L, Yuan W et al (2018) Investigation on pyrolysis mechanism of guaiacol as lignin model compound at atmospheric pressure. Fuel 232:632–638. https://doi.org/10.1016/J.FUEL.2018.05.162
123. Huang J, Li X, Wu D et al (2013) Theoretical studies on pyrolysis mechanism of guaiacol as lignin model compound. J Renew Sustain Energy 5:043112. https://doi.org/10.1063/1.4816497
124. Yang H, Norinaga K, Li J et al (2018) Effects of HZSM-5 on volatile products obtained from the fast pyrolysis of lignin and model compounds. Fuel Process Technol 181:207–214. https://doi.org/10.1016/j.fuproc.2018.09.022
125. Yang H, Furutani Y, Kudo S et al (2016) Experimental investigation of thermal decomposition of dihydroxybenzene isomers: catechol, hydroquinone, and resorcinol. J Anal Appl Pyrolysis 120:321–329. https://doi.org/10.1016/J.JAAP.2016.05.019
126. Asmadi M, Kawamoto H, Saka S (2011) Thermal reactivities of catechols/pyrogallols and cresols/xylenols as lignin pyrolysis intermediates. J Anal Appl Pyrolysis 92:76–87. https://doi.org/10.1016/J.JAAP.2011.04.012
127. Scheer AM, Mukarakate C, Robichaud DJ et al (2012) Unimolecular thermal decomposition of phenol and d 5-phenol: direct observation of cyclopentadiene formation via cyclohexadienone. J Chem Phys. https://doi.org/10.1063/1.3675902
128. Moldoveanu SC (2019) Pyrolysis of alcohols and phenols. Pyrolysis of organic molecules. Elsevier, Amsterdam, pp 207–278
129. Jarvis MW, Daily JW, Carstensen HH et al (2011) Direct detection of products from the pyrolysis of 2-phenethyl phenyl ether. J Phys Chem A 115:428–438. https://doi.org/10.1021/jp1076356
130. Huang X, Liu C, Huang J, Li H (2011) Theory studies on pyrolysis mechanism of phenethyl phenyl ether. Comput Theor Chem 976:51–59. https://doi.org/10.1016/J.COMPTC.2011.08.001
131. Beste A, Buchanan AC (2011) Kinetic analysis of the phenyl-shift reaction in β-O-4 lignin model compounds: a computational study. J Org Chem 76:2195–2203. https://doi.org/10.1021/jo2000385
132. Beste A, Buchanan AC (2012) Role of carbon-carbon phenyl migration in the pyrolysis mechanism of β-O-4 lignin model compounds: phenethyl phenyl ether and α-hydroxy phenethyl phenyl ether. J Phys Chem A 116:12242–12248. https://doi.org/10.1021/jp3104694
133. Beste A, Buchanan AC (2012) Kinetic simulation of the thermal degradation of phenethyl phenyl ether, a model compound for the β-O-4 linkage in lignin. Chem Phys Lett 550:19–24. https://doi.org/10.1016/J.CPLETT.2012.08.040
134. Chen L, Ye X, Luo F et al (2015) Pyrolysis mechanism of βO_4 type lignin model dimer. J Anal Appl Pyrolysis 115:103–111. https://doi.org/10.1016/J.JAAP.2015.07.009

135. Akazawa M, Kojima Y, Kato Y (2016) Effect of pyrolysis temperature on the pyrolytic degradation mechanism of β-aryl ether linkages. J Anal Appl Pyrolysis 118:164–174. https://doi.org/10.1016/j.jaap.2016.02.001
136. Wang S, Ru B, Dai G et al (2017) Mechanism study on the pyrolysis of a synthetic β-O-4 dimer as lignin model compound. Proc Combust Inst 36:2225–2233. https://doi.org/10.1016/J.PROCI.2016.07.129
137. Shen D, Jin W, Gu S (2017) Enhancement of aromatic monomer production from pyrolysis of lignin-related β-O-4 contained model compound. J Anal Appl Pyrolysis 127:176–182. https://doi.org/10.1016/J.JAAP.2017.08.010
138. Wang M, Liu C (2016) Theoretic studies on decomposition mechanism of O-methoxy phenethyl phenyl ether: primary and secondary reactions. J Anal Appl Pyrolysis 117:325–333. https://doi.org/10.1016/J.JAAP.2015.10.016
139. Moldoveanu SC (2019) Pyrolysis of alcohols and phenols. Pyrolysis of organic molecules. Elsevier, Amsterdam, pp 207–278
140. Park HW, Kim JK, Hong UG et al (2013) Catalytic decomposition of lignin model compounds to aromatics over acidic catalysts. Catal Surv from Asia 17:119–131. https://doi.org/10.1007/s10563-013-9158-3
141. Huang J, He C (2015) Pyrolysis mechanism of α-O-4 linkage lignin dimer: a theoretical study. J Anal Appl Pyrolysis 113:655–664. https://doi.org/10.1016/J.JAAP.2015.04.012
142. Huang J, He C, Pan G, Tong H (2016) A theoretical research on pyrolysis reactions mechanism of coumarone-contained lignin model compound. Comput Theor Chem 1091:92–98. https://doi.org/10.1016/J.COMPTC.2016.07.016
143. Valle B, García-Gómez N, Arandia A et al (2019) Effect of phenols extraction on the behavior of Ni-spinel derived catalyst for raw bio-oil steam reforming. Int J Hydrogen Energy. https://doi.org/10.1016/J.IJHYDENE.2018.12.057
144. Yu N, Rahman MM, Chen J et al (2019) Steam reforming of simulated bio-oil on K–Ni–Cu–Mg–Ce–O/Al_2O_3: the effect of K. Catal Today 323:183–190. https://doi.org/10.1016/J.CATTOD.2018.04.010
145. Spragg J, Mahmud T, Dupont V (2018) Hydrogen production from bio-oil: a thermodynamic analysis of sorption-enhanced chemical looping steam reforming. Int J Hydrogen Energy. https://doi.org/10.1016/j.ijhydene.2018.10.068
146. Bizkarra K, Bermudez JM, Arcelus-Arrillaga P et al (2018) Nickel based monometallic and bimetallic catalysts for synthetic and real bio-oil steam reforming. Int J Hydrogen Energy. https://doi.org/10.1016/j.ijhydene.2018.03.049
147. Bizkarra K, Barrio VL, Gartzia-Rivero L et al (2019) Hydrogen production from a model bio-oil/bio-glycerol mixture through steam reforming using Zeolite L supported catalysts. Int J Hydrogen Energy 44:1492–1504. https://doi.org/10.1016/J.IJHYDENE.2018.11.079
148. Sun Z, Fridrich B, De Santi A, et al (2018) Bright side of lignin depolymerization: toward new platform chemicals. Chem. Rev. https://doi.org/10.1021/acs.chemrev.7b00588
149. Sarvaramini A, Larachi F (2012) Catalytic oxygenless steam cracking of syngas-containing benzene model tar compound over natural Fe-bearing silicate minerals. Fuel. https://doi.org/10.1016/j.fuel.2012.02.039
150. Kaisalo N, Simell P, Lehtonen J (2016) Benzene steam reforming kinetics in biomass gasification gas cleaning. Fuel. https://doi.org/10.1016/j.fuel.2016.06.042
151. Nam H, Wang Z, Shanmugam SR et al (2018) Chemical looping dry reforming of benzene as a gasification tar model compound with Ni- and Fe-based oxygen carriers in a fluidized bed reactor. Int J Hydrogen Energy. https://doi.org/10.1016/j.ijhydene.2018.08.103
152. Sarıoğlan A (2012) Tar removal on dolomite and steam reforming catalyst: benzene, toluene and xylene reforming. Int J Hydrogen Energy 37:8133–8142. https://doi.org/10.1016/j.ijhydene.2012.02.045
153. Keller M, Leion H, Mattisson T (2016) Chemical looping tar reforming using La/Sr/Fe-containing mixed oxides supported on ZrO2. Appl Catal B Environ. https://doi.org/10.1016/j.apcatb.2015.10.047
154. Josuinkas FM, Quitete CPB, Ribeiro NFP, Souza MMVM (2014) Steam reforming of model gasification tar compounds over nickel catalysts prepared from hydrotalcite precursors. Fuel Process Technol. https://doi.org/10.1016/j.fuproc.2014.01.007

155. Gao N, Wang X, Li A et al (2016) Hydrogen production from catalytic steam reforming of benzene as tar model compound of biomass gasification. Fuel Process Technol 148:380–387. https://doi.org/10.1016/J.FUPROC.2016.03.019
156. Virginie M, Courson C, Niznansky D et al (2010) Characterization and reactivity in toluene reforming of a Fe/olivine catalyst designed for gas cleanup in biomass gasification. Appl Catal B Environ 101:90–100. https://doi.org/10.1016/J.APCATB.2010.09.011
157. Li L, Song Z, Zhao X et al (2016) Microwave-induced cracking and CO_2 reforming of toluene on biomass derived char. Chem Eng J 284:1308–1316. https://doi.org/10.1016/J.CEJ.2015.09.040
158. Kim S, Chun D, Rhim Y et al (2014) Catalytic reforming of toluene using a nickel ion-exchanged coal catalyst. Int J Hydrogen Energy. https://doi.org/10.1016/j.ijhydene.2015.06.103
159. Liu X, Yang X, Liu C et al (2016) Low-temperature catalytic steam reforming of toluene over activated carbon supported nickel catalysts. J Taiwan Inst Chem Eng. https://doi.org/10.1016/j.jtice.2016.05.006
160. Yamaguchi S, Ozaki T, Suyama T et al (2019) Evaluation of regenerative function and activity of reforming toluene by composite catalyst containing spinel oxide. Int J Hydrogen Energy 44:9338–9347. https://doi.org/10.1016/J.IJHYDENE.2019.02.101
161. Zhang R, Wang H, Hou X (2014) Catalytic reforming of toluene as tar model compound: effect of Ce and Ce–Mg promoter using Ni/olivine catalyst. Chemosphere 97:40–46. https://doi.org/10.1016/J.CHEMOSPHERE.2013.10.087
162. Oh G, Park SY, Seo MW et al (2016) Ni/Ru–Mn/Al_2O_3 catalysts for steam reforming of toluene as model biomass tar. Renew Energy 86:841–847. https://doi.org/10.1016/J.RENENE.2015.09.013
163. Xu M, Hu H, Yang F et al (2018) Novel findings in conversion mechanism of toluene as model compound of biomass waste tar in molten salt. J Anal Appl Pyrolysis 134:274–280. https://doi.org/10.1016/J.JAAP.2018.06.017
164. Świerczyński D, Libs S, Courson C, Kiennemann A (2007) Steam reforming of tar from a biomass gasification process over Ni/olivine catalyst using toluene as a model compound. Appl Catal B Environ 74:211–222. https://doi.org/10.1016/J.APCATB.2007.01.017
165. Li Q, Wang Q, Kayamori A, Zhang J (2018) Experimental study and modeling of heavy tar steam reforming. Fuel Process Technol 178:180–188. https://doi.org/10.1016/J.FUPROC.2018.05.020
166. Bizkarra K, Barrio VL, Arias PL, Cambra JF (2016) Sustainable hydrogen production from bio-oil model compounds (meta-xylene) and mixtures (1-butanol, meta-xylene and furfural). Bioresour Technol. https://doi.org/10.1016/j.biortech.2016.05.038
167. Hu X, Lu G (2009) Investigation of the steam reforming of a series of model compounds derived from bio-oil for hydrogen production. Appl Catal B Environ. https://doi.org/10.1016/j.apcatb.2008.10.021
168. Boldrin P, Ruiz-Trejo E, Mermelstein J, et al (2016) Strategies for carbon and sulfur tolerant solid oxide fuel cell materials, incorporating lessons from heterogeneous catalysis. Chem. Rev. https://doi.org/10.1021/acs.chemrev.6b00284
169. Nabgan W, Tuan Abdullah TA, Mat R et al (2017) Renewable hydrogen production from bio-oil derivative via catalytic steam reforming: an overview. Renew Sustain Energy Rev. https://doi.org/10.1016/j.rser.2017.05.069
170. García-García I, Acha E, Bizkarra K et al (2015) Hydrogen production by steam reforming of *m*-cresol, a bio-oil model compound, using catalysts supported on conventional and unconventional supports. Int J Hydrogen Energy 40:14445–14455. https://doi.org/10.1016/j.ijhydene.2015.07.155
171. Wu C, Liu R (2010) Hydrogen production from steam reforming of *m*-cresol, a model compound derived from bio-oil: green process evaluation based on liquid condensate recycling. Energy Fuels 24:5139–5147. https://doi.org/10.1021/ef100369g
172. Trane-Restrup R, Jensen AD (2015) Steam reforming of cyclic model compounds of bio-oil over Ni-based catalysts: product distribution and carbon formation. Appl Catal B Environ 165:117–127. https://doi.org/10.1016/j.apcatb.2014.09.026
173. Zamboni I, Courson C, Kiennemann A (2017) Fe–Ca interactions in Fe-based/CaO catalyst/sorbent for CO_2 sorption and hydrogen production from toluene steam reforming. Appl Catal B Environ 203:154–165. https://doi.org/10.1016/J.APCATB.2016.10.024
174. Nabgan W, Tuan Abdullah TA, Mat R et al (2016) Influence of Ni to Co ratio supported on ZrO_2 catalysts in phenol steam reforming for hydrogen production. Int J Hydrogen Energy 41:22922–22931. https://doi.org/10.1016/J.IJHYDENE.2016.10.055

175. Nabgan W, Tuan Abdullah TA, Mat R et al (2016) Hydrogen production from catalytic steam reforming of phenol with bimetallic nickel-cobalt catalyst on various supports. Appl Catal A Gen 527:161–170. https://doi.org/10.1016/J.APCATA.2016.08.033
176. Meng J, Zhao Z, Wang X et al (2018) Comparative study on phenol and naphthalene steam reforming over Ni–Fe alloy catalysts supported on olivine synthesized by different methods. Energy Convers Manag 168:60–73. https://doi.org/10.1016/J.ENCONMAN.2018.04.112
177. Artetxe M, Nahil MA, Olazar M, Williams PT (2016) Steam reforming of phenol as biomass tar model compound over Ni/Al_2O_3 catalyst. Fuel 184:629–636. https://doi.org/10.1016/j.fuel.2016.07.036
178. Remiro A, Arandia A, Oar-Arteta L et al (2018) Regeneration of $NiAl_2O_4$ spinel type catalysts used in the reforming of raw bio-oil. Appl Catal B Environ 237:353–365. https://doi.org/10.1016/J.APCATB.2018.06.005
179. Remiro A, Ochoa A, Arandia A et al (2019) On the dynamics and reversibility of the deactivation of a Rh/CeO_2ZrO_2 catalyst in raw bio-oil steam reforming. Int J Hydrogen Energy 44:2620–2632. https://doi.org/10.1016/J.IJHYDENE.2018.12.073
180. Esteban-Díez G, Gil MV, Pevida C et al (2016) Effect of operating conditions on the sorption enhanced steam reforming of blends of acetic acid and acetone as bio-oil model compounds. Appl Energy. https://doi.org/10.1016/j.apenergy.2016.05.149
181. Vanga G, Gattia DM, Stendardo S, Scaccia S (2019) Novel synthesis of combined CaO-$Ca1_2Al1_4O_33$-Ni sorbent-catalyst material for sorption enhanced steam reforming processes. Ceram Int 45:7594–7605. https://doi.org/10.1016/J.CERAMINT.2019.01.054
182. Yu Z, Yang Y, Yang S et al (2018) Iron-based oxygen carriers in chemical looping conversions: a review. Carbon Resour Convers. https://doi.org/10.1016/j.crcon.2018.11.004
183. De Vos Y, Jacobs M, Van Der Voort P et al (2019) Sustainable iron-based oxygen carriers for chemical looping for hydrogen generation. Int J Hydrogen Energy 44:1374–1391. https://doi.org/10.1016/J.IJHYDENE.2018.11.099
184. Udomchoke T, Wongsakulphasatch S, Kiatkittipong W et al (2016) Performance evaluation of sorption enhanced chemical-looping reforming for hydrogen production from biomass with modification of catalyst and sorbent regeneration. Chem Eng J. https://doi.org/10.1016/j.cej.2016.05.115
185. Saisu M, Sato T, Watanabe M et al (2003) Conversion of lignin with supercritical water-phenol mixtures. Energy Fuels. https://doi.org/10.1021/ef0202844
186. Wahyudiono Sasaki M, Goto M (2011) Thermal decomposition of guaiacol in sub- and supercritical water and its kinetic analysis. J Mater Cycles Waste Manag 13:68–79. https://doi.org/10.1007/s10163-010-0309-6
187. Yong TLK, Yukihiko M (2013) Kinetic analysis of guaiacol conversion in sub- and supercritical water. Ind Eng Chem Res 52:9048–9059. https://doi.org/10.1021/ie4009748
188. Yang L, Li Y, Savage PE (2014) Hydrolytic cleavage of C–O linkages in lignin model compounds catalyzed by water-tolerant Lewis acids. Ind Eng Chem Res 53:2633–2639. https://doi.org/10.1021/ie403545n
189. Rebacz NA, Savage PE (2013) Anisole hydrolysis in high temperature water. Phys Chem Chem Phys 15:3562–3569. https://doi.org/10.1039/c3cp43877e
190. Penninger JML, Kersten RJA, Baur HCL (1999) Reactions of diphenylether in supercritical water—mechanism and kinetics. J Supercrit Fluids 16:119–132. https://doi.org/10.1016/S0896-8446(99)00024-8
191. Roberts V, Fendt S, Lemonidou AA et al (2010) Influence of alkali carbonates on benzyl phenyl ether cleavage pathways in superheated water. Appl Catal B Environ 95:71–77. https://doi.org/10.1016/j.apcatb.2009.12.010
192. Siskin M, Katritzky AR, Balasubramanian M (1991) Aqueous organic chemistry. 4. Cleavage of diaryl ethers. Energy Fuels. https://doi.org/10.1021/ef00029a028
193. Chatterjee M, Chatterjee A, Ishizaka T, Kawanami H (2015) Rhodium-mediated hydrogenolysis/hydrolysis of the aryl ether bond in supercritical carbon dioxide/water: an experimental and theoretical approach. Catal Sci Technol 5:1532–1539. https://doi.org/10.1039/c4cy01417k
194. He J, Zhao C, Lercher JA (2012) Ni-catalyzed cleavage of aryl ethers in the aqueous phase. J Am Chem Soc 134:20768–20775. https://doi.org/10.1021/ja309915e
195. Zhao C, Kou Y, Lemonidou AA et al (2009) Highly selective catalytic conversion of phenolic bio-oil to alkanes. Angew Chemie Int Ed 48:3987–3990. https://doi.org/10.1002/anie.200900404

196. Abd Hamid SB, Ambursa MM, Sudarsanam P et al (2017) Effect of Ti loading on structure-activity properties of Cu–Ni/Ti-MCM-41 catalysts in hydrodeoxygenation of guaiacol. Catal Commun. https://doi.org/10.1016/j.catcom.2017.02.006
197. Resende KA, Braga AH, Noronha FB, Hori CE (2019) Hydrodeoxygenation of phenol over Ni/Ce1–xNbxO_2 catalysts. Appl Catal B Environ. https://doi.org/10.1016/j.apcatb.2018.12.040

Affiliations

Mikel Oregui-Bengoechea[1] · Ion Agirre[1] · Aitziber Iriondo[1] · Alexander Lopez-Urionabarrenechea[1] · Jesus M. Requies[1] · Iker Agirrezabal-Telleria[1] · Kepa Bizkarra[1] · V. Laura Barrio[1] · Jose F. Cambra[1]

[1] Department of Chemical and Environmental Engineering, School of Engineering, University of the Basque Country EHU/UPV, Plaza Ingeniero Torres Quevedo 1, 48013 Bilbao, Spain

The manufacturer's authorised representative in the EU is Springer Nature Customer Service Centre GmbH, Europaplatz 3, 69115 Heidelberg, Germany. If you have any concerns regarding our products, please contact ProductSafety@springernature.com

Printed and bound by CPI Group (UK) Ltd, Croydon, CR0 4YY
05/10/2026
02228123-0009